Intrinsically Biocompatible Polymer Systems

Intrinsically Biocompatible Polymer Systems

Special Issue Editor

Marek Kowalczuk

MDPI • Basel • Beijing • Wuhan • Barcelona • Belgrade • Manchester • Tokyo • Cluj • Tianjin

Special Issue Editor
Marek Kowalczuk
Polish Academy of Sciences
Poland

Editorial Office
MDPI
St. Alban-Anlage 66
4052 Basel, Switzerland

This is a reprint of articles from the Special Issue published online in the open access journal *Polymers* (ISSN 2073-4360) (available at: https://www.mdpi.com/journal/polymers/special_issues/biocompatible_polymers).

For citation purposes, cite each article independently as indicated on the article page online and as indicated below:

LastName, A.A.; LastName, B.B.; LastName, C.C. Article Title. *Journal Name* **Year**, *Article Number*, Page Range.

ISBN 978-3-03928-420-7 (Pbk)
ISBN 978-3-03928-421-4 (PDF)

Contents

About the Special Issue Editor

Marek Kowalczuk is a professor at the Centre of Polymer and Carbon Materials, Polish Academy of Sciences, Zabrze, Poland, and the head of the Group of Innovation, Technology, and Analysis Service. He received his Ph.D. in 1984 from the Faculty of Chemistry, Silesian University of Technology, Gliwice, Poland, and D.S. degree in 1994 at the same university. Since 2010, he has been a professor of chemistry, nominated by the President of Poland. He was a visiting lecturer at the University of Massachusetts in Amherst, MA, USA; Marie Curie EU fellow at the University of Bologna, Italy; and professor in synthetic/polymer chemistry at the University of Wolverhampton, UK. Recently, he was elected a member of Chemistry Committee of Polish Academy of Sciences. He is the author and co-author of over 200 scientific papers and a score of patents. His main scientific interests are: biodegradable and functional polymer biomaterials, novel initiators and mechanisms of anionic polymerization related to the synthesis of biodegradable polymers possessing desired architecture, biodegradation of synthetic and natural polymers, polymer mass spectrometry, and forensic engineering of advanced polymeric materials.

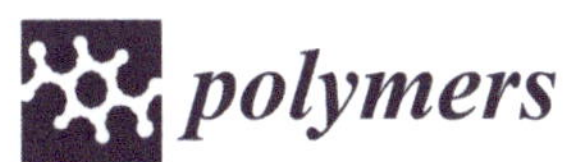 

Editorial

Intrinsically Biocompatible Polymer Systems

Marek Kowalczuk

Centre of Polymer and Carbon Materials Polish Academy of Sciences, 34 M. Curie-Sklodowska St., 41-800 Zabrze, Poland; marek.kowalczuk@cmpw-pan.edu.pl; Tel.: +48-322-716-077-(225)

Received: 16 January 2020; Accepted: 17 January 2020; Published: 29 January 2020

Polymers are everywhere, even inside of the human body. Polymers can be produced by living organisms, in which case they are called biopolymers, while polymers which possess the ability to be in contact with a living system without producing any adverse effect are referred to as polymeric biomaterials [1]. Polymeric biomaterials may be of natural or synthetic origin.

The term "biocompatibility" reflects the ability of a polymer material to perform with an appropriate host response in a specific application. Thus, biocompatibility is a property of a polymeric biomaterial–host system [2]. The definition of biocompatibility was redefined over ten years ago, and is now accepted to refer to: "the ability of a biomaterial to perform its desired function with respect to a medical therapy, without eliciting any undesirable local or systemic effects in the recipient or beneficiary of that therapy, but generating the most appropriate beneficial cellular or tissue response in that specific situation, and optimising the clinically relevant performance of that therapy" [3]. The mechanism of biocompatibility is not yet well understood; however, countless attempts have been made to elucidate the framework of mechanisms that controls the events that occur when a biomaterial is exposed to tissue in a human body [4].

The range of biological hazards of polymeric biomaterials is wide and complex. The ISO 10993-1 addresses the determination of the effects of medical devices on tissues, mostly in a general way, rather than in terms of specific devices [5]. Therefore, the biosafety of polymeric biomaterials needs to be predictable, in order to undertake assessments of the potential complications arising from their usage and on the formation of their degradation products. Forensic engineering of advanced polymeric materials (FEAPM) deals with the evaluation and understanding of the relationships between their structure, properties, and behavior, before, during, and after their practical application. Both *ex ante* investigations and *ex post* studies are needed in order to define and minimize any potential failure of novel polymeric biomaterials in specific applications. These elements in the FEAPM methodology are currently being studied in the area of polymeric biomaterials [6].

This book comprises 15 chapters, each of which was published previously as original research contributions of the Polymers Special Issue devoted to biocompatible polymer systems: https://www.mdpi.com/journal/polymers/special_issues/biocompatible_polymers?view=compact&listby=date.

Conflicts of Interest: Author declare no conflict of Interest.

References

1. Vert, M.; Doi, Y.; Hellwich, K.-H.; Hess, M.; Hodge, P.; Kubisa, P.; Rinaudo, M.; Schué, F. Terminology for biorelated polymers and applications (IUPAC Recommendations 2012). *Pure Appl. Chem.* **2012**, *84*, 377–410. [CrossRef]
2. Williams, D.F. There is no such thing as a biocompatible material. *Biomaterials* **2014**, *35*, 10009–10014. [CrossRef] [PubMed]
3. Williams, D.F. On the mechanisms of biocompatibility. *Biomaterials* **2008**, *29*, 2941–2953. [CrossRef] [PubMed]
4. Williams, D.F. Biocompatibility pathways: Biomaterials-induced sterile inflammation, mechanotransduction and principles of biocompatibility control. *ACS Biomater. Sci. Eng.* **2017**, *3*, 2–35. [CrossRef]

5. Bernard, M.; Jubeli, E.; Pungente, M.D.; Yagoubi, N. Biocompatibility of polymer-based biomaterials and medical devices — Regulations, in vitro screening and risk-management. *Biomater. Sci.* **2018**, *6*, 2025–2053. [CrossRef] [PubMed]
6. Kowalczuk, M. Forensic engineering of advanced polymeric materials. *M J Foren.* **2017**, *1*, 1–2.

 *polymers*

Article

Bioresorbable Stent in Anterior Cruciate Ligament Reconstruction

Krzysztof Ficek [1,2], Jolanta Rajca [1,*], Mateusz Stolarz [1,3], Ewa Stodolak-Zych [4], Jarosław Wieczorek [5], Małgorzata Muzalewska [6], Marek Wyleżoł [6], Zygmunt Wróbel [7], Marcin Binkowski [8] and Stanisław Błażewicz [4]

[1] Department of Science, Innovation and Development, Galen-Orthopaedics, 43-150 Bierun, Poland; krzysztof.ficek@galen.pl (K.F.); matstolarz@gmail.com (M.S.)
[2] Department of Physiotherapy, Academy of Physical Education, 40-065 Katowice, Poland
[3] Department of Orthopedics and Traumatology, City Hospital in Zabrze, 41-803 Zabrze, Poland
[4] Department of Biomaterials and Composites, Faculty of Materials Science and Ceramics, AGH University of Science and Technology, 30-059 Krakow, Poland; stodolak@agh.edu.pl (E.S.-Z.); blazew@agh.edu.pl (S.B.)
[5] University Center of Veterinary Medicine UJ-UR, University of Agriculture in Krakow, 30-059 Krakow, Poland; jaroslaw.wieczorek@urk.edu.pl
[6] Institute of Fundamentals of Machinery Design, Faculty of Mechanical Engineering, Silesian University of Technology, 44-100 Gliwice, Poland; muzalewska.malgosia@gmail.com (M.M.); marek.wylezol@polsl.pl (M.W.)
[7] Institute of Biomedical Engineering, Faculty of Science and Technology, University of Silesia, 41-205 Sosnowiec, Poland; zygmunt.wrobel@us.edu.pl
[8] X-ray Microtomography Lab, Department of Computer Biomedical Systems, Institute of Computer Science, Faculty of Computer and Materials Science, University of Silesia, 41-200 Sosnowiec, Poland; binkowski.marcin@gmail.com
* Correspondence: jolanta.rajca@galen.pl

Received: 18 October 2019; Accepted: 25 November 2019; Published: 29 November 2019

Abstract: The exact causes of failure of anterior cruciate ligament (ACL) reconstruction are still unknown. A key to successful ACL reconstruction is the prevention of bone tunnel enlargement (BTE). In this study, a new strategy to improve the outcome of ACL reconstruction was analyzed using a bioresorbable polylactide (PLA) stent as a catalyst for the healing process. The study included 24 sheep with 12 months of age. The animals were randomized to the PLA group (n = 16) and control group (n = 8), subjected to the ACL reconstruction with and without the implantation of the PLA tube, respectively. The sheep were sacrificed 6 or 12 weeks post-procedure, and their knee joints were evaluated by X-ray microcomputed tomography with a 50 μm resolution. While the analysis of tibial and femoral tunnel diameters and volumes demonstrated the presence of BTE in both groups, the enlargement was less evident in the PLA group. Also, the microstructural parameters of the bone adjacent to the tunnels tended to be better in the PLA group. This suggested that the implantation of a bioresorbable PLA tube might facilitate osteointegration of the tendon graft after the ACL reconstruction. The beneficial effects of the stent were likely associated with osteogenic and osteoconductive properties of polylactide.

Keywords: anterior cruciate ligament reconstruction; bone tunnel enlargement; X-ray microtomography; polylactide

---

1. Introduction

The causes of failure of anterior cruciate ligament (ACL) reconstruction are still a matter of debate. While many various ACL reconstruction techniques have been proposed thus far, none of them seems to be optimal and complication-free [1–3]. The failures of ACL reconstruction might be associated with an inappropriate orientation of bone tunnels, use of improper fixation methods and materials, and inadequate rehabilitation, as well as with mechanical behavior of the bone and biological processes that occur during remodeling, maturation, and incorporation of the graft [4,5]. The healing potential of a newly implanted graft is relatively low [6–8] and is primarily determined by conditions within proximity of the bone tunnel and soft tissue of the graft, including the intra-articular environment. Osteointegration of the tendon grafts used for ACL reconstruction is still far from satisfactory, although several strategies have been postulated to improve the process [9–19].

Another critical determinant of successful ACL reconstruction is the prevention of bone tunnel enlargement (BTE), a phenomenon of mechanical and biological etiology. The mechanical causes of BTE might be related to the tunnel drilling technique, graft fixation technique, vibrations at the tunnel entry, and movements of the graft referred to as "bungee effect" and "windshield wiper effect" [20–25]. The biological mechanisms involved in the BTE include accumulation of intra-articular fluid, which penetrates to the space between the graft and the wall of the bone tunnel. The sites in which the graft is not adjacent closely to the bone tunnel wall, the so-called "dead space", are particularly prone to fluid accumulation. The intra-articular fluid that accumulates after the ACL rupture contains proinflammatory cytokines, which are responsible for local osteolysis [26–29]. Another biological mechanism implicated in BTE pathogenesis might be the so-called "synovial bathing effect"; due to its excessive accumulation after ACL reconstruction, synovial fluid might be pressed into the bone tunnel and cause enlargement thereof [23,30,31].

Polylactide (PLA) is a polymer used in surgical practice for bone anastomosis implants (screws, plates, rods). PLA owes its popularity not only to its good biocompatibility (the best among polymers) but also to the ease of implant formation (injection, extrusion, printing). Polylactide is a biodegradable polymer, the durability of which can be determined in vivo by in vitro degradation tests. Degradation of polylactide was a subject of many published studies [32–34]. According to some authors, the degradation of PLA in strongly hydrated environments occurs through the penetration of water molecules and hydrolysis of ester bonds, which results in an increase in the concentration of terminal carboxylic groups [35]. Other researchers have suggested that the PLA polymer chain is autocatalyzed due to acidification of the environment (dissociation effect of carboxyl groups) [36]. The result of material degradation is a decrease in molecular weight and a larger dispersion of average molecular weight. The kinetics of the process depends on the structure of the polymer chain, especially the presence of L or D enantiomer and their ratio [34]. Many methods can be used to control the degradation process in order to reduce the negative effect of PLA autocatalysis. One of them is the addition of modifiers capable of connecting hydronium ions to the polymeric matrix [37], as it is the case with PLA-based nanocomposites modified with ceramic particles. Unfortunately, in the case of new materials, many other parameters need to be tested, and validated in vitro and in vivo tests are required. Therefore, it seems easier to use a minimum amount of pure PLA, e.g., in the form of membranes or highly porous materials, especially when their task is not limited to the transfer of stress. An example of such a solution is presented below. The perforated microscopic material also has pores at the microstructural level, thus reducing the total amount of material used to produce the implant. The presented solutions significantly affect the durability of the material in vitro, which allows approximating the time of degradation in vivo.

In our present study, we analyzed another strategy to improve the outcome of ACL reconstruction, using a bioresorbable polylactide (PLA) stent as a catalyst for the healing process. The stent was produced from poly(L/DL-lactide) 80/20 (80:20 PL/DLA is name of molar ratio the combination of L-lactate (80%) and DL-lactate (20%)) [38–42]. Bioresorbable PLA polymers, in the form of powder, beads, or paste, have already been tested in sheep [39,41], rabbits [42,43], and humans [41,43]. However, applied in a non-solid form, PLA did not provide sufficient mechanical support, which eventually contributed to BTE. Thus, in this study, we verified whether PLA in another form, a tube-shaped perforated stent with a porosity of 45%, improved graft-bone integration after ACL reconstruction. We hypothesized that implantation of the stent could accelerate the bone healing process and prevent the accumulation of non-uniform forces inside the bone tunnel, reducing the risk of BTE. This hypothesis was first verified in a virtual environment, based on finite element analysis of the stress-strain response from the bone, graft, and PLA stent (see: Appendix A, Figures A2–A5). The results constituted the basis for the proper animal experiment, the results of which are described below. The microarchitecture of bone tunnels and adjacent bone was analyzed based on high-resolution, high-quality images obtained during X-ray microtomography (micro-CT). The study involved sheep, as this animal was previously shown to be a good model for orthopedic research [44,45].

2. Materials and Methods

2.1. Animals and Specimen Preparation

The study included 24 male sheep with 12 months of age and body weights ranging between 35 and 40 kg. The protocol of the study was approved by the Animal Ethics Committee at the Institute of Pharmacology, Polish Academy of Sciences in Krakow (decisions no. 820/2011 and 836/2011 of 27 January 2011 and 19 May 2011, respectively). The animals were randomized into two groups: the PLA group (n = 16) and the control group (n = 8). The sheep were kept under standard husbandry conditions (stalls with straw bedding, temperature 10–25 °C, natural light/dark cycle according to the season, with 8 to 16 h lights on), two animals per stall, matched according to the experimental group. The animals were fed with hay (ad libitum) and pasture (0.3 kg/animal/day) and had unlimited access to water. The food was restricted one day before the ACL reconstruction procedure.

The reconstruction procedure was carried out under general anesthesia by inhalation nitrous oxide 25–40% (Linde Gas, Krakow, Poland), isoflurane 0.6–3% (Aerrane 250 mL, Baxter, Warsaw, Poland), and oxygen 30–40% (Medical Oxygen, Linde Gas, Warsaw, Poland). First, the animal's ACL was cut and removed from the joint. Then, tibial and femoral bone tunnels, 4.5 mm in diameter, were drilled using the ACL insertions as the landmarks (Figure 1). The tunnels were drilled in the medial aspect of the proximal tibial metaphysis and distal femoral metaphysis lateral to the condyles. The autograft was harvested from the Achilles tendon. In the control group, the autograft, 4.5 mm in diameter, was pulled through both bone tunnels, and then, its ends were fixed extracortically to the femur and tibia with an EndoButton and button (Figure 1). In the PLA group, PLA tubes (3.5 mm in diameter) were first placed in the tunnels, and then, the autograft, also 3.5 mm in diameter, was inserted into the tubes, so it adhered tightly to their inner walls. Finally, the graft's ends were fixed analogically, as in the control group.

After the procedure, the sheep were placed in stalls and allowed to move freely. The animals were controlled on a daily basis for 6–12 weeks. Twelve sheep, four from the control group and eight from the PLA group, were sacrificed at 6 weeks post-reconstruction, followed by another 12 sheep, four from the control group and eight from the PLA group, at 12 weeks after the procedure. The animals were euthanized with pentobarbital sodium (Morbital, Biovet, Pulawy, Poland) overdose (50–80 mg/kg), and the previously operated knee joints were dissected to enable accurate examination of the bone tunnels.

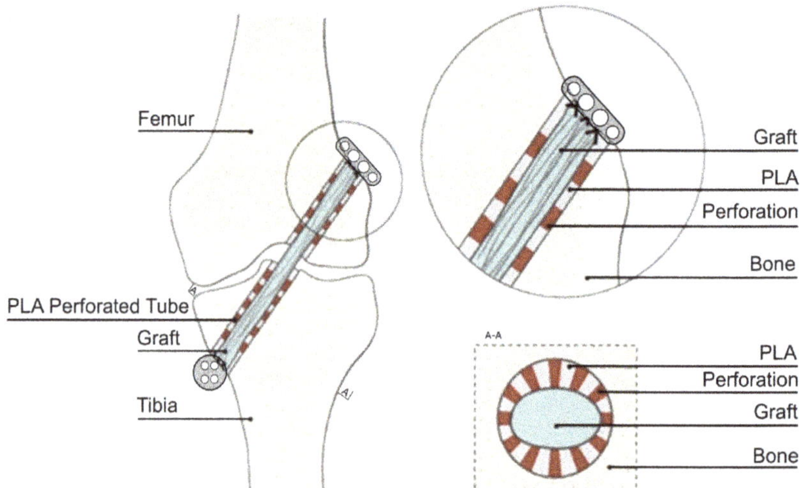

Figure 1. Scheme of the ACL reconstruction, type of graft fixation, and the implantation of polylactide (PLA) perforated tube.

2.2. Polylactide Tubes

The PLA stents were produced from bioresorbable poly(L/DL-lactide) 80/20 (PURAC Biochem, Gorinchem, The Netherlands). The tubes were prepared using a thermal method. Briefly, the polymer pellets were heated to 156 °C, which is close to the melting point for PLA, and shaped by compression in a cylindrical metallic form to obtain a polymer film with approximately 500 µm thickness. Each PLA tube, 3.5 mm in diameter and 25 mm in length, had twenty circular-shaped perforations (0.5 mm in diameter) per square cm (Figure 2). The perforations were made with a laser device. Upon manufacturing, the PLA tubes underwent low-temperature plasma sterilization (H_2O_2, 40 °C).

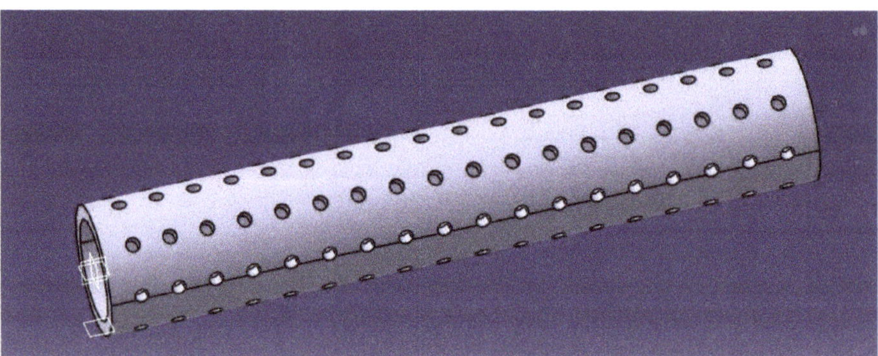

Figure 2. PLA perforated tube.

Detailed methodology of polymer membrane production and the results of functional testing can be found elsewhere [46,47]. Briefly, polymer membranes were produced from a synthetic PL/DLA polymer, using a phase inversion method. The flat membrane used later for the production of the stent was manufactured by casting. A combination of tetrahydrofuran and acetone in a 10:1 ratio was used as a solvent. The polymer was homogenized for 3 h to obtain a homogeneous solution (3% *w/v*). Then, a porogenic agent, dimethylsulfoxide (DMSO), and pure water (UHQ) were added in a 1:1 ratio.

The suspension was homogenized for a few minutes, poured onto Petri plates, dried, and conditioned. Then, the membranes obtained, as described above, were bathed in ethyl alcohol to remove the remains of DMSO. The final porosity of the membrane was 45%. The pore distribution in the membrane was binomodal, proving the existence of two-pore populations: the first one, more numerous with the size of about 26 μm (85%), and the second one with the average size of about 7 μm (10%). These values were determined on the basis of microscopic image analysis obtained with a scanning electron microscope (Nova NanoSEM, FEI, Hillsboro, OR, USA)) (Figure 3). The physicochemical properties of the membranes were determined based on their roughness and wettability tests. The influence of the in vitro environment on the durability of the membrane's material was monitored based on pH alterations in water and phosphate-buffered saline extracts. In line with the standard requirements for degradability testing, the membranes were kept in an incubator for 4 weeks. The molecular weight of the membrane at the end of the experiment was determined with an Ubbelohde viscometer, with norm DIN 51562, SI Analytics, Germany (measuring liquid tethrahydrofurane, $K = 4.85 \times 10^{-4}$ dL/g and $a = 0.68$). The degradability of the membrane was also monitored *in vitro* (three months/H_2O/ 37 °C/5% CO_2 incubation); over the three-month monitoring period, the mean molecular weight of the membrane determined with an Ubbelohde viscometer decreased from 200 to 145 kDa. The pH of the immersion medium decreased slightly during the monitoring period, down to 6.2.

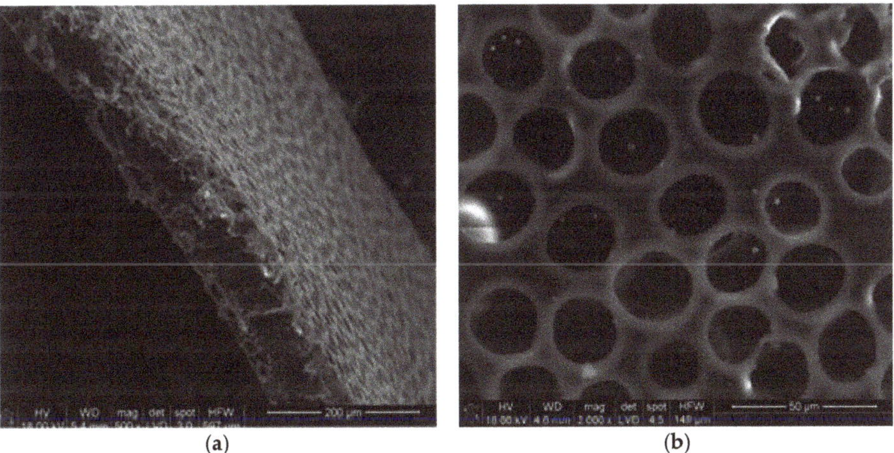

(a)   (b)

Figure 3. The microstructure of the polymer membranes used to form implants in the form of tubes. (**a**) membrane cross-section, (**b**) membrane surface.

2.3. Micro-CT Scanning

The knee joints from 24 sheep were scanned at the X-ray Microtomography Lab (XML), Institute of Computer Science, University of Silesia (Sosnowiec, Poland), using an XMT scanner (v|tome|x s, GE Sensing and Inspection Technologies, Phoenix|x-ray, Wunstorf, Germany). The X-ray images were acquired using a 140 kV voltage, 350 mA current, and 50 μm resolution. To accurately analyze the regeneration of the tibial and femoral tunnels on micro-CT, the images underwent reslicing so that the long central axis of each image corresponded to the exact center of the bone tunnel. An example of reslicing and image transformation, with axial cross-sections perpendicular to the bone tunnel's axis of symmetry, is shown in Figure 4. All advanced image analyses were carried out with ImageJ, an open-source image enumeration software package (US National Institute of Health, Bethesda, MD, USA) [48].

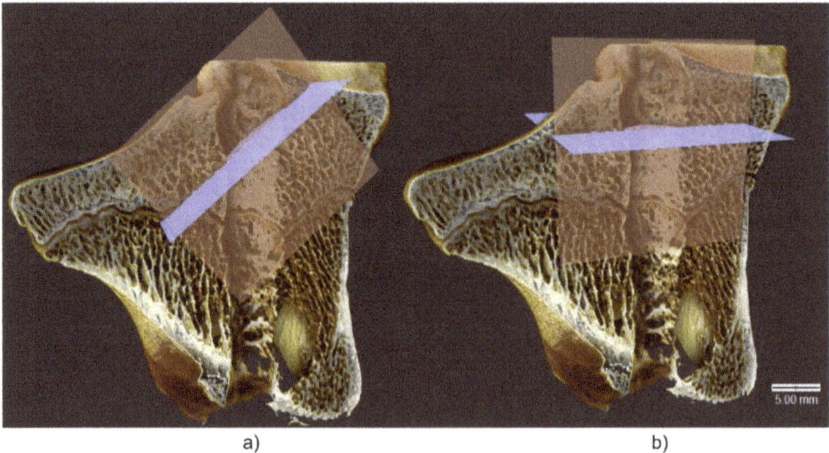

a)   b)

Figure 4. Three-dimensional visualization of the tibia. Axial cross-section—blue, sagittal cross-section—orange; (**a**) before reslicing, (**b**) after reslicing.

The analysis began with the selection of the tunnel's edge on each axial cross-section of the tibia and femur (Figure 5). The boundary between the intra-tunnel space and the bone was chosen manually, to obtain an optimal area in the variable portion of the bone tunnel. To avoid a time-consuming selection of edges on each slice, the operator marked the edges on several slices, and then, the regions of interest (ROIs) were interpolated onto the remaining slices.

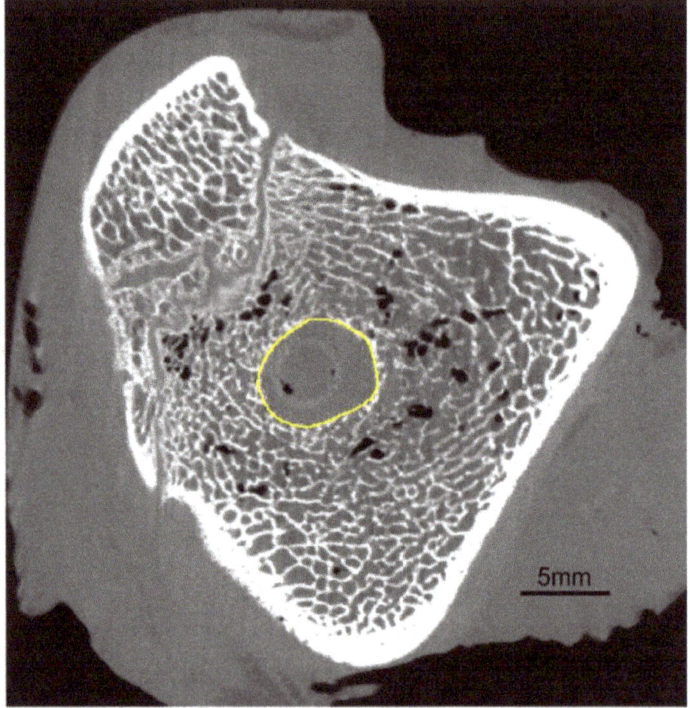

Figure 5. Axial cross-section of the tibia; yellow—a selection of ROI (regions of interest).

2.4. Analysis of the Tunnel Diameter After the ACL Reconstruction

The diameter of the bone tunnel in the axial cross-section (Figure 5) was estimated by computing the diameter of the circle fitted automatically into the selected ROI of each slice, using ImageJ 1.49b software (Wayne Rasband National Institutes of Health, Bethesda, MD, USA) [48]. Then, mean diameters were calculated for three segments of the tunnel: entry, midportion, and exit (Figure 6) using pre-specified ROIs. The three segments of the tunnel were defined by dividing its entire length into three equal parts.

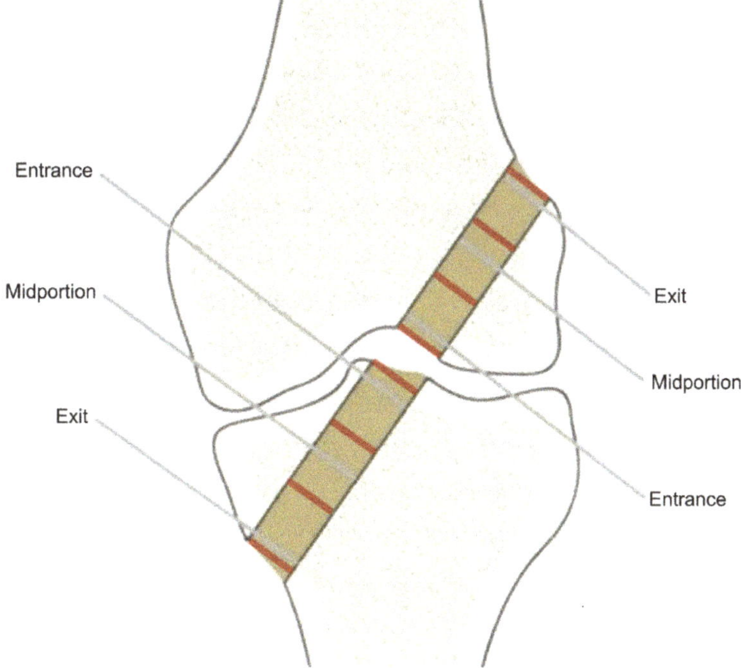

Figure 6. Division of bone tunnels for the analysis.

2.5. Measurement of Tunnel Volume and Determination of Histomorphometric Parameters

To examine the entry of bone tunnel in more detail, its volume and histomorphometric parameters of the adjacent bone were determined. The measurements were taken inside the tunnel, 5 mm from its entry. Tunnel volume was calculated using the previously defined ROIs. Histomorphometric parameters of the adjacent bone, such as bone volume fraction (BV/TV), trabecular thickness (Tb.Th), and connectivity density (Conn.D), were determined based on micro-CT images. To obtain the measurements of trabecular structures, the area inside each ROI (Figure 5) was increased three times (Figure 7a,b). The maximum entropy method, an automated global thresholding method, was proposed for the segmentation of bone samples in our study; in this method, the threshold was chosen based on maximizing the inter-class entropy. Then, different automated algorithms were tested to select the most appropriate method for the segmentation of trabecular bone in sheep. Bone volume fraction (BV/TV), calculated as bone volume divided by total volume, corresponds to the percentage of mineralized bone located within the volume of interest (VOI). Trabecular thickness (Tb.Th) provides information about the thickness of trabecular structures. Connectivity is defined as the maximum number of trabecular connections that can be severed before the structure is split into two separate parts [49]. All three parameters were calculated using the BoneJ plugin [50] within the ImageJ software.

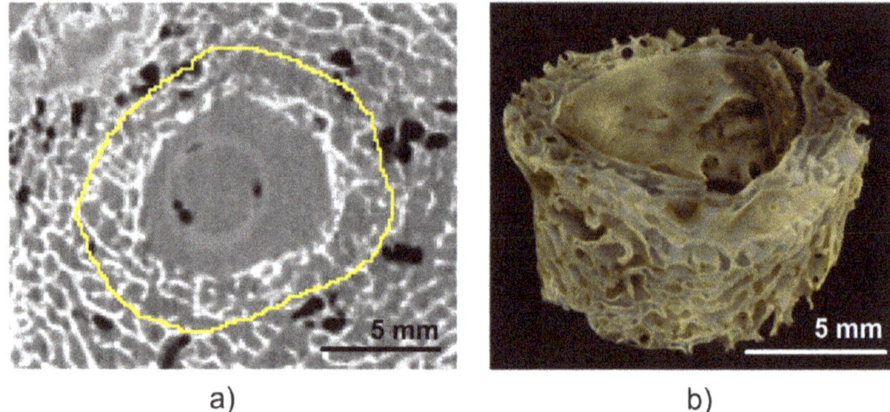

Figure 7. Region of interest for the analysis of histomorphometric parameters. (**a**) axial cross-section with the boundary of the ROI magnified three times—yellow line, (**b**) visualization of the segmented bone tissue.

2.6. Three-Dimensional Visualization

The changes in histomorphometric parameters of adjacent bone were visualized using Drishti open-source software [51]. On three-dimensional images (Figure 8), the bone tunnels were presented in sagittal cross-section, with the extracortical button placed at the bottom left side of the image.

2.7. Statistical Analysis

Normal distribution of the study variables was verified with the Shapiro–Wilk test, and the equality of their variances was checked with Levene's test. The significance of intergroup differences was verified with non-parametric Mann–Whitney U-test. All calculations were carried out with Statistica 10 (StatSoft, Tulsa, OK, USA), with the threshold of statistical significance set at $p < 0.05$.

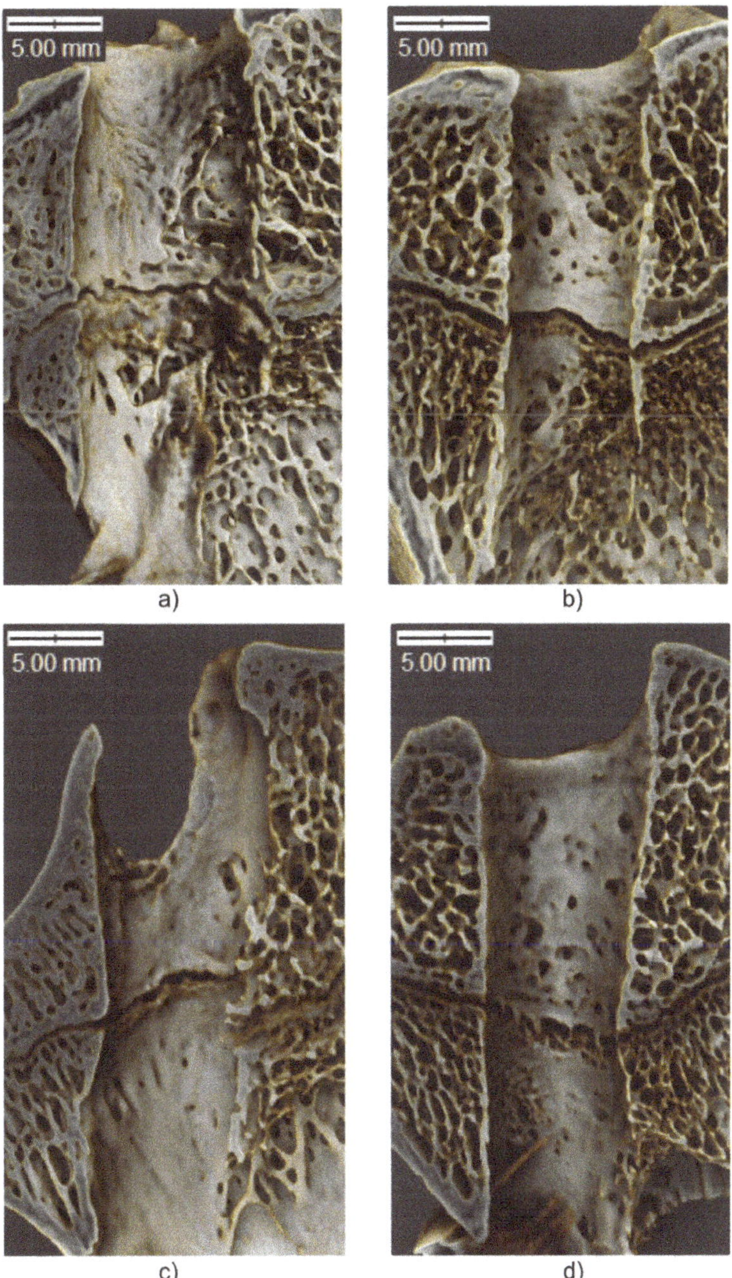

Figure 8. Three-dimensional visualization of the tibial tunnel in sagittal cross-section. (**a**) control at 6 weeks, (**b**) PLA at 6 weeks, (**c**) control at 12 weeks, (**d**) PLA at 12 weeks.

3. Results

3.1. Bone Tunnel Measurements

Diameters of the bone tunnels after various healing time (6 and 12 weeks) are presented in Figure 8 and Table 1. At 6 weeks, the diameters of the tunnels in all segments except the midportion and exit of the femoral tunnel turned out to be significantly smaller in the PLA group than in the controls. At 12 weeks, the statistically significant between-group differences were observed solely for the midportion and exit of the tibial tunnel. While the diameters of most segments at 6 and 12 weeks did not differ significantly in the control group, a significant increase in all three diameters of the femoral tunnel was observed in the PLA group (Figure 9).

Table 1. Morphometric parameters of tibial and femoral tunnels at 6 and 12 weeks post-ACL (anterior cruciate ligament).

Variable	Tibia			Femur		
	6 Weeks	12 Weeks	*p*-Value	6 Weeks	12 Weeks	*p*-Value
Tunnel Entry Diameter						
Controls	8.1 ± 1.3	7.7 ± 1.2	0.665006	8.3 ± 0.9	8.8 ± 1.5	0.665006
PLA group	5.7 ± 0.6	6.8 ± 1.3	0.156255	6.1 ± 0.9	7.6 ± 1.1	0.017673
p-value	0.008475	0.350238		0.008475	0.298618	
Tunnel Midportion Diameter						
Controls	8.1 ± 1.8	8.5 ± 1.5	0.885234	6.2 ± 1.2	8.0 ± 1.4	0.193932
PLA group	5.7 ± 0.7	6.3 ± 1.0	0.270149	5.8 ± 0.5	7.1 ± 0.4	0.003167
p-value	0.021857	0.021857		0.552215	0.219304	
Tunnel Exit Diameter						
Controls	7.4 ± 1.4	9.2 ± 2.0	0.193932	4.4 ± 0.3	6.3 ± 1.7	0.112352
PLA group	5.1 ± 0.8	5.8 ± 0.8	0.103563	5.6 ± 0.5	6.4 ± 0.5	0.032278
p-value	0.008475	0.033753		0.008475	0.924719	
Tunnel Volume						
Controls	277.6 ± 80.3	235.5 ± 75.0	0.470487	257.2 ± 95.5	311.5 ± 104.2	0.470487
PLA group	133.9 ± 31.1	200.4 ± 75.6	0.083124	142.6 ± 44.4	213.8 ± 53.9	0.033161
p-value	0.008475	0.444697		0.033753	0.240956	

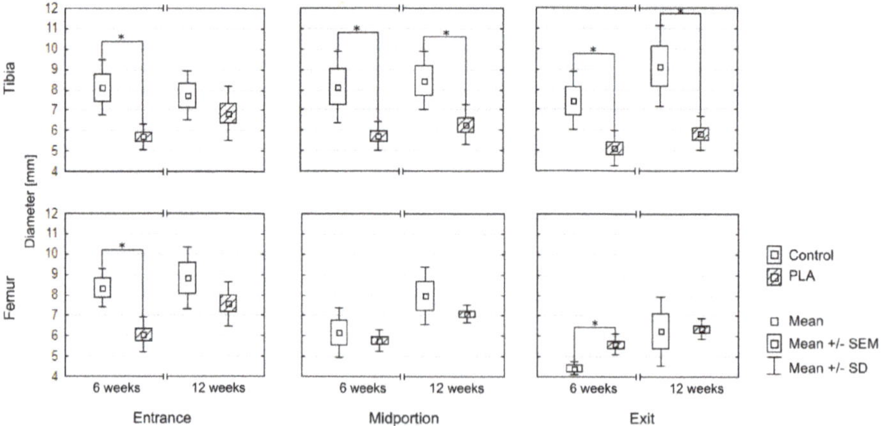

Figure 9. Comparison of bone tunnel diameters in the controls and PLA group; * statistically significant difference at $p < 0.05$.

The results for the tunnel volume were consistent with those for the tunnel diameters (Figure 10, Table 1). In both tibia and femur, mean tunnel volumes at 6 weeks, but not at 12 weeks, were significantly smaller in the PLA group than in the controls. In line with those findings, a significant increase in the femoral tunnel volume and a tendency to increase in the tibial tunnel volume were observed at 12 weeks in the PLA group but not in the control group.

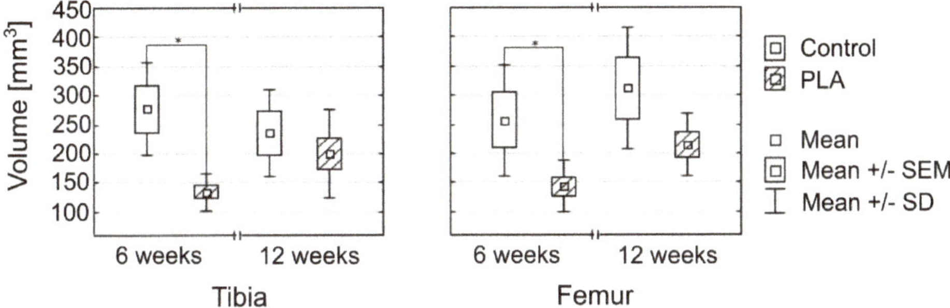

Figure 10. Comparison of bone tunnel volumes in the controls and PLA group at 6 and 12 weeks; * statistically significant difference at $p < 0.05$.

3.2. Bone Microstructure

While no statistically significant differences were found in the microstructure of adjacent bone in the PLA and control group ($p < 0.05$), a tendency to better outcomes was observed in the former. Specifically, BV/TV values in the control group tended to be lower than in the PLA group (Figure 11, Table 2). In both groups, the values of the BV/TV tended to increase with the time elapsed since the reconstruction. However, considerable variance in BV/TV values was observed, especially in the controls, as shown by relatively high standard deviations.

Table 2. Microstructural parameters of tibial and femoral tunnels at 6 and 12 weeks post-ACL.

Variable	Tibia			Femur		
	6 Weeks	12 Weeks	*p*-Value	6 Weeks	12 Weeks	*p*-Value
			BV/TV			
Controls	20.63 ± 8.04	24.98 ± 6.50	0.470487	19.73 ± 4.90	21.90 ± 4.22	0.470487
PLA group	22.30 ± 3.42	26.51 ± 4.81	0.103563	21.15 ± 4.36	23.22 ± 2.73	0.245279
p-value	0.671126	0.552215		0.932324	0.749119	
			Tb.Th			
Controls	0.44 ± 0.16	0.50 ± 0.37	0.665006	0.37±0.10	0.34 ± 0.07	0.885234
PLA group	0.44 ± 0.09	0.45 ± 0.05	0.494837	0.34±0.04	0.35 ± 0.04	0.332922
p-value	0.798907	0.932324		0.671126	0.594033	
			Conn.D			
Controls	1.61 ± 0.81	1.63 ± 0.14	0.885234	1.27 ± 0.44	1.72 ± 0.45	0.312322
PLA group	1.80 ± 0.46	1.57 ± 0.25	0.156255	1.95 ± 0.61	2.05 ± 0.16	0.948533
p-value	0.671126	0.932324		0.074532	0.240956	

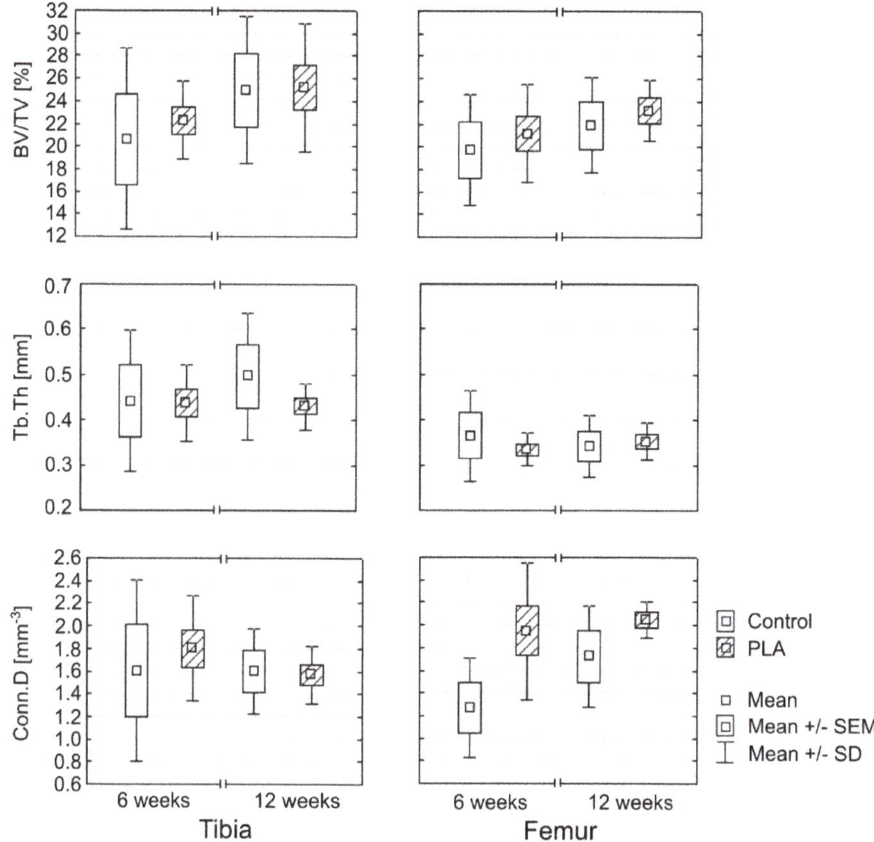

Figure 11. Comparison of histomorphometric parameters.

While in the PLA group, Tb.Th values for the bone adjacent to the tibial and femoral tunnels were essentially the same at 6 and 12 weeks post-reconstruction, higher variance in the trabecular thickness at various study time points was observed in the control group (Table 2).

At 6 weeks after the reconstruction, Conn.D values for the tibial tunnel in the PLA group tended to be higher than in the controls, whereas virtually no difference in Conn.D values for the tibial tunnel in the PLA and control group was observed at 12 weeks. Regardless of the study timepoint, a trend towards a between-group difference in Conn.D values for the femoral tunnel was observed in favor of the PLA group (Table 2).

3.3. Three-Dimensional Visualization

The microstructure of bone remodeling is depicted in Figure 8. As shown in the figure, the edges of the bone tunnel in the PLA group were smooth, and no evidence of a non-uniform tunnel enlargement was observed. Also, the formation of the callus at both study timepoints seemed to be better in the PLA group than in the controls (Figure 8a,b). Regardless of the study group, the remodeling process was more advanced at 12 weeks post-reconstruction, which is consistent with the increase in BV/TV values.

4. Discussion

Early osteointegration of the tendon graft is crucial for successful rehabilitation and return to normal activities of daily living. The aim of this study was to improve the quality of tendon-bone fusion after ACL reconstruction. However, rather than focusing on soft tissues, such as the tendon, or mechanical properties of the tendon-stent-bone system, the study centered around interactions between the stent and the bone, especially bone remodeling after implantation of the stent. In the earlier animal study using a finite element analysis, we demonstrated that placement of the stent between the bone and the graft had a beneficial effect on the stress-strain response (See: Appendix A). The PLA tube was designated to act as a "shock absorber and stabilizer" during the ligament movements, preventing hypermobility of the graft within the bone tunnel. The numerical analysis demonstrated that the insertion of the stent contributed to a more favorable distribution of forces and lesser strain within the bone tunnel and created uniform conditions at the whole tunnel length. Based on those findings, we hypothesized that the implant not only acted as a scaffold for bone cells but might also protect the graft during the implantation and prevent the undesirable tunneling phenomenon. The aim of our present study was to verify this hypothesis in an animal model.

In our present study, we analyzed the influence of the PLA tube on bone tunnel remodeling using high-resolution micro-CT imaging; this enabled us to determine a set of objective parameters, such as tunnel diameter and volume, as well as histomorphometric characteristics of adjacent bone, such as BV/TV, Tb.Th., and Conn.D. The three-dimensionally molded stent was used as a buffer to minimize or eliminate undesired effects associated with post-traumatic or post-operative bone damage. Moreover, the stent was intended to act as a "sealant" filling the space between the tendon graft and bone tunnel wall, and hence, preventing penetration of synovial fluid into the tunnel. This would minimize the unfavorable effects of proinflammatory cytokines contained in the synovial fluid on bone remodeling, especially the BTE phenomenon.

We analyzed the diameters of the bone tunnels as a marker of potential BTE and a measure of the PLA tube effect on bone healing. Our findings suggested that the use of the PLA implant might not prevent the BTE completely, as the diameters of the femoral tunnel increased with the time elapsed since the reconstruction procedure. However, it needs to be stressed that no significant changes in the diameters of the tibial tunnel were observed over time in the PLA group. BTE might be a consequence of too early load of the operated limb, as the study animals did not undergo any controlled post-procedure rehabilitation, and the only factor limiting their activity was postoperative pain. However, it should be stressed that according to some rehabilitation protocols, human patients after ACL reconstruction are expected to weight-bear the operated extremity early after the procedure and to walk without crutches or braces. Nevertheless, our study showed that implantation of the PLA tube prevented the excessive widening of the bone tunnels.

Furthermore, a comparative analysis of femoral tunnel diameters in three segments (entry, midportion, and exit) demonstrated that in the control group, the tunnel tended to be wider at its entry. This might be a consequence of synovial fluid penetration to this segment of the tunnel and the detrimental effect of proinflammatory cytokines contained in the fluid on bone remodeling. In turn, the diameters of the three segments of the femoral tunnel in the PLA group were essentially the same, implying that insertion of the PLA tube might prevent the excessive inflow of the synovial fluid and minimize the harmful effects of proinflammatory cytokines.

To provide a better insight into the role of the PLA stent in bone remodeling after ACL reconstruction, we also analyzed the microstructure of adjacent bone. While the PLA group and the controls did not differ significantly in terms of their BV/TV values, the analysis of the latter in conjunction with Conn.D suggested that the use of the PLA stent might promote the faster formation of new bone. An increase in trabecular thickness and connectivity constitutes a response to mechanical overload [52]. After insertion of the PLA tube into the bone tunnel, some mechanical loads might be absorbed by the stent, rather than being directly delivered to the bone. Due to the resultant lesser mechanical load, the trabecular thickness in the PLA group did not increase significantly over time and was essentially the same at both 6 and 12 weeks post-reconstruction. This implied that the PLA stent prevented a non-uniform distribution of forces within the bone tunnel, creating a more stable environment for new bone formation. In contrast, variance in the trabecular thickness in the control group increased considerably with the time elapsed since the reconstruction procedure. The beneficial effect of the PLA stent on the remodeling of the adjacent bone was also confirmed by a tendency to higher connectivity density, especially around the femoral tunnel. Higher connectivity density reflected a larger number of connections between the trabeculae in the PLA group. Furthermore, the porosity of 45% and perforations of the stent could promote the accumulation of biological material from the bone marrow blood in the form of incubation over the entire surface of the stent and support bone tissue overgrowth.

Strengths and Limitations

Our present study has several strengths and limitations. We proposed a new, accurate method to analyze bone remodeling after ACL reconstruction. Specifically, we resliced high-resolution micro-CT images to obtain accurate measurements in axial cross-sections. The results of previous studies, in which the images were not adjusted by reslicing [53,54], might have been biased due to a slight axial displacement of the examined cross-sections. In our opinion, this could have a considerable impact on the results, as the measurements of cross-sections that are not perfectly perpendicular might be slightly inaccurate. To improve the reproducibility of the measurements, we applied automatic thresholding (maximum entropy method); as a result, segmentation of the images was independent of the observer, making the results more reliable. One potential limitation of this study might be a too small number of examined sheep and a disproportion in the number of animals included in the PLA and control group (16 and 8, respectively). However, even in such a small sample, we found significant intergroup differences, confirming the beneficial effect of the PLA tube on bone healing. Another potential limitation of this study might be a relatively short follow-up. According to literature, full integration of the graft after ACL reconstruction takes longer than 12 weeks [55,56], and an increase in bone tunnel diameter and BTE phenomenon might be observed even up to 12 months post-reconstruction [57,58]. Nevertheless, the aim of our study was to analyze the changes that occurred early after the reconstruction rather than the long-term outcomes of the procedure. Finally, interpreting the hereby presented findings, one should remember that the sheep included in this study weight-bore their operated limbs as soon as they recovered from anesthesia, whereas human patients after ACL reconstruction are mobilized gradually and fully weight-bear their extremities even several weeks after the procedure.

5. Conclusions

The results of this study suggested that the application of a bioresorbable PLA tube might facilitate osteointegration of the tendon graft after the ACL reconstruction. The use of the bioresorbable PLA stent might partially prevent BTE and stimulate new bone formation, as shown by changes in BV/TV and Conn.D values. The beneficial effects of the PLA tube were likely associated with osteogenic and osteoconductive properties of polylactide. Our findings implied that the PLA tube might positively affect the quality of the graft-bone interface within the tibial and femoral tunnels, eventually resulting in better osteointegration of the ACL graft.

Author Contributions: Conceptualization, K.F.; methodology, K.F., S.B., and E.S.-Z.; validation, K.F., M.B., and Z.W.; formal analysis, M.S., J.R., M.M., and M.W.; investigation, K.F., M.S., J.R., and E.S.-Z.; resources, K.F., J.R., and J.W.; data curation, J.R. and M.B.; writing—original draft preparation, K.F. and M.S.; visualization, M.S. and J.R.; project administration, K.F. and J.R.

Funding: This research received no external funding.

Acknowledgments: The authors would like to express their gratitude to Dr. Szymon Bruzewicz (SciencePro, http://www.sciencepro.co.kr) for this linguistic and editorial assistance in the preparation of the manuscript.

Conflicts of Interest: The authors declare no conflict of interest.

Appendix A

Appendix A.1 Construction of a Three-Dimensional Finite Element Model

The geometry of sheep femoral bone was modeled by the use of the advanced integrated CAx system (CATIA v5). During the numerical experiment, the levels of stress and strain were calculated for the two configurations: 1) cortical bone-cancellous bone stent-graft, and 2) cortical bone-cancellous bone graft.

To perform numerical analysis, a model of sheep posterior femur was created and used to simulate the postoperative load of the lower limb. The three-dimensional model, including both cortical bone (approximately 2.4 mm thick) and cancellous bone, was designed with CATIA v5 software. The model had a bone tunnel for graft implantation (with 4.5 mm in diameter, which corresponded to the maximum outer diameter of the stent used in the proper experiment). The model was cut, so the cortical bone was exposed, and the force corresponding to 1/4 of the sheep's body weight was applied, with a resultant load of 100 N. During the experiment, the model was fixed on its medial and lateral condyle (Figure A1).

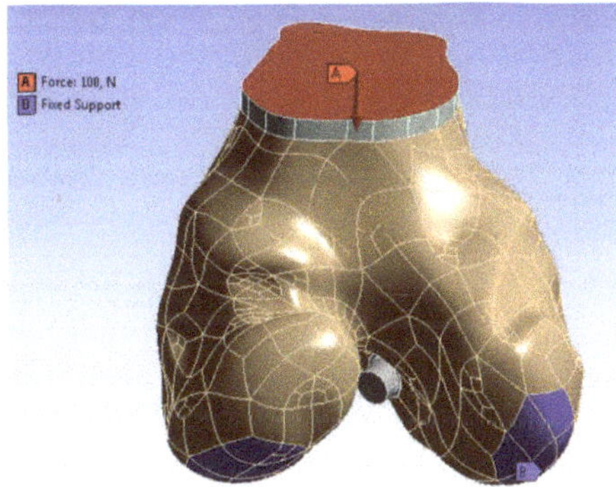

Figure A1. Application of force and fixed support to the femur model.

Material properties of the blood, cortical, and cancellous bone tissue used during the numerical study of fluid flow are summarized in Table A1. The calculations were carried out for the following mechanical properties of polylactide (LATI Latigea B01 F1 Bio-Polymer): density = 1.26 g/cm^3, tensile strength = 55 MPa, modulus of elasticity = 3 GPa.

Table A1. Material properties of the blood, cortical, and cancellous bone.

Material Properties	Blood	Cortical Bone	Cancellous Bone
Density (kg/cm^3)	1060	-	-
Proper heat (J/kg*K)	1006.43	-	-
Thermal conductivity (W/m*K)	0.0242	-	-
Viscosity (kg/m*s)	0.003	-	-
Input speed (mm/s)	1	-	-
Young's modulus (MPA)	-	17,000	300
Density (g/cm^3)	-	1.9	0.46
Poisson's ratio	-	0.3	0.3
Compressive strength (MPa)	-	200	6

Appendix A.2 Stress-Strain Analysis

The numerical simulation demonstrated that the stress was transmitted primarily through the cortical layer of the bone, which is consistent with the real-life conditions (Figure A2). The levels of stress in the graft and stent were small, no greater than several tens of Pa. Hence, to better visualize the stress levels, the scale was changed, so the red color corresponded to >100 Pa tension. The stress-strain analysis showed that implantation of the stent had a particular effect on stress distribution within the cortical bone and stent/graft interface (Figure A3). The analysis demonstrated that the distribution of the strain within the graft was more favorable after implantation of the stent (Figures A4 and A5). This implied that implantation of the stent was not associated with additional stress for the graft. Theoretically, a decrease in the strain at the bone/graft interface observed after implantation of the stent should contribute to better ingrowth of the graft and reduce the likelihood of tunnel widening.

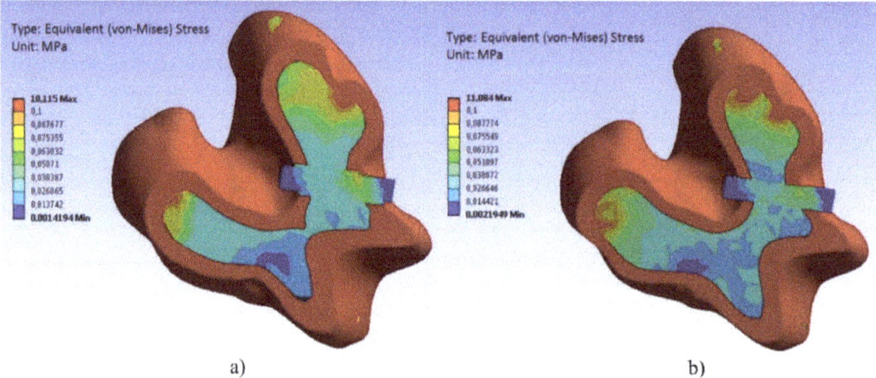

a)   b)

Figure A2. Stress distribution within the femur: (**a**) with a stent, (**b**) without a stent.

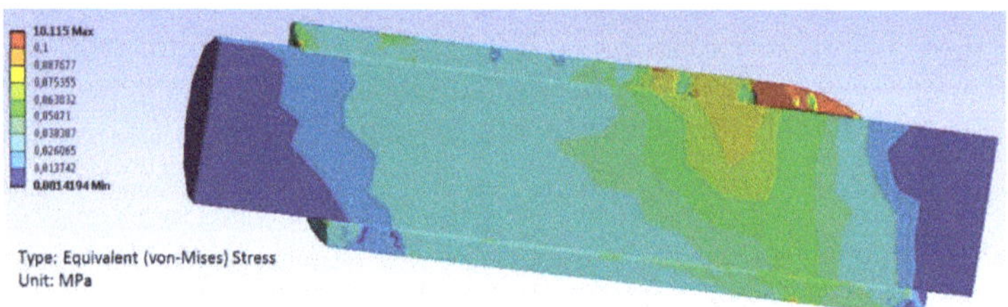

Figure A3. Stress distribution within the graft and stent.

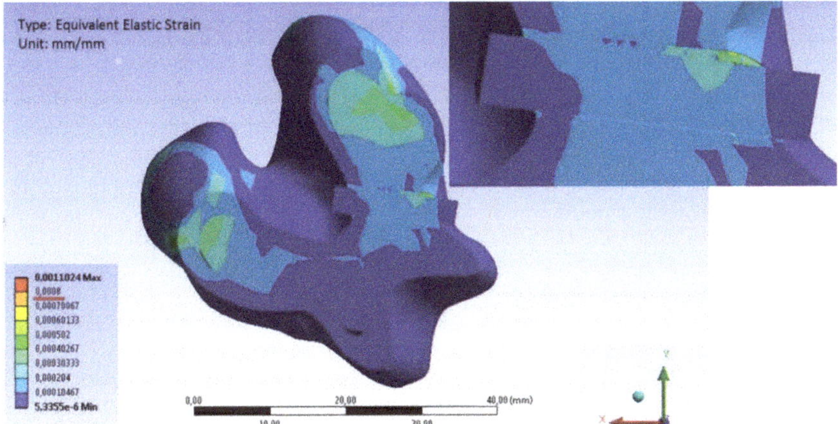

Figure A4. Strain distribution within the femur with the stent.

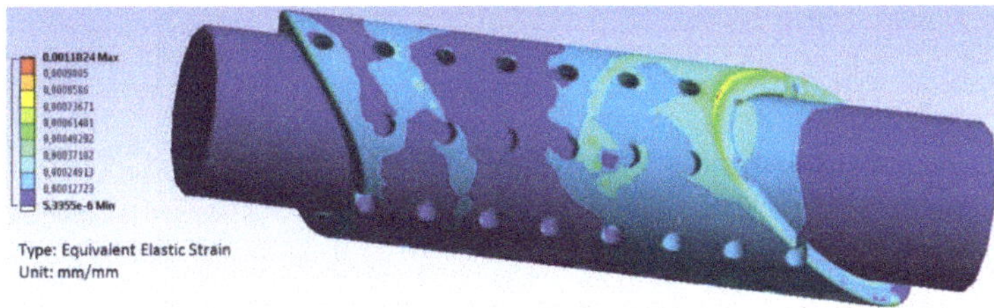

Figure A5. Strain distribution within the graft and stent.

References

1. Kuskucu, S.M. Comparison of short-term results of bone tunnel enlargement between EndoButton CL and cross-pin fixation systems after chronic anterior cruciate ligament reconstruction with autologous quadrupled hamstring tendons. *J. Int. Med. Res.* **2008**, *36*, 23–30. [CrossRef]
2. Moisala, A.S.; Järvela, T.; Paakkala, A.; Paakkala, T.; Kannus, P.; Järvinen, M. Comparison of the bioabsorbable and metal screw fixation after ACL reconstruction with a hamstring autograft in MRI and clinical outcome: A prospective randomized study. *Knee Surg. Sports Traumatol. Arthrosc.* **2008**, *16*, 1080–1086. [CrossRef] [PubMed]
3. Siebold, R.; Kiss, Z.S.; Morris, H.G. Effect of compaction drilling during ACL reconstruction with hamstrings on postoperative tunnel widening. *Arch. Orthop. Trauma Surg.* **2008**, *128*, 461–468. [CrossRef] [PubMed]
4. Jaureguito, J.W.; Paulos, L.E. Why grafts fail. *Clin. Orthop. Relat. Res.* **1996**, *325*, 25–41. [CrossRef] [PubMed]
5. Meyers, A.B.; Haims, A.H.; Menn, K.; Moukaddam, H. Imaging of anterior cruciate ligament repair and its complications. *AJR Am. J. Roentgenol.* **2010**, *194*, 476–484. [CrossRef] [PubMed]
6. Beynnon, B.D.; Johnson, R.J.; Abate, J.A.; Fleming, B.C.; Nichols, C.E. Treatment of anterior cruciate ligament injuries, part 2. *Am. J. Sports Med.* **2005**, *33*, 1751–1767. [CrossRef]
7. Beynnon, B.D.; Johnson, R.J.; Abate, J.A.; Fleming, B.C.; Nichols, C.E. Treatment of anterior cruciate ligament injuries, part I. *Am. J. Sports Med.* **2005**, *33*, 1579–1602. [CrossRef]
8. Sundar, S.; Pendegrass, C.J.; Blunn, G.W. Tendon bone healing can be enhanced by demineralized bone matrix: A functional and histological study. *J. Biomed. Mater. Res. B Appl. Biomater.* **2009**, *88*, 115–122. [CrossRef]
9. Chen, C.H. Strategies to enhance tendon graft-bone healing in anterior cruciate ligament reconstruction. *Chang Gung Med. J.* **2009**, *32*, 483–493.

10. Chen, C.H.; Chang, C.H.; Su, C.I.; Wang, K.C.; Liu, H.T.; Yu, C.M.; Wong, C.B.; Wang, I.C. Arthroscopic single-bundle anterior cruciate ligament reconstruction with periosteum-enveloping hamstring tendon graft: Clinical outcome at 2 to 7 years. *Arthroscopy* **2010**, *26*, 907–917. [CrossRef]

11. Chen, C.H.; Chen, W.J.; Shih, C.H.; Chou, S.W. Arthroscopic anterior cruciate ligament reconstruction with periosteum-enveloping hamstring tendon graft. *Knee Surg. Sports Traumatol. Arthrosc.* **2004**, *12*, 398–405. [CrossRef] [PubMed]

12. Karaoglu, S.; Celik, C.; Korkusuz, P. The effects of bone marrow or periosteum on tendon-to-bone tunnel healing in a rabbit model. *Knee Surg. Sports Traumatol. Arthrosc.* **2009**, *17*, 170–178. [CrossRef] [PubMed]

13. Rodeo, S.A.; Suzuki, K.; Deng, X.H.; Wozney, J.; Warren, R.F. Use of recombinant human bone morphogenetic protein-2 to enhance tendon healing in a bone tunnel. *Am. J. Sports Med.* **1999**, *27*, 476–488. [CrossRef]

14. Baxter, F.R.; Bach, J.S.; Detrez, F.; Cantournet, S.; Corté, L.; Cherkaoui, M.; Ku, D.N. Augmentation of bone tunnel healing in anterior cruciate ligament grafts: Application of calcium phosphates and other materials. *J. Tissue Eng.* **2010**, *2010*, 712370. [CrossRef] [PubMed]

15. Soon, M.Y.; Hassan, A.; Hui, J.H.; Goh, J.C.; Lee, E.H. An analysis of soft tissue allograft anterior cruciate ligament reconstruction in a rabbit model: A short-term study of the use of mesenchymal stem cells to enhance tendon osteointegration. *Am. J. Sports Med.* **2007**, *35*, 962–971. [CrossRef] [PubMed]

16. Yeh, W.L.; Lin, S.S.; Yuan, L.J.; Lee, K.F.; Lee, M.Y.; Ueng, S.W. Effects of hyperbaric oxygen treatment on tendon graft and tendon-bone integration in bone tunnel: Biochemical and histological analysis in rabbits. *J. Orthop. Res.* **2007**, *25*, 636–645. [CrossRef]

17. Yamazaki, S.; Yasuda, K.; Tomita, F.; Tohyama, H.; Minami, A. The effect of transforming growth factor-beta1 on intraosseous healing of flexor tendon autograft replacement of anterior cruciate ligament in dogs. *Arthroscopy* **2005**, *21*, 1034–1041. [CrossRef]

18. Sasaki, K.; Kuroda, R.; Ishida, K.; Kubo, S.; Matsumoto, T.; Mifune, Y.; Kinoshita, K.; Tei, K.; Akisue, T.; Tabata, Y.; et al. Enhancement of tendon-bone osteointegration of anterior cruciate ligament graft using granulocyte colony-stimulating factor. *Am. J. Sports Med.* **2008**, *36*, 1519–1527. [CrossRef]

19. Darabos, N.; Haspl, M.; Moser, C.; Darabos, A.; Bartolek, D.; Groenemeyer, D. Intraarticular application of autologous conditioned serum (ACS) reduces bone tunnel widening after ACL reconstructive surgery in a randomized controlled trial. *Knee Surg. Sports Traumatol. Arthrosc.* **2011**, *19* (Suppl. S1), S36–S46. [CrossRef]

20. Gokce, A.; Beyzadeoglu, T.; Ozyer, F.; Bekler, H.; Erdogan, F. Does bone impaction technique reduce tunnel enlargement in ACL reconstruction? *Int. Orthop.* **2009**, *33*, 407–412. [CrossRef]

21. Höher, J.; Livesay, G.A.; Ma, C.B.; Withrow, J.D.; Fu, F.H.; Woo, S.L. Hamstring graft motion in the femoral bone tunnel when using titanium button/polyester tape fixation. *Knee Surg. Sports Traumatol. Arthrosc.* **1999**, *7*, 215–219. [CrossRef] [PubMed]

22. Höher, J.; Scheffler, S.U.; Withrow, J.D.; Livesay, G.A.; Debski, R.E.; Fu, F.H.; Woo, S.L. Mechanical behavior of two hamstring graft constructs for reconstruction of the anterior cruciate ligament. *J. Orthop. Res.* **2000**, *18*, 456–461. [CrossRef] [PubMed]

23. Iorio, R.; Vadalà, A.; Argento, G.; Di Sanzo, V.; Ferretti, A. Bone tunnel enlargement after ACL reconstruction using autologous hamstring tendons: A CT study. *Int. Orthop.* **2007**, *31*, 49–55. [CrossRef] [PubMed]

24. Jagodzinski, M.; Foerstemann, T.; Mall, G.; Krettek, C.; Bosch, U.; Paessler, H.H. Analysis of forces of ACL reconstructions at the tunnel entrance: Is tunnel enlargement a biomechanical problem? *J. Biomech.* **2005**, *38*, 23–31. [CrossRef]

25. L'Insalata, J.C.; Klatt, B.; Fu, F.H.; Harner, C.D. Tunnel expansion following anterior cruciate ligament reconstruction: A comparison of hamstring and patellar tendon autografts. *Knee Surg. Sports Traumatol. Arthrosc.* **1997**, *5*, 234–238. [CrossRef]

26. Agarwal, S. Osteolysis—Basic science, incidence and diagnosis. *Curr. Orthop.* **2004**, *18*, 220–231. [CrossRef]

27. Irie, K.; Uchiyama, E.; Iwaso, H. Intraarticular inflammatory cytokines in acute anterior cruciate ligament injured knee. *Knee* **2003**, *10*, 93–96. [CrossRef]

28. Zysk, S.P.; Fraunberger, P.; Veihelmann, A.; Dörger, M.; Kalteis, T.; Maier, M.; Pellengahr, C.; Refior, H.J. Tunnel enlargement and changes in synovial fluid cytokine profile following anterior cruciate ligament reconstruction with patellar tendon and hamstring tendon autografts. *Knee Surg. Sports Traumatol. Arthrosc.* **2004**, *12*, 98–103. [CrossRef]

29. Darabos, N.; Hundric-Haspl, Z.; Haspl, M.; Markotic, A.; Darabos, A.; Moser, C. Correlation between synovial fluid and serum IL-1beta levels after ACL surgery-preliminary report. *Int. Orthop.* **2009**, *33*, 413–418. [CrossRef]

30. Clatworthy, M.G.; Annear, P.; Bulow, J.U.; Bartlett, R.J. Tunnel widening in anterior cruciate ligament reconstruction: A prospective evaluation of hamstring and patella tendon grafts. *Knee Surg. Sports Traumatol. Arthrosc.* **1999**, *7*, 138–145. [CrossRef]

31. Iorio, R.; Vadalà, A.; Di Vavo, I.; De Carli, A.; Conteduca, F.; Argento, G.; Ferretti, A. Tunnel enlargement after anterior cruciate ligament reconstruction in patients with post-operative septic arthritis. *Knee Surg. Sports Traumatol. Arthrosc.* **2008**, *16*, 921–927. [CrossRef] [PubMed]

32. Södergård, A.; Stolt, M. Industrial Production of High Molecular Weight Poly (Lactic Acid). In *Poly (Lactic Acid). Synthesis, Structures, Properties, Processing, and Application*; Auras, R., Lim, L.T., Selke, S.E.M., Tsuji, H., Eds.; John Wiley & Sons: Hoboken, NJ, USA, 2010; pp. 27–41.

33. Li, S.M.; Garreau, H.; Vert, M. Structure Property Relationships in the Case of the Degradation of Massive Aliphatic Poly-(Alpha-Hydroxy Acids) in Aqueous-Media. *J. Mater. Sci. Mater. Med.* **1990**, *1*, 123–130. [CrossRef]

34. Gorrasi, G.; Pantani, R. Effect of PLA grades and morphologies on hydrolytic degradation at composting temperature: Assessment of structural modification and kinetic parameters. *Polym. Degrad. Stab.* **2013**, *98*, 1006–1014. [CrossRef]

35. Hakim, R.H.; Cailloux, J.; Santana, O.O.; Bou, J.; Sánchez-Soto, M.; Odent, J.; Raquez, J.M.; Dubois, P.; Carrasco, F.; Maspoch, M.L. PLA/SiO$_2$ composites: Influence of the filler modifications on the morphology, crystallization behavior, and mechanical properties. *J. Appl. Polym. Sci.* **2017**, *134*, 45367. [CrossRef]

36. Gleadall, A.; Pan, J.; Kruft, M.A.; Kellomäki, M. Degradation mechanisms of bioresorbable polyesters. Part 1. Effects of random scission, end scission and autocatalysis. *Acta Biomater.* **2014**, *10*, 2223–2232. [CrossRef]

37. Rapacz-Kmita, K.; Stodolak-Zych, E.; Szaraniec, B.; Gajek, M.; Dudek, P. Effect of clay mineral on the accelerated hydrolytic degradation of polylactide in the polymer/clay nanocomposites. *Mater. Lett.* **2015**, *146*, 73–76. [CrossRef]

38. Ficek, K.; Stodolak, E.; Tomczak, A.; Stolarz, M. Bioresorbable Polylactide Implant used in orthopaedic surgery—Case report. *PrzypadkiMedyczne.pl* **2012**, *22*, 84–88.

39. Gugala, Z.; Gogolewski, S. The in vitro growth and activity of sheep osteoblasts on three-dimensional scaffolds from poly(L/DL-lactide) 80/20%. *J. Biomed. Mater. Res. A* **2005**, *75*, 702–709. [CrossRef]

40. Gugala, Z.; Gogolewski, S. Differentiation, growth and activity of rat bone marrow stromal cells on resorbable poly(L/DL-lactide) membranes. *Biomaterials* **2004**, *25*, 2299–2307. [CrossRef]

41. Gugala, Z.; Lindsey, R.W.; Gogolewski, S. New approaches in the treatment of critical-size segmental defects in long bones. *Macromol. Symp.* **2007**, *253*, 147–161. [CrossRef]

42. Leiggener, C.S.; Curtis, R.; Müller, A.A.; Pfluger, D.; Gogolewski, S.; Rahn, B.A. Influence of copolymer composition of polylactide implants on cranial bone regeneration. *Biomaterials* **2006**, *27*, 202–207. [CrossRef] [PubMed]

43. Ficek, K.; Wieczorek, J.; Stodolak-Zych, E.; Kosenyuk, Y. A revised surgical concept of anterior cruciate ligament replacement in a rabbit model. *Eng. Biomater.* **2012**, *113*, 33–34.

44. Nuss, K.M.; Auer, J.A.; Boos, A.; von Rechenberg, B. An animal model in sheep for biocompatibility testing of biomaterials in cancellous bones. *BMC Musculoskelet. Disord.* **2006**, *7*, 67. [CrossRef] [PubMed]

45. Turner, A.S. The sheep as a model for osteoporosis in humans. *Vet. J.* **2002**, *163*, 232–239. [CrossRef] [PubMed]

46. Stodolak-Zych, E.; Szumera, M.; Błażewicz, M. Osteoconductive nanocomposite materials for bone regeneration. *Mater. Sci. Forum* **2013**, *730–732*, 38–43. [CrossRef]

47. Stodolak-Zych, E.; Łuszcz, A.; Menaszek, E.; Ścisłowska-Czarencka, A. Resorbable polymer membranes for medical applications. *J. Biomim. Biomater. Tissue Eng.* **2014**, *19*, 99–108. [CrossRef]

48. Schneider, C.A.; Rasband, W.S.; Eliceiri, K.W. NIH Image to ImageJ: 25 years of image analysis. *Nat. Methods* **2012**, *9*, 671–675. [CrossRef]

49. Skyscan, N.V. *Structural Parameters Measured by the Skyscan CT-Analyser Software*; Yumpu: Diepoldsau, Switzerland, 1987; pp. 1–15.

50. Doube, M.; Kłosowski, M.M.; Arganda-Carreras, I.; Cordelières, F.P.; Dougherty, R.P.; Jackson, J.S.; Schmid, B.; Hutchinson, J.R.; Shefelbine, S.J. BoneJ: Free and extensible bone image analysis in ImageJ. *Bone* **2010**, *47*, 1076–1079. [CrossRef]

51. Limaye, A. Drishti: A volume exploration and presentation tool. In *Developments in X-ray Tomography VIII, Proceedings of the SPIE Optical Engineering + Applications, San Diego, CA, USA, 12–16 August 2012*; Stock, S.R., Ed.; Society of Photo-Optical Instrumentation Engineers: Bellingham, WA, USA, 2012.

52. Ruimerman, R.; Hilbers, P.; van Rietbergen, B.; Huiskes, R. A theoretical framework for strain-related trabecular bone maintenance and adaptation. *J. Biomech.* **2005**, *38*, 931–941. [CrossRef]

53. Iorio, R.; Di Sanzo, V.; Vadalà, A.; Conteduca, J.; Mazza, D.; Redler, A.; Bolle, G.; Conteduca, F.; Ferretti, A. ACL reconstruction with hamstrings: How different technique and fixation devices influence bone tunnel enlargement. *Eur. Rev. Med. Pharmacol. Sci.* **2013**, *17*, 2956–2961.

54. Silva, A.; Sampaio, R.; Pinto, E. Femoral tunnel enlargement after anatomic ACL reconstruction: A biological problem? *Knee Surg. Sports Traumatol. Arthrosc.* **2010**, *18*, 1189–1194. [CrossRef] [PubMed]

55. Weiler, A.; Hoffmann, R.F.; Bail, H.J.; Rehm, O.; Südkamp, N.P. Tendon healing in a bone tunnel. Part II: Histologic analysis after biodegradable interference fit fixation in a model of anterior cruciate ligament reconstruction in sheep. *Arthroscopy* **2002**, *18*, 124–135. [CrossRef] [PubMed]

56. Weiler, A.; Peine, R.; Pashmineh-Azar, A.; Abel, C.; Südkamp, N.P.; Hoffmann, R.F. Tendon healing in a bone tunnel. Part I: Biomechanical results after biodegradable interference fit fixation in a model of anterior cruciate ligament reconstruction in sheep. *Arthroscopy* **2002**, *18*, 113–123. [CrossRef] [PubMed]

57. Buelow, J.U.; Siebold, R.; Ellermann, A. A prospective evaluation of tunnel enlargement in anterior cruciate ligament reconstruction with hamstrings: Extracortical versus anatomical fixation. *Knee Surg. Sports Traumatol. Arthrosc.* **2002**, *10*, 80–85. [CrossRef] [PubMed]

58. Vadalà, A.; Iorio, R.; De Carli, A.; Argento, G.; Di Sanzo, V.; Conteduca, F.; Ferretti, A. The effect of accelerated, brace free, rehabilitation on bone tunnel enlargement after ACL reconstruction using hamstring tendons: A CT study. *Knee Surg. Sports Traumatol. Arthrosc.* **2007**, *15*, 365–371. [CrossRef]

 *polymers*

Article

The Influence of Mucin-Based Artificial Saliva on Properties of Polycaprolactone and Polylactide

Dawid Łysik [1,*], Joanna Mystkowska [1,*], Grzegorz Markiewicz [1], Piotr Deptuła [2] and Robert Bucki [2]

[1]   Institute of Biomedical Engineering, Bialystok University of Technology, Wiejska 45C, 15-351 Bialystok, Poland; g.markiewicz@doktoranci.pb.edu.pl
[2]   Department of Microbiological and Nanobiomedical Engineering, Medical University of Bialystok, Mickiewicza 2C, 15-222 Bialystok, Poland; piotr.deptula@umb.edu.pl (P.D.); buckirobert@gmail.com (R.B.)
*   Correspondence: d.lysik@pb.edu.pl (D.Ł.); j.mystkowska@pb.edu.pl (J.M.)

Received: 25 October 2019; Accepted: 13 November 2019; Published: 14 November 2019

Abstract: Polycaprolactone (PCL) and polylactide (PLA) are the two most common biodegradable polymers with potential use in oral applications. Both polymers undergo mainly slow hydrolytic degradation in the human body. However, specific conditions of the oral cavity, like elevated temperature, low pH, and presence of saliva affect the rate of hydrolysis. The study examined the properties of solid samples of PCL and PLA subjected to degradation in phosphate buffered saline (PBS) and artificial saliva (AS) at temperatures of 37 or 42 °C, and pH values 2 or 7.4. A number of tests were performed, including measurement of the degree of swelling, weight loss, molecular weight, differential scanning calorimetry, and thermogravimetry of polymers, as well as hardness and tensile strength. Additionally, topography and stiffness of surfaces using atomic force microscopy are presented. It has been noticed that in the artificial saliva, the processes of polymer degradation occur slightly more slowly, and the effects of temperature and pH are less pronounced. We believe that a layer of porcine gastric mucin from artificial saliva that adsorbed on the surface of polymers may have a key role in the observed differences; this layer resembles protective mucin coating tissues in the oral cavity.

Keywords: degradation; saliva; mechanical properties; molecular weight; thermal properties; activation energy of thermal decomposition

---

1. Introduction

Biomaterials are a wide group of materials used to evaluate, treat or replace tissues, organs or functions in the human body. In many areas of medicine, a paradigm shift from biostable materials to biodegradable materials has been observed in recent years. In dentistry, some of the permanent prosthetic devices used for temporary applications have been or will be changed to biodegradable products. The main reason for this situation is the biostability requirement of long-term implants and the need for revision surgery. The most important group of this type of applications are polymers [1–4]. Depending on the intended use, polymeric biomaterials may require a specific stability/degradation time. The time for which a polymer should retain the designed functions in the tissue environment, the so-called functional time, is the most important measure of its properties. Another, also important, parameter is the disappearance time, which determines the time for the total removal of material from the body. Between the functional time and the disappearance time, the biomaterial does not fulfill its functions. However, the polymer still releases degradation products, which, depending on the release rate and the physical and chemical properties, can trigger the body's reactions [2]. The design of the polymer-based medical device requires determining the behavior of the material in the tissue environment. This is a difficult task because there are many factors affecting the degradation. In the bioactive environment of the human body, polymers can be decomposed during physicochemical and biological processes. This group contains hydrolysis, enzymatic or chemical (e.g., oxidation) reactions, and physical factors (including water absorption and swelling, fatigue by mechanical stresses, and wear). Usually, most of them act simultaneously and it is difficult to predict the behavior of the material *in vivo* [3,5]. Polymer degradation can be analyzed using complex mathematical models based on experimental data [6–8]. These models, however, are not able to predict the properties of the material during degradation, i.e., mechanical changes. If for various reasons, it is not possible to conduct *in vivo* experiments, then it is necessary to perform *in vitro* tests in conditions as close as possible to the properties of the implantation site.

The most promising group of biodegradable polymers are aliphatic polyesters, among which the most extensively investigated are polycaprolactone (PCL) and polylactide (PLA). PCL is a semi-crystalline polymer with a glass transition temperature from −60 to −55 °C and a low melting point of about 60 °C. PCL has low mechanical strength, but very high elongation at break (4700%), which makes it extremely flexible [9]. PCL is mainly used in tissue engineering for the regeneration of bone tissues [10], ligaments [11], cartilage [12], skin [13], nerves [14], and vessels [15]. Excellent processability of PCL allows scaffolding by electrospinning or by forming porous structures by treatment with a porogen. Due to slow degradation time, it is also used in long-term drug delivery systems, such as contraceptives [16].

Due to the fact that lactide is a chiral molecule and exist in two optically active forms—L-lactide and D-lactide—the polymerization of these monomers leads to the formation of semi-crystalline polymers: poly(L-lactide) (PLLA) and poly(D-lactide) (PDLA). The polymerization of racemic (D,L)-lactide and meso-lactide results in the formation of amorphous polymers poly(D,L-lactide) (PDLLA) and meso-polylactide [17]. Among them, PLLA and PDLLA are used for biomedical applications [18]. PLLA has good tensile strength and stiffness, its elastic modulus is 2.7 Gpa. The glass transition temperature is 60–65 °C and the melting point is 173–178 °C [19]. The time of complete resorption of PLLA is above 24 months, therefore it is most often used as a material for bone fixators [19]. It is also used in tissue engineering for the regeneration of bones [20,21], tendons [22], cartilage [23], nerves [24] and vessels [25,26]. Due to the slow degradation time, PLLA is not often used alone as a drug delivery system [27]. Sometimes, however, by modifying, mixing or copolymerization with other polymers (for example PGA), its degradation time is shortened [28–31]. PDLLA is an amorphous polymer due to the random positions of its two isomeric monomers in the chain, whereby the glass transition temperature is about 55–60 °C and the Young modulus is 1.9 Gpa [19]. The PDLLA full resorption time is shorter than PLLA resorption time and ranges from 12 to 16 months, therefore it is more often used as a drug delivery system [32] and in tissue engineering [33].

Due to their properties, such as biodegradability and ease of processing (common use in 3D printing [34]), PCL and PLA are used for oral cavity applications. According to the literature [35–41], they are used for scaffolds for trachea, teeth, and tooth–ligament interfaces, and as carriers for oral insulin delivery.

Both polymers in the human body undergo mainly slow hydrolytic degradation [5]. From a chemical point of view, PCL and PLA hydrolysis is a bimolecular nucleophilic substitution reaction that can be catalyzed by the presence of acids or bases [42]. The form of solid samples in which long chains are condensed means that the behavior of these polymers during degradation is more complicated than in other forms (microspheres, films). This is due to the fact that the reactivity in the crystal domains may be different than in amorphous ones [43,44]; the rate of degradation near chain ends may differ in comparison to other regions of the chain [45]. In addition, depending on the reaction rate and speed of molecule transport, surface erosion or bulk erosion may occur [46].

It is assumed that the kinetics of aliphatic polyesters' degradation is the third order reaction and depends on the concentration of polymer bonds, water, and acid hydrolysis products [42]. According to this theory, there is a linear relationship between the logarithm of the polymer molecular weight and the degradation time, which was confirmed by the results of a series of experiments carried out by

Pitt et al. [47,48] in which PCL and PLA degradation was examined *in vivo*. In the same studies, it was also observed that the weight loss of the samples was negligible until the molecular weight of PCL and PLA reached 5000 and 15,000 g/mol, respectively. Another feature of PCL and PLA hydrolysis is autocatalysis, whereby the degradation of solid samples is heterogeneous and occurs faster inside the sample than in the outer parts [49,50]. This feature is associated with the formation of the outer layer of slowly degrading polymer, whose macromolecules are trapped in the surface layer. Only oligomers with sufficiently low molecular weight are able to diffuse and dissolve. Inside the sample, however, the rapidly growing number of carboxyl groups accelerates the cleavage of ester bonds leading to auto-acceleration. This phenomenon was not observed in the pH buffered medium, which was thought to suggest that auto-acceleration was associated with low pH, although studies carried out in low-pH buffer solutions did not show enhancement of auto-acceleration [42]. Therefore, the effect of pH on degradation is still an open question, especially when we consider oral conditions in which pH variability is common. An important factor during degradation is also the temperature of the contact environment. In a few studies [51,52], it was confirmed according to the Arrhenius equation that with higher contact temperature, faster degradation process in the material is observed.

There are also studies on accelerated hydrolysis in the presence of some enzymes like pronase, proteinase K, or lipase [53–57]. However, specific conditions of the oral cavity, such as variability of pH and temperature, bacteria, fungi, and the presence of various types of chemical substances, affect the rate of degradation. On the other hand, biomaterials in direct contact with the oral environment coincide with a film with specific properties, mainly composed of macromolecules, such as mucins derived from saliva [58–61]. We believe that the conditions created by the saliva environment can affect the degradation of polymers in the oral cavity. In this work, we presented the influence of a mucin-based artificial saliva (AS) environment on degradation processes of PCL and PLA at temperatures of 37 and 42 °C, and pH values 2 and 7.4.

2. Materials and Methods

2.1. PCL and PLA Preparation for Degradation Studies

PCL (Sigma-Aldrich, Saint Louis, MO, USA) and PLA (3001D, D content 1.6%) (NatureWorks, Minnetonka, MN, USA) in the form of pellets were dried for 3 h. Then, on the Borche BS60 device (Borche, Rancho Cucamonga, CA, USA), samples of 30 mm × 5 mm × 4 mm dimensions were processed by injection molding. Next, obtained samples were conditioned at room temperature for 24 h (21 °C, 40% humidity). Two media were prepared for degradation studies: phosphate-buffered saline (PBS) (pH 7.4) and artificial saliva (AS) based on PBS, porcine gastric mucin (type III) (PGM), and xanthan gum (Sigma-Aldrich, Saint Louis, MO, USA) (pH 7.4), tested earlier [62,63]. Subsequently, the pH of half of the prepared medium was reduced to pH 2 by adding the appropriate volume of hydrochloric acid. The PCL and PLA samples were separately immersed in the medium in sealed, plastic containers and placed in two incubators, the internal temperatures of which were 37 ± 0.5 °C and 42 ± 0.5 °C. Designations adopted in the article are presented in Table 1. The whole cycle of tests lasted 8 weeks; every 5 days the medium was changed for a fresh one for each sample (n = 5). The polymers were tested for time intervals of 1, 2, 4, and 8 weeks.

Table 1. Designations used in the work.

Designation	Material	Medium	Temperature (°C)	pH
PCL/PBS/37/7	PCL	PBS	37	7.4
PCL/PBS/37/2	PCL	PBS	37	2
PCL/PBS/42/7	PCL	PBS	42	7.4
PCL/PBS/42/2	PCL	PBS	42	2
PCL/AS/37/7	PCL	Artificial saliva	37	7.4
PCL/AS/37/2	PCL	Artificial saliva	37	2
PCL/AS/42/7	PCL	Artificial saliva	42	7.4
PCL/AS/42/2	PCL	Artificial saliva	42	2
PLA/PBS/37/7	PLA	PBS	37	7.4
PLA/PBS/37/2	PLA	PBS	37	2
PLA/PBS/42/7	PLA	PBS	42	7.4
PLA/PBS/42/2	PLA	PBS	42	2
PLA/AS/37/7	PLA	Artificial saliva	37	7.4
PLA/AS/37/2	PLA	Artificial saliva	37	2
PLA/AS/42/7	PLA	Artificial saliva	42	7.4
PLA/AS/42/2	PLA	Artificial saliva	42	2

2.2. Degree of Swelling, Weight Loss, and CLSM Observations

The degradation process of PCL and PLA was followed by determining weight loss and water absorption of the materials. Samples were washed with distilled water and gently wiped with paper. The degree of swelling was determined by comparing the wet weight (w_w) with dry weight (w_d) according to Equation (1):

$$degree\ of\ swelling(\%) = \frac{w_w - w_d}{w_d} \times 100\% \tag{1}$$

The percentage of weight loss was determined after drying the samples (24 h, room temperature) by comparing dry weight (w_d) with the initial weight (w_0) according to Equation (2):

$$weight\ loss(\%) = \frac{w_0 - w_d}{w_0} \times 100\% \tag{2}$$

A balance (Mettler Toledo, Columbus, OH, USA) with a sensitivity of 0.01 mg was used to weigh the samples.

The surfaces of the samples were observed using a non-destructive real-time imaging technique: confocal laser scanning microscopy (CLSM) LEXT OLS 4000 (Olympus, Tokyo, Japan). For each test, five raw samples were tested. Samples were observed just after being rinsed three times in pure water to remove free molecules from polymer surface.

2.3. Molecular Weight

The molecular weight of polymers was estimated based on intrinsic viscosity measurements of polymer solutions in chloroform using iVisc Capillary Viscometer (Lauda-Brinkmann LP, Delran, NJ, USA) (Ubbelohde, capillary constant 0.003 mm^2/s^2). The Mark–Houwink equation, Equation (3), was used in the calculations:

$$[\eta] = KM_\eta^a \tag{3}$$

where [η] is intrinsic viscosity, M_η is the estimated polymer molecular weight, K and a are constants of the equation which depend on polymer type, solvent, and temperature (for PCL-chloroform at 30 °C: K = 1.298·10^{-2} g/cm^3, a = 0.828 [64]; for PLA-chloroform at 25 °C: K = 6.06·10^{-2} g/cm^3, a = 0.64 [65]).

2.4. Differential Scanning Calorimetry (DSC) and Degree of Crystallinity

DSC studies were carried out using a DSC Discovery apparatus (TA Instruments, New Castle, DE, USA). The measurements were conducted in three cycles (heat–cool–heat), in a temperature range from −90 to 320 °C with a heating and cooling rate of 10 °C/min for PCL, and in a temperature range from −30 to 300 °C with a heating and cooling rate of 10°C/min for PLA. Results discussed in work were taken from second heating curves (first heating and cooling were performed to reduce the thermal history of tested samples). Three samples of each polymer were analyzed. Obtained DSC curves were used to analyze the glass transition temperature (T_g), crystallization temperature (T_c), cold crystallization enthalpy (ΔH_c), melting temperature I, and fusion enthalpy (ΔH_m). The degree of crystallinity (X_c) for the samples was determined according to the following Equation (4):

$$X_c = \frac{\Delta H_m - \Delta H_c}{\Delta H_m^{100}} \times 100\% \tag{4}$$

where ΔH_m^{100} is fusion enthalpy of 100% crystalline PLA/PCL (93.7 J/g for PLA [66] and 135.3 J/g for PCL [67]).

2.5. Thermogravimetry (TG) and Activation Energy

TG tests were carried out using the Q500 thermogravimetric analyzer (TA Instruments, New Castle, DE, USA) in a temperature range from 30 to 500 °C and nitrogen atmosphere with heating rates (k) values of 5, 10, 20 °C/min. Special attention was paid to temperature at the start of thermal decomposition (T_{DS}) (taken as 2% of the weight loss of the sample) and the end of thermal decomposition (T_{DE}) (taken as 95% weight loss of the sample), and temperature at which the rate of weight loss was the highest (T_{max}).

The Kissinger method was used to determine the activation energy of thermal decomposition of the tested materials. This method is based on the dependence of the temperature value T_{max} (corresponding to the maximum of the DTG signal) to the heating rate k (Equation (5)):

$$ln\left(\frac{k}{T_{max}^2}\right) = -\frac{E_a}{R} \times \frac{1}{T_{max}} + const. \tag{5}$$

As the heating rate increases, the temperature of the maximum intensity of the DTG signal also increases. The Kissinger method is based on presenting the obtained T_{max} values in the configuration $\ln(k/T_{max}^2) \sim 1/T_{max}$. The directional coefficient of the obtained straight line corresponds to the value E_a/R (where E_a is activation energy, R is a gas constant equal to 8.31 J/mol·K).

2.6. Hardness and Tensile Strength

Mechanical tests were carried out for wet samples, only washed with distilled water and wiped with paper. Hardness measurements of materials were performed using a Shore durometer (Type D) according to ASTM D2240. Tensile strength tests were carried out using a universal testing machine Zwick/Roell Z010 (Zwick Roell Group, Ulm, Germany) in accordance with ISO 527. Five samples of each polymer were analyzed for each test.

2.7. Atomic Force Microscopy Measurments

Topography and mechanical property measurements were recorded using an atomic force microscope (AFM) NanoWizard 4 BioScience AFM (JPK Instruments, Bruker, MA, USA) equipped with a liquid cell setup. Triangular-shaped cantilevers (AppNano NITRA-TALL-V-G) characterized by a spring constant of 0.37 N/m were used. Due to the lateral forces during contact mode scanning, a force curves-based imaging mode was used (JPK QI mode), with the resolution of 128 pixels per line, to show topography of the samples. QI maps also served as data for surface profiles and roughness examinations. Elastic modulus (i.e., the Young's modulus) of the polymers was calculated based on force indentation curves from AFM, collected using the same cantilever. Elasticity maps size of 10 μm × 10 μm corresponding to a grid of 8 × 8 pixels and 25 μm × 25 μm areas corresponding to a grid of 16 × 16 pixels were carried out. Elasticity maps were collected from various samples areas. Young's modulus was derived from the Hertz–Sneddon model applied to force-indentation curves.

3. Results and Discussion

At work, 8-week *in vitro* degradation studies of PCL and PLA samples, processed by injection molding were carried out. PBS or artificial saliva (AS, whose composition was based on mucin and xanthan gum) were used as the contacting environment. The main aim of this work was to verify whether the presence of macromolecules such as mucin and xanthan gum affects the process of polymer degradation. It was also examined whether the changes in pH and temperature of the environment, which reflect similar changes in the oral cavity and throat, will affect the degradation process. For this purpose, a number of physicochemical and mechanical tests were carried out, including evaluation of the material degree of swelling, weight and molecular weight change during degradation, hardness, tensile strength measurements, and DSC and TG thermal studies with the evaluation of the crystallinity of polymers and activation energy observed for polymers' thermal decomposition. In addition, the layer resulting from the adsorption of macromolecules from the saliva was examined by atomic force microscopy to assess topography and mechanical properties.

3.1. Degree of Swelling, Weight Loss, and CLSM Observations

Figure 1 presents the degree of swelling of PCL and PLA as a function of incubation time. In case of artificial saliva, the absorption of PCL was 5–7% and was 3.5–4% for PLA after 8 weeks. For both materials, the degree of swelling was higher for samples incubated in AS. Higher swelling degree probably mainly results from the presence of the adsorbed layer of macromolecules on the surface of samples, which has not been washed away, than from the penetration of mucin and xanthan gum molecules into the bulk of the material. In both materials, the degree of swelling was higher at 42 °C than at 37 °C, which results from faster diffusion of water into the material at a higher temperature. In addition, it was observed that the highest increase of absorption occurred at the beginning of the degradation period (2–4 weeks), in subsequent time intervals a plateau in level was noticed.

Figure 2 shows the weight loss as a function of degradation time. After 8 weeks the weight loss of PCL incubated in PBS reached 0.5–0.7% and reached 0.3–0.5% for PCL incubated in the AS. In case of PLA, weight loss after 8 weeks reached 0.75–0.95% in PBS and reached 0.45–0.7% in AS. It was observed that there is a correlation between swelling degree and weight loss (in frame of each polymer and type of environment). Hydrolysis is the major mechanism of aliphatic polyester degradation; PCL and PLA belong to this group. The rate of its hydrolytic degradation is determined by the water concentration, and thus the degree of swelling. The polyesters can absorb up to 1% water, however, due to hydrolysis, more hydrophilic chain ends are formed, and polymers may absorb increasing amounts of water. The kinetics of hydrolytic degradation, therefore, depend on the diffusion coefficient of water to the polymer, which for polyesters, regardless of the state, is about 10^{-8} cm^2/s (at 37 °C), which can be compared to penetrating 1 mm of the sample in a few days [5]. The coefficient of water diffusion (and the diffusion coefficient of chain fragments within the polymer, and the solubility of degradation products) through the polymer increases with temperature, which agrees with the observations obtained in this study. At 42 °C, the weight loss was higher than at 37 °C.

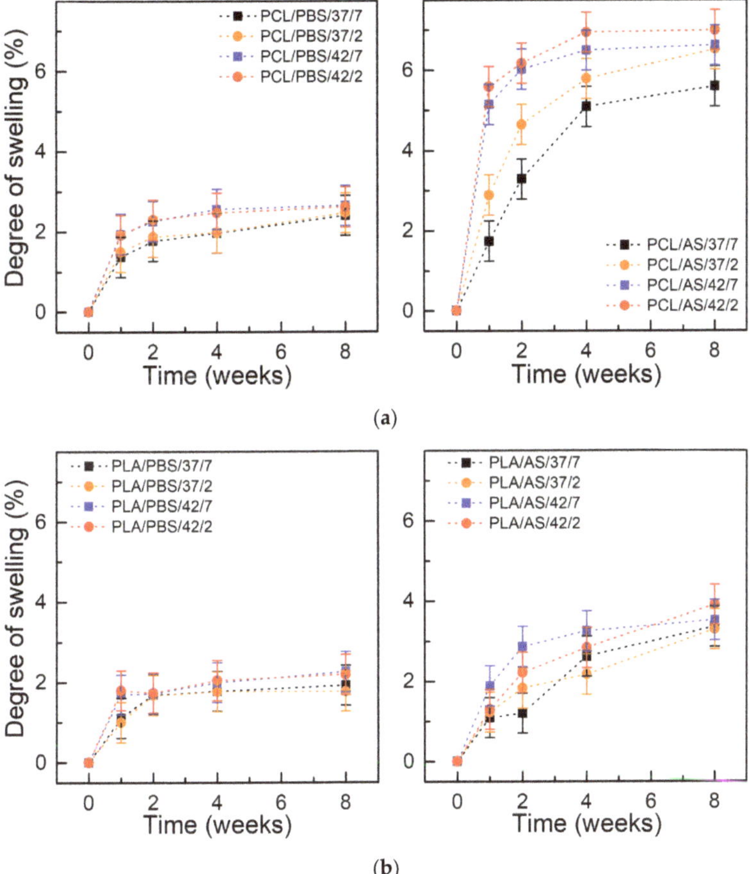

Figure 1. Swelling degree of materials during degradation: (**a**) polycaprolactone (PCL); (**b**) polylactide (PLA); (±SD, n = 5).

The weight loss of both materials was lower in the artificial saliva when compared to PBS. We believe that the mucin layer formed on polymers' surface, like in the oral cavity (or the respiratory/gastrointestinal tract) [68], covers the surfaces of materials and creates a protective layer [61,62]. It can act as a barrier and finally slow down the degradation of polymers.

For materials incubated in PBS at different pH values, 2 and 7.4, small differences were observed either in the degree of swelling or in weight loss. The influence of pH in aqueous solutions on the rate of degradation has been studied in several works [5,69–72]. Studies of PLA dissolved in tetrahydrofuran with a pH in the range of 0–14 showed the slowest degradation at pH 4 [5]. Other studies [69] on the degradation of PLA polymer chains attached to the surface (brush structure) in the pH range 3–8 showed the slowest degradation at pH 3. PLA degradation is the slowest in solutions with pH 4, because pKa of lactic acid is 3.84. In solutions with pH > 4, lactic acid is in dissociated form, which favors faster hydrolysis. In solutions with pH < 4, lactic acids at the chain ends prefer the associated acid form, which also accelerates the hydrolysis [5]. However, in solid polymers, the effect of pH on the rate of degradation is often negligible [5,71]. This is due to the low solubility of H^+ and OH^- ions in polymers so that they cannot effectively affect degradation.

For both materials incubated in artificial saliva at different pH levels, the differences were visible, and the weight loss of polymers contacted with solutions with pH 2 was lower. This may be due to the specific properties of mucin, which under the influence of low pH, can change their conformation, making the created layer a stronger barrier.

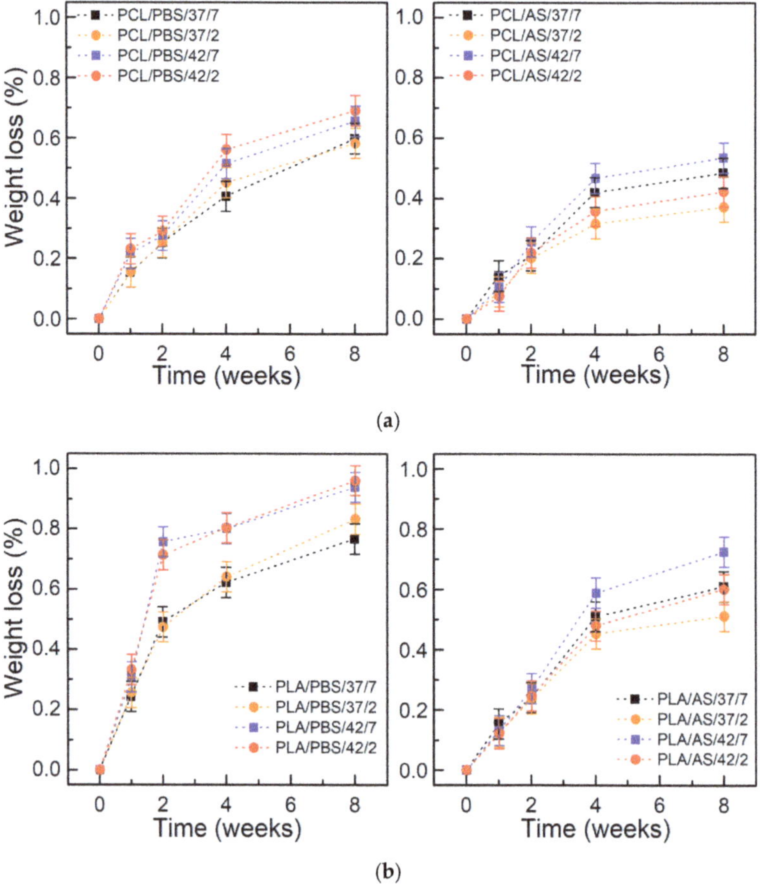

Figure 2. Weight loss of materials during degradation: (**a**) PCL; (**b**) PLA; (±SD, n = 5).

Figure 3a shows the surfaces of PCL samples before (panel A) and after 8 weeks of degradation in PBS (panels B–E) and artificial saliva (panels F–I). CLSM observations were performed on identical parts of the samples. Before degradation tests (Figure 3a, panel A) a characteristic texture is visible for elements prepared by the injection molding method. After 8 weeks of degradation, the surface texture is less visible, which may indicate that surface erosion has occurred. In samples F and H, which present samples after degradation in artificial saliva at pH = 7.4, there are probably residues on the macromolecular layer adsorbed on the surface. Figure 3b presents the surface of the PLA samples before (panel A) and after 8 weeks of degradation in PBS (panels B–E) and artificial saliva (panels F–I) solutions. Significant differences are visible between the surfaces of the tested polymer samples. In samples B and C, the surface morphology is more complex than in sample A, the characteristic striated texture has been enhanced. The surfaces for samples E, F, H, and I after degradation at 42 °C have changed to "milky" color and the surface is smoother due to erosion; these results correlate with weight loss. However, in samples F and G there are visible residues from the adsorbed layer of macromolecules (which correlates with results of swelling).

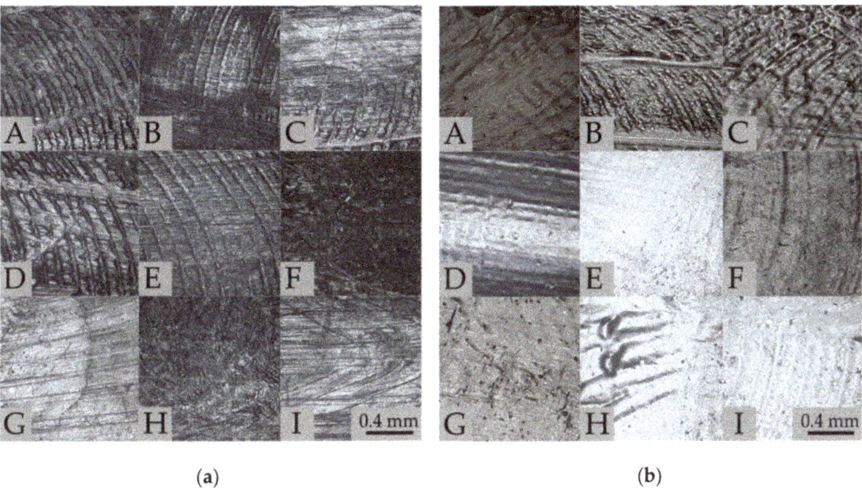

Figure 3. Surfaces of tested polymers before and after 8 weeks degradation imaged using CLSM:
(**a**) PCL—before degradation (A); PBS/37/7 (B); PBS/37/2 (C); PBS/42/7 (D); PBS/42/2 (E); AS/37/7 (F);
AS/37/2 (G); AS/42/7 (H); AS/42/2 (I). (**b**) PLA—before degradation (A); PBS/37/7 (B); PBS/37/2 (C);
PBS/42/7 (D); PBS/42/2 (E); AS/37/7 (F); AS/37/2 (G); AS/42/7 (H); AS/42/2 (I).

3.2. Molecular Weight

Figure 4 presents the results of molecular weight measurements of polymers PCL and PLA
(estimated on the basis of intrinsic viscosity measurements) after 8 weeks of degradation in the PBS and
artificial saliva in relation to the molecular weight of non-degraded polymer samples. The molecular
weight of PCL samples kept in PBS solution at 37 °C decreased after 8 weeks by about 13%; at 42 °C
the decrease was much higher—about 30–32%. The artificial saliva slightly affected the decrease of
molecular weight in relation to the PBS. At 37 °C about 12–13% of molecular weight loss was observed;
at 42 °C about 24–26% of molecular weight loss was observed. Similar trends were observed for PLA.
In the case of PBS environment at 37 °C the decrease in molecular weight was about 11–13%, and at
42 °C, the decrease was about 38–40%. For the artificial saliva at 37 °C, molecular weight decrease
was 10–12%; at 42 °C it was 36–37%. For both polymers and both contact solutions, no significant
differences between low and neutral pH were observed (these differences were a maximum of 2% of
molecular weight loss). The observed trends correlate with the results of weight loss, where a lower
decrease of mass in the environment of artificial saliva was also observed. In case of this medium,
an increase of temperature had a much greater impact on the degradation process of polymers in
comparison to the influence of the pH of the contact environment.

3.3. Thermal Studies

Thermal tests using DSC and TG are very helpful in assessing changes in the material, especially in
the initial stage of degradation, which are difficult to investigate by other methods. There are few works
in which the glass transition temperature, melting temperature, or the degree of crystallinity calculated
on the basis of the enthalpy of physical transformations were used to monitor the degradation process.
Reich [73] observed that the glass transition temperature and the melting temperature of the PLA
microparticles decreases due to degradation process. In the work of Paul et al. [74], it has been shown
that the decrease in molecular weight during hydrolysis is accompanied by decreases in the glass
transition and crystallization temperatures, as well as an increase in the degree of crystallinity.

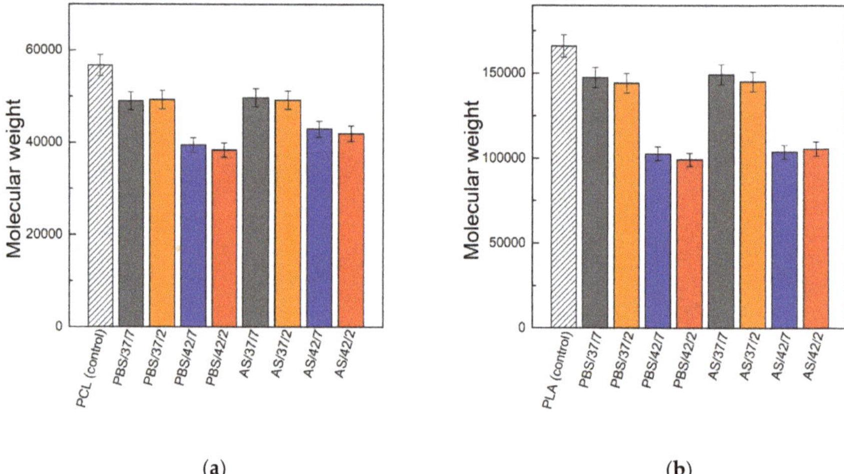

(**a**)   (**b**)

Figure 4. Molecular weight of polymers before (control) and after 8-weeks of degradation: (**a**) PCL; (**b**) PLA.

The thermal and crystalline properties of the PCL and PLA before and after the degradation period obtained from the DSC and TG curves are shown in Table 2. For both materials, changes in thermal properties were observed. For PCL, there was a significant reduction in glass transition temperature from the initial value of −65.2 °C to about −80 °C for samples kept in the AS. In the case of PLA, a significant decrease in the glass transition temperature was observed for samples PLA/PBS/42/7 and PCL/AS/42/2 (from initial value 48.5 °C to below 42 °C). The decrease in T_g is related to both the reduction of molecular weight and the plasticizing effect by acid oligomers formed during degradation.

Table 2. Glass transition temperature (T_g), crystallization temperature (T_c), melting temperature (T_m), and crystallinity (X_c) obtained from DSC curves, along with starting thermal decomposition temperature (TDS), end of thermal decomposition temperature (T_{DE}), and activation energy of thermal decomposition (E_a) obtained from TG of PCL and PLA after 8 weeks of degradation.

Sample	T_g (°C)	T_c (°C)	T_m (°C)	X_c (%)	T_{DS} (°C)	T_{DE} (°C)	E_a (kJ/mol)
PCL	−65.2	37.5	54	59.3	360.6	450.9	221.2
PCL/PBS/37/7	−70.6	36.4	54.8	59.6	361.4	450.6	218.8
PCL/PBS/37/2	−65.3	36.1	53.5	59.9	365.2	436.7	174.4
PCL/PBS/42/7	−65.3	35.5	53.2	65.5	365.6	433.3	183.8
PCL/PBS/42/2	−69.95	35.1	53.5	63.6	367.1	433.2	160.3
PCL/AS/37/7	−78.3	36.7	53.8	57.8	364.3	426.8	162.7
PCL/AS/37/2	−79.7	36.8	53.8	59.0	366.4	434.9	181.3
PCL/AS/42/7	−79.4	36.4	53.4	60.0	365.3	434.1	177.6
PCL/AS/42/2	−82.2	36.7	53.1	61.5	366.4	434.4	177.9
PLA	48.5	107.7	146.7	2.6	324.2	388.7	190.2
PLA/PBS/37/7	46.8	96.9	143.7	3.2	319.8	385.9	163.0
PLA/PBS/37/2	47.1	97.6	145.6	3.1	319.7	385.2	140.4
PLA/PBS/42/7	41.1	103.5	134.5	2.8	316.9	384.7	154.2
PLA/PBS/42/2	41.6	100	135.5	2.6	316.6	384.4	144.0
PLA/AS/37/7	47.3	96.6	144.2	3.1	308.9	378.5	158.2
PLA/AS/37/2	46.5	99.2	141.7	3.5	318.0	385.1	170.1
PLA/AS/42/7	45.7	99.3	140.2	4.2	314.1	382.1	157.1
PLA/AS/42/2	47.5	95	145.3	4.6	310.7	383.0	177.2

In both materials, degradation decreased the value of crystallization temperature. In case of PCL, this was from the initial value of 37.5 °C to below 36 °C for samples PCL/PBS/42/7 and PCL/PBS/42/2, and for PLA with an initial value of 107.7 °C, this decreased to even 95 °C for PLA/AS/42/2. The decrease in the crystallization temperature can be interpreted as a decrease in molecular weight since shorter polymer chains tend to crystallize at lower temperature [74].

The melting temperature was also generally reduced after degradation. The highest changes were observed for PLA samples aged in PBS at 42 °C (from initial 146.7 °C to about 135 °C). For PCL, changes in the melting temperature were lower than for PLA. Thus, it can be concluded that the observed changes were irrelevant (melting point for all samples was in the range 53–54 °C).

The evolution of the global crystallinity X_c of the PLA and PCL during hydrolysis can be deduced through the evolution of both ΔH_c and ΔH_m. PLA has a generally low crystallinity which indicates a quasi-amorphous structure. As a result of degradation performed in this study, its crystallinity increases, but these are rather small changes (from the initial 2.6% to a maximum of 4.6%). PCL is a semi-crystalline polymer and due to degradation its crystallinity also increases, from the initial 59.3% to a maximum of 65.5%. Due to the fact that the range of changes in the weight of samples during degradation was low (up to 1%), the increase in crystallinity cannot be attributed to the change of the ratio of the crystalline phase to amorphous phase (the amorphous phase is generally decomposed faster than the crystalline one). This increase has to be attributed to an effective crystallization occurring during degradation. A possible mechanism of such a process may be an increase of the polymer chains' mobility due to the action of water, PBS medium, saliva macromolecules, and resulting degradation products (by the decrease of the chains' molecular weight and plasticization), which could result in structure reorganization and crystallization [74].

Hydrolytic degradation affects the thermal stability of polymers, often determined by the temperature of the start (T_{DS}) and end (T_{DE}) of thermal decomposition. In the case of untreated PCL, the temperatures were 360.6 °C and 450.9 °C, respectively. After 8 weeks of degradation, PCL had a higher T_{DS} (361.4–367.1 °C) and lower T_{DE} (426.8–450.6 °C) in all tested conditions. However, there were no differences between individual environments. In the case of PLA, for an unaged sample, T_{DS} was 324.2 °C and T_{DE} was 388.7 °C. It may be concluded that after degradation both T_{DS} (308.9–319.8 °C) and T_{DE} (378.5–385.9 °C) were decreased. Samples incubated in artificial saliva were characterized by a lower T_{DS}. Decrease of both temperatures is a result of molecular mass decrease. When considering the type of environment, the surface mucin-based saliva layer may decrease the rate of degradation.

The kinetics of the thermal decomposition process depend on the structure of the polymer and molecular weight. Therefore, structural changes due to hydrolysis may be indirectly observed by determining the activation energy of thermal decomposition. The activation energies of polymers measured before and after incubation in contact solutions were 221.2 kJ/mol (before) and 160.3–218.8 kJ/mol (after incubation in PBS and AS) for PCL; and 190.2 kJ/mol (before) and 144.0–177.2 kJ/mol (after incubation in PBS and AS) for PLA. After degradation in various tested environments, the activation energy generally decreased, which means that less energy is required to initiate thermal decomposition of tested materials. In the case of degradation performed in PBS, the greatest decreases of activation energy were noted for the environment at pH 2 (samples PCL/PBS/42/2, PLA/PBS/37/2, and PLA/PBS/42/2) in comparison to pH 7.4. The increase of contact environment temperature reduced the activation energy of the thermal decomposition process for both polymers.

3.4. Mechanical Tests

Water absorption, weight loss, and molecular weight loss as a function of time, either *in vivo* or *in vitro* studies in simulated body fluids, are useful parameters for assessing the behavior of biodegradable materials. Implants in the body environment often require the preservation of their mechanical properties, such as hardness, stiffness, tensile, compression, or bending strength. While the increase in water sorption, the weight loss, and the decrease of polymer molecular weight are obvious, and the conducted research gives information about the rate of these processes, changes in mechanical properties are more difficult to predict. In addition, the loss of weight and molecular weight does not always correlate with the decrease in mechanical properties—especially in the initial stages of the degradation process.

Figure 5 shows the Shore hardness of PCL and PLA as a function of degradation time. For both materials incubated in PBS and AS solutions, a slight increase of hardness during degradation up to 4 weeks was observed, followed by a slight drop to the initial or lower values at the end of the considered period. The higher hardness was observed in the case of materials kept in the environment of artificial saliva, which correlates with lower weight loss.

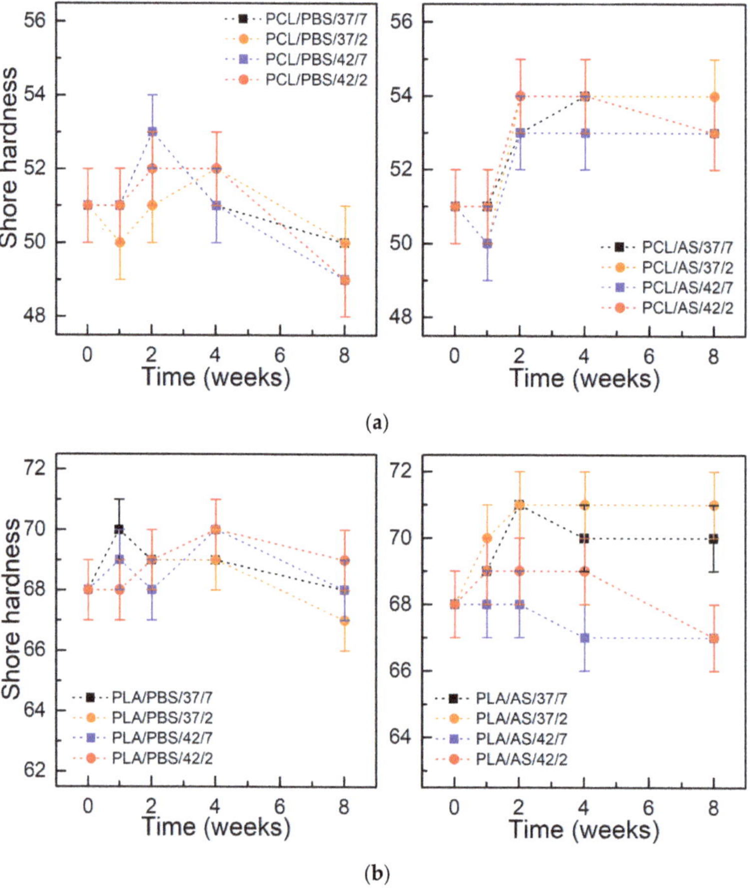

Figure 5. Shore hardness of materials during degradation: (**a**) PCL; (**b**) PLA; (±SD, n = 5).

Figure 6 presents the tensile strength of PCL and PLA samples as a function of degradation time. Similarly to the case of hardness tests, a slight increase in tensile strength was observed over the first 4 weeks of testing, then a gentle decrease was observed. In case of PCL immersed in PBS, the value of the tensile strength after 8 weeks did not change and, depending on the pH and temperature of the environment, it was within 18.25–18.75 MPa (initial value was 18.5 MPa). For PCL kept in AS, the tensile strength after 8 weeks was 19 MPa for a medium with pH 7.4 and temperature 37 °C, and about 18.5 MPa for the remaining conditions. For PLA immersed in PBS, a slight increase of tensile strength was observed after the first week of degradation from an initial value of 65 MPa to about

68 MPa for the PLA/PBS/42/2 sample. In the next weeks, a continuous decrease in the value of up to 62 MPA for the PLA/PBS/37/7 sample and around 58 MPa for the PLA/PBS/37/2 sample were observed. For the PLA kept in the AS environment, a decrease in the tensile strength value up to 61–62 MPa was also observed, however, preceded by an increase in the 4th week for samples kept at 42 °C.

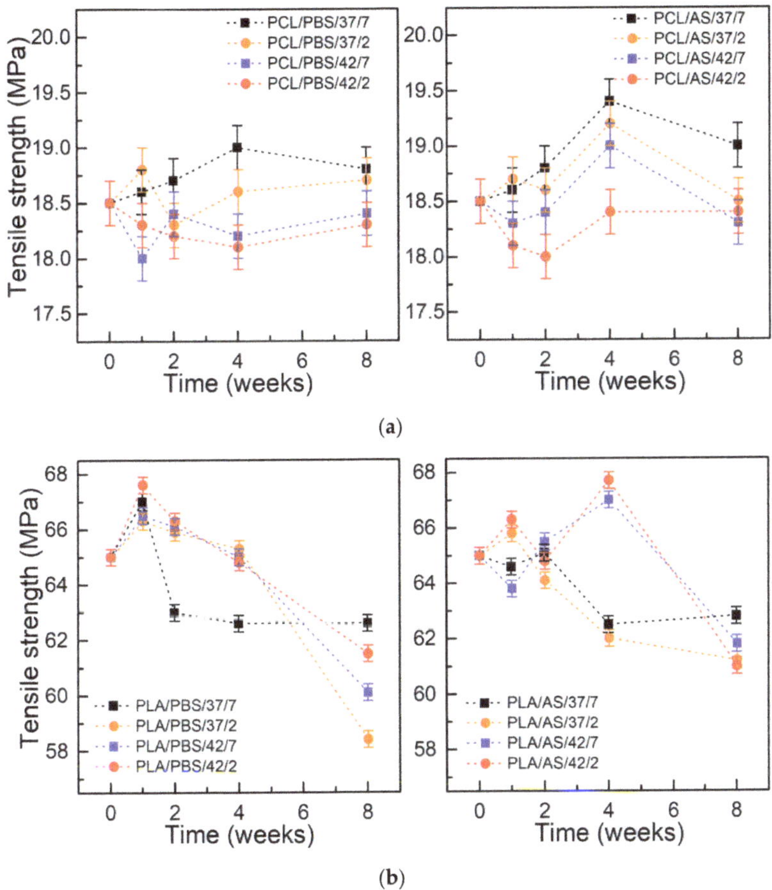

Figure 6. Tensile strength of materials during degradation: (**a**) PCL; (**b**) PLA; (±SD, n = 5).

During the degradation process of both PCL and PLA, an increase in the mechanical strength was initially perceived, which is caused by the absorption of water. An increase of hydrogen bond formation restricts the movement of chains past each other, and a consequential increase of the tensile strength during first 4 weeks of testing was observed. After strengthening the material, further absorption of water leads to polymer chains shortening. As a result, a decrease of mechanical properties (tensile strength, hardness) occurred. A similar character of mechanical changes was noticed in a few other works. In [75], after 30 days of hydrolytic degradation, a 6% increase in tensile strength and a 3% increase in Young's modulus were observed, after which a significant decrease in mechanical strength was noticed. In [76], the *in vivo* degradation of PLA implants for bone fixation was investigated. Until the 12th week, a slight increase in the mechanical strength of the examined implants was observed; 36 weeks after implantation, the mechanical strength was reduced by 23%.

3.5. AFM Measurments

To assess the properties of the thin layer of mucin adsorbed on the surface of samples during incubation in artificial saliva (pH 7.4 and pH 2) for 7 days, AFM measurements of topography (Figure 7) and mechanical properties (Figure 8) were carried out.

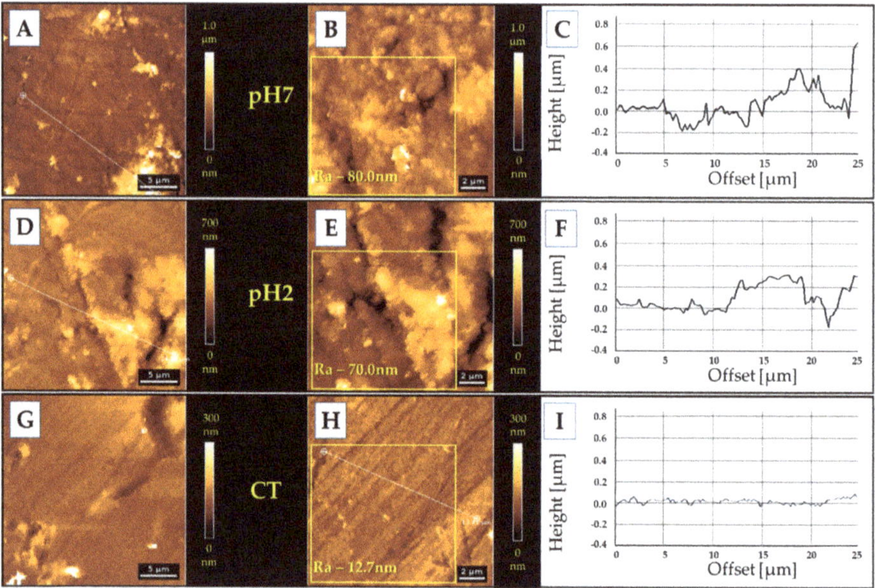

Figure 7. Topography of polymer surfaces: (**A–C**) the samples kept in pH 7 artificial saliva; (**D–F**) samples kept in pH 2 artificial saliva; (**G–I**) the control sample.

An adsorbed layer (Figure 7A,B,D,E) is visible on the surface of samples incubated in an artificial saliva, which changes the surface topography to a more rough one—from Ra = 12.7 nm (control sample) to Ra = 80 nm and Ra = 70 nm, for pH 7 and pH 2, respectively. The height of the layer formed at pH 7 is about 400 nm; at pH 2 the height is about 300 nm.

The adsorbed layer of mucin is definitely less stiff (80–133 MPa) than the polymer surface, whose Young's modulus was estimated at 3 GPa. However, it is so stiff and compact that it can be a barrier similar to that created by mucus on the tissues/biomaterials. Moreover, the layer formed on polymers incubated at lower pH is stiffer than that created in a neutral pH environment. Similar trends were observed ex vivo via AFM on fresh mucus-covered tissues [77]. The mucus tested at lower pH was not only stiffer, but also less adhesive. Using dynamic light scattering studies [78] it was demonstrated that pH lowering results in conformational porcine gastric mucin transition from the isotropic random coil at pH 7 to the anisotropic extended random coil at pH 2. Lowering pH results in a significant increase of viscosity and viscoelasticity of solutions as well as concentrated PGM [79], [80]. This transition is driven by a decrease in electrostatic repulsion between anionic fragments of mucin molecules, which allows interaction and aggregation of their hydrophobic residues [77].

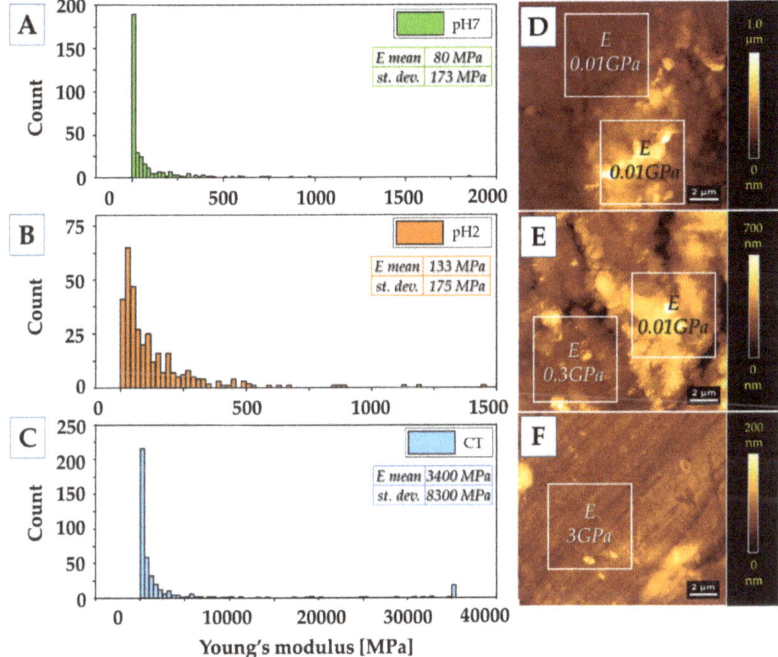

Figure 8. Mechanical properties of the polymer surfaces. (**A–C**) Young's modulus values distributions. (**A**) pH 7 sample; (**B**) pH 2 sample; (**C**) control sample. Inside the panels, mean values of Young's modulus (E) and standard deviations (st. dev.) are presented. (**D–F**) Topography of the samples' surfaces with values of Young's modulus from two areas—flat surface and high additional layer. (**D**) pH 7 sample; (**E**) pH 2 sample; (**F**) control sample.

4. Summary and Conclusions

The study investigated the PCL and PLA—two most common biodegradable polymers with potential use in oral applications. Degradation was carried out in two contact solutions—PBS and artificial saliva based on porcine gastric mucin and xanthan gum. The results of sample weight loss and molecular weight loss indicate a greater degree of polymer degradation in the PBS medium than in the artificial saliva environment. It may be concluded that the layer of mucin-based saliva acts as a barrier, and finally, a decrease in the degradation of polymers is observed. In addition, a significant effect of elevated temperature on the rate of degradation was observed, particularly evident in the results of molecular weight studies, in which a greater decrease was observed in case of higher temperature. However, lowering the pH of the environment had a lesser impact on the degradation process. An increase of temperature had a much greater impact on the degradation process of polymers in comparison to the influence of the pH of the contact environment. The increase of contact solution temperature reduced the activation energy of the thermal decomposition process for the tested polymers. The observed changes in physicochemical and thermal tests reflect the results of tests of mechanical properties. An increase of mechanical properties during the first 4 weeks of testing was observed. Further water absorption leads to the decrease of tested mechanical parameters.

The macromolecule layer formed on the surface of polymers can affect the rate of degradation. On the other hand, in applications such as drug carriers, the resulting macromolecule layer may impede the release of the substance from polymer.

Polymers **2019**, *11*, 1880

Author Contributions: D.Ł. and J.M. conceived the concept for this paper; D.Ł. wrote the original version of the manuscript; D.Ł., G.M., P.D. and J.M. performed an experimental part; D.Ł., J.M., G.M., P.D. and R.B. contributed to the writing of this paper.

Funding: This scientific work was realized in the frame of work, No. S/WM/2/2017 and financed from research funds of the Ministry of Science and Higher Education, Poland. The paper was financed from the Minister of Science and Higher Education program "Regional Initiative of Excellence" in 2019–2022 project number 011/RID/2018/19 (amount of funding 12,000,000 PLN).

Conflicts of Interest: The authors declare no conflict of interest.

References

1. Tian, H.; Tang, Z.; Zhuang, X.; Chen, X.; Jing, X. Biodegradable synthetic polymers: Preparation, functionalization and biomedical application. *Prog. Polym. Sci.* **2012**, *37*, 237–280. [CrossRef]
2. Nair, L.S.; Laurencin, C.T. Biodegradable polymers as biomaterials. *Prog. Polym. Sci.* **2007**, *32*, 762–798. [CrossRef]
3. Teo, A.J.T.; Mishra, A.; Park, I.; Kim, Y.J.; Park, W.T.; Yoon, Y.J. Polymeric Biomaterials for Medical Implants and Devices. *ACS Biomater. Sci. Eng.* **2016**, *2*, 454–472. [CrossRef]
4. Mystkowska, J.; Mazurek-Budzyńska, M.; Piktel, E.; Niemirowicz, K.; Karalus, W.; Deptuła, P.; Pogoda, K.; Łysik, D.; Dąbrowski, J.R.; Rokicki, G.; et al. Assessment of aliphatic poly(ester-carbonate-urea-urethane)s potential as materials for biomedical application. *J. Polym. Res.* **2017**, *24*, 144. [CrossRef]
5. Lyu, S.; Untereker, D. Degradability of polymers for implantable biomedical devices. *Int. J. Mol. Sci.* **2009**, *10*, 4033–4065. [CrossRef]
6. Han, X.; Pan, J. A model for simultaneous crystallisation and biodegradation of biodegradable polymers. *Biomaterials* **2009**, *30*, 423–430. [CrossRef]
7. Bikiaris, D.N.; Papageorgiou, G.Z.; Achilias, D.S.; Pavlidou, E.; Stergiou, A. Miscibility and enzymatic degradation studies of poly(ε-caprolactone)/poly(propylene succinate) blends. *Eur. Polym. J.* **2007**, *43*, 2491–2503. [CrossRef]
8. Metzmacher, I.; Radu, F.; Bause, M.; Knabner, P.; Friess, W. A model describing the effect of enzymatic degradation on drug release from collagen minirods. *Eur. J. Pharm. Biopharm.* **2007**, *67*, 349–360. [CrossRef]
9. Woodruff, M.A.; Hutmacher, D.W. The return of a forgotten polymer—Polycaprolactone in the 21st century. *Prog. Polym. Sci.* **2010**, *35*, 1217–1256. [CrossRef]
10. Ren, K.; Wang, Y.; Sun, T.; Yue, W.; Zhang, H. Electrospun PCL/gelatin composite nanofiber structures for effective guided bone regeneration membranes. *Mater. Sci. Eng. C* **2017**, *78*, 324–332. [CrossRef]
11. Hayami, J.W.S.; Surrao, D.C.; Waldman, S.D.; Amsden, B.G. Design and characterization of a biodegradable composite scaffold for ligament tissue engineering. *J. Biomed. Mater. Res. Part A* **2010**, *92*, 1407–1420. [CrossRef] [PubMed]
12. Li, W.J.; Tuli, R.; Okafor, C.; Derfoul, A.; Danielson, K.G.; Hall, D.J.; Tuan, R.S. A three-dimensional nanofibrous scaffold for cartilage tissue engineering using human mesenchymal stem cells. *Biomaterials* **2005**, *26*, 599–609. [CrossRef] [PubMed]
13. Chaudhari, A.A.; Vig, K.; Baganizi, D.R.; Sahu, R.; Dixit, S.; Dennis, V.; Singh, S.R.; Pillai, S.R. Future prospects for scaffolding methods and biomaterials in skin tissue engineering: A review. *Int. J. Mol. Sci.* **2016**, *17*, 1974. [CrossRef] [PubMed]
14. Ghasemi-Mobarakeh, L.; Prabhakaran, M.P.; Morshed, M.; Nasr-Esfahani, M.H.; Ramakrishna, S. Electrospun poly(ε-caprolactone)/gelatin nanofibrous scaffolds for nerve tissue engineering. *Biomaterials* **2008**, *29*, 4532–4539. [CrossRef]
15. Ju, Y.M.; Choi, J.S.; Atala, A.; Yoo, J.J.; Lee, S.J. Bilayered scaffold for engineering cellularized blood vessels. *Biomaterials* **2010**, *31*, 4313–4321. [CrossRef]
16. Dash, T.K.; Konkimalla, V.B. Poly-ε-caprolactone based formulations for drug delivery and tissue engineering: A review. *J. Control. Release* **2012**, *158*, 15–33. [CrossRef]
17. Tsuji, H. Poly(lactide) stereocomplexes: Formation, structure, properties, degradation, and applications. *Macromol. Biosci.* **2005**, *5*, 569–597. [CrossRef]
18. Ulery, B.D.; Nair, L.S.; Laurencin, C.T. Biomedical applications of biodegradable polymers. *J. Polym. Sci. Part B Polym. Phys.* **2011**, *49*, 832–864. [CrossRef]

19. Middleton, J.C.; Tipton, A.J. Synthetic biodegradable polymers as orthopedic devices. *Biomaterials* **2000**, *21*, 2335–2346. [CrossRef]
20. Shim, I.K.; Jung, M.R.; Kim, K.H.; Seol, Y.J.; Park, Y.J.; Park, W.H.; Lee, S.J. Novel three-dimensional scaffolds of poly(L -lactic acid) microfibers using electrospinning and mechanical expansion: Fabrication and bone regeneration. *J. Biomed. Mater. Res. Part B Appl. Biomater.* **2010**, *95*, 150–160. [CrossRef]
21. Cui, F.Z.; Li, Y.; Ge, J. Self-assembly of mineralized collagen composites. *Mater. Sci. Eng. R Rep.* **2007**, *57*, 1–27. [CrossRef]
22. Barber, J.G.; Handorf, A.M.; Allee, T.J.; Li, W.J. Braided nanofibrous scaffold for tendon and ligament tissue engineering. *Tissue Eng. Part A* **2013**, *19*, 1265–1274. [CrossRef] [PubMed]
23. Tanaka, Y.; Yamaoka, H.; Nishizawa, S.; Nagata, S.; Ogasawara, T.; Asawa, Y.; Fujihara, Y.; Takato, T.; Hoshi, K. The optimization of porous polymeric scaffolds for chondrocyte/atelocollagen based tissue-engineered cartilage. *Biomaterials* **2010**, *31*, 4506–4516. [CrossRef] [PubMed]
24. Yang, F.; Murugan, R.; Ramakrishna, S.; Wang, X.; Ma, Y.X.; Wang, S. Fabrication of nano-structured porous PLLA scaffold intended for nerve tissue engineering. *Biomaterials* **2004**, *25*, 1891–1900. [CrossRef]
25. Hu, J.; Sun, X.; Ma, H.; Xie, C.; Chen, Y.E.; Ma, P.X. Porous nanofibrous PLLA scaffolds for vascular tissue engineering. *Biomaterials* **2010**, *31*, 7971–7977. [CrossRef]
26. Ma, H.; Hu, J.; Ma, P.X. Polymer scaffolds for small-diameter vascular tissue engineering. *Adv. Funct. Mater.* **2010**, *20*, 2833–2841. [CrossRef]
27. Farah, S.; Anderson, D.G.; Langer, R. Physical and mechanical properties of PLA, and their functions in widespread applications—A comprehensive review. *Adv. Drug Deliv. Rev.* **2016**, *107*, 367–392. [CrossRef]
28. Rasal, R.M.; Janorkar, A.V.; Hirt, D.E. Poly(lactic acid) modifications. *Prog. Polym. Sci.* **2010**, *35*, 338–356. [CrossRef]
29. Lu, L.; Garcia, C.A.; Mikos, A.G. *In vitro* degradation of thin poly(DL-lactic-co-glycolic acid) films. *J. Biomed. Mater. Res.* **1999**, *46*, 236–244. [CrossRef]
30. McDonald, P.F.; Lyons, J.G.; Geever, L.M.; Higginbotham, C.L. *In vitro* degradation and drug release from polymer blends based on poly(dl-lactide), poly(l-lactide-glycolide) and poly(ε-caprolactone). *J. Mater. Sci.* **2010**, *45*, 1284–1292. [CrossRef]
31. Vey, E.; Rodger, C.; Booth, J.; Claybourn, M.; Miller, A.F.; Saiani, A. Degradation kinetics of poly(lactic-co-glycolic) acid block copolymer cast films in phosphate buffer solution as revealed by infrared and Raman spectroscopies. *Polym. Degrad. Stab.* **2011**, *96*, 1882–1889. [CrossRef]
32. Xue, Y.N.; Huang, Z.Z.; Zhang, J.T.; Liu, M.; Zhang, M.; Huang, S.W.; Zhuo, R.X. Synthesis and self-assembly of amphiphilic poly(acrylic acid-b-dl-lactide) to form micelles for pH-responsive drug delivery. *Polymer* **2009**, *50*, 3706–3713. [CrossRef]
33. Bao, M.; Lou, X.; Zhou, Q.; Dong, W.; Yuan, H.; Zhang, Y. Electrospun biomimetic fibrous scaffold from shape memory polymer of PDLLA-co-TMC for bone tissue engineering. *ACS Appl. Mater. Interfaces* **2014**, *6*, 2611–2621. [CrossRef] [PubMed]
34. Goyanes, A.; Det-Amornrat, U.; Wang, J.; Basit, A.W.; Gaisford, S. 3D scanning and 3D printing as innovative technologies for fabricating personalized topical drug delivery systems. *J. Control. Release* **2016**, *234*, 41–48. [CrossRef] [PubMed]
35. Park, C.H.; Rios, H.F.; Jin, Q.; Bland, M.E.; Flanagan, C.L.; Hollister, S.J.; Giannobile, W.V. Biomimetic hybrid scaffolds for engineering human tooth-ligament interfaces. *Biomaterials* **2010**, *31*, 5945–5952. [CrossRef] [PubMed]
36. Park, J.S.; Lee, S.J.; Jo, H.H.; Lee, J.H.; Kim, W.D.; Lee, J.Y.; Park, S.A. Fabrication and characterization of 3D-printed bone-like β-tricalcium phosphate/polycaprolactone scaffolds for dental tissue engineering. *J. Ind. Eng. Chem.* **2017**, *46*, 175–181. [CrossRef]
37. Xiong, X.Y.; Li, Q.H.; Li, Y.P.; Guo, L.; Li, Z.L.; Gong, Y.C. Pluronic P85/poly(lactic acid) vesicles as novel carrier for oral insulin delivery. *Colloids Surf. B Biointerfaces* **2013**, *111*, 282–288. [CrossRef]
38. Lin, C.H.; Su, J.M.; Hsu, S.H. Evaluation of type II collagen scaffolds reinforced by poly(ε- caprolactone) as tissue-engineered trachea. *Tissue Eng. Part C Methods* **2008**, *14*, 69–77. [CrossRef]
39. Mäkitie, A.A.; Korpela, J.; Elomaa, L.; Reivonen, M.; Kokkari, A.; Malin, M.; Korhonen, H.; Wang, X.; Salo, J.; Sihvo, E.; et al. Novel additive manufactured scaffolds for tissue engineered trachea research. *Acta Otolaryngol.* **2013**, *133*, 412–417. [CrossRef]

40. Gao, M.; Zhang, H.; Dong, W.; Bai, J.; Gao, B.; Xia, D.; Feng, B.; Chen, M.; He, X.; Yin, M.; et al. Tissue-engineered trachea from a 3D-printed scaffold enhances whole-segment tracheal repair. *Sci. Rep.* **2017**, *7*. [CrossRef]

41. Luo, Y.Y.; Xiong, X.Y.; Tian, Y.; Li, Z.L.; Gong, Y.C.; Li, Y.P. A review of biodegradable polymeric systems for oral insulin delivery. *Drug Deliv.* **2016**, *23*, 1882–1891. [CrossRef] [PubMed]

42. Lyu, S.; Schley, J.; Loy, B.; Lind, D.; Hobot, C.; Sparer, R.; Untereker, D. Kinetics and time-temperature equivalence of polymer degradation. *Biomacromolecules* **2007**, *8*, 2301–2310. [CrossRef] [PubMed]

43. Mathisen, T.; Lewis, M.; Albertsson, A.-C. Hydrolytic degradation of nonoriented poly(β-propiolactone). *J. Appl. Polym. Sci.* **1991**, *42*, 2365–2370. [CrossRef]

44. Tsuji, H.; Mizuno, A.; Ikada, Y. Properties and morphology of poly(L-lactide). III. Effects of initial crystallinity on long-term *in vitro* hydrolysis of high molecular weight poly(L-lactide) film in phosphate-buffered solution. *J. Appl. Polym. Sci.* **2000**, *77*, 1452–1464. [CrossRef]

45. Wiggins, J.S.; Hassan, M.K.; Mauritz, K.A.; Storey, R.F. Hydrolytic degradation of poly(d,l-lactide) as a function of end group: Carboxylic acid vs. hydroxyl. *Polymer* **2006**, *47*, 1960–1969. [CrossRef]

46. Lyu, S.; Sparer, R.; Untereker, D. Analytical solutions to mathematical models of the surface and bulk erosion of solid polymers. *J. Polym. Sci. Part B Polym. Phys.* **2005**, *43*, 383–397. [CrossRef]

47. Pitt, C.G.; Chasalow, F.I.; Hibionada, Y.M.; Klimas, D.M.; Schindler, A. Aliphatic polyesters. I. The degradation of poly(ε-caprolactone) *in vivo*. *J. Appl. Polym. Sci.* **1981**, *26*, 3779–3787. [CrossRef]

48. Pitt, G.G.; Gratzl, M.M.; Kimmel, G.L.; Surles, J.; Sohindler, A. Aliphatic polyesters II. The degradation of poly (DL-lactide), poly (ε-caprolactone), and their copolymers *in vivo*. *Biomaterials* **1981**, *2*, 215–220. [CrossRef]

49. Vert, M.; Li, S.; Garreau, H. More about the degradation of LA/GA-derived matrices in aqueous media. *J. Control. Release* **1991**, *16*, 15–26. [CrossRef]

50. Göpferich, A. Polymer bulk erosion. *Macromolecules* **1997**, *30*, 2598–2604. [CrossRef]

51. Tsuji, H.; Nakahara, K.; Ikarashi, K. Poly(L-lactide), 8 high-temperature hydrolysis of poly(L-lactide) films with different crystallinities and crystalline thicknesses in phosphate-buffered solution. *Macromol. Mater. Eng.* **2001**, *286*, 398–406. [CrossRef]

52. Deng, M.; Zhou, J.; Chen, G.; Burkley, D.; Xu, Y.; Jamiolkowski, D.; Barbolt, T. Effect of load and temperature on *in vitro* degradation of poly(glycolide-co-L-lactide) multifilament braids. *Biomaterials* **2005**, *26*, 4327–4336. [CrossRef] [PubMed]

53. Chen, D.R.; Bei, J.Z.; Wang, S.G. Polycaprolactone microparticles and their biodegradation. *Polym. Degrad. Stab.* **2000**, *67*, 455–459. [CrossRef]

54. Sekosan, G.; Vasanthan, N. Morphological changes of annealed poly-ε-caprolactone by enzymatic degradation with lipase. *J. Polym. Sci. Part B Polym. Phys.* **2010**, *48*, 202–211. [CrossRef]

55. Castilla-Cortázar, I.; Más-Estellés, J.; Meseguer-Dueñas, J.M.; Ivirico, J.L.E.; Marí, B.; Vidaurre, A. Hydrolytic and enzymatic degradation of a poly(ε-caprolactone) network. *Polym. Degrad. Stab.* **2012**, *97*, 1241–1248. [CrossRef]

56. Reeve, M.S.; McCarthy, S.P.; Downey, M.J.; Gross, R.A. Polylactide Stereochemistry: Effect on Enzymic Degradability. *Macromolecules* **1994**, *27*, 825–831. [CrossRef]

57. Gan, Z.; Yu, D.; Zhong, Z.; Liang, Q.; Jing, X. Enzymatic degradation of poly(ε-caprolactone)/poly(DL-lactide) blends in phosphate buffer solution. *Polymer* **1999**, *40*, 2859–2862. [CrossRef]

58. Gibbins, H.L.; Yakubov, G.E.; Proctor, G.B.; Wilson, S.; Carpenter, G.H. What interactions drive the salivary mucosal pellicle formation? *Colloids Surf. B Biointerfaces* **2014**, *120*, 184–192. [CrossRef]

59. Mystkowska, J.; Niemirowicz-Laskowska, K.; Łysik, D.; Tokajuk, G.; Dąbrowski, J.R.; Bucki, R. The role of oral cavity biofilm on metallic biomaterial surface destruction–corrosion and friction aspects. *Int. J. Mol. Sci.* **2018**, *19*, 743. [CrossRef]

60. Łysik, D.; Niemirowicz-Laskowska, K.; Bucki, R.; Tokajuk, G.; Mystkowska, J. Artificial Saliva: Challenges and Future Perspectives for the Treatment of Xerostomia. *Int. J. Mol. Sci.* **2019**, *20*, 3199. [CrossRef]

61. Mystkowska, J.; Dabrowski, J.R.; Romanowska, J.; Klekotka, M.; Car, H.; Milewska, A.J. Artificial mucin-based saliva preparations—Physicochemical and tribological properties. *Oral Health Prev. Dent.* **2018**, *16*, 183–193. [PubMed]

62. Mystkowska, J.; Łysik, D.; Klekotka, M. Effect of saliva and mucin-based saliva substitutes on fretting processes of 316 austenitic stainless steel. *Metals* **2019**, *9*, 178. [CrossRef]

63. Mystkowska, J.; Jałbrzykowski, M.; Dabrowski, J.R. Tribological properties of selected self-made solutions of synthetic saliva. *Solid State Phenom.* **2013**, *199*, 567–572. [CrossRef]

64. Kouparitsas, I.K.; Mele, E.; Ronca, S. Synthesis and electrospinning of polycaprolactone from an aluminium-based catalyst: Influence of the ancillary ligand and initiators on catalytic efficiency and fibre structure. *Polymers* **2019**, *11*, 677. [CrossRef]

65. Rak, J.; Ford, J.L.; Rostron, C.; Walters, V. The preparation and characterization of poly(D,L-lactic acid) for use as a biodegradable drug carrier. *Pharm. Acta Helv.* **1985**, *60*, 162–169.

66. Jia, S.; Yu, D.; Zhu, Y.; Wang, Z.; Chen, L.; Fu, L. Morphology, crystallization and thermal behaviors of PLA-based composites: Wonderful effects of hybrid GO/PEG via dynamic impregnating. *Polymers* **2017**, *9*, 528. [CrossRef]

67. Liang, J.Z.; Zhou, L.; Tang, C.Y.; Tsui, C.P. Crystallization properties of polycaprolactone composites filled with nanometer calcium carbonate. *J. Appl. Polym. Sci.* **2013**, *128*, 2940–2944. [CrossRef]

68. Bansil, R.; Turner, B.S. Mucin structure, aggregation, physiological functions and biomedical applications. *Curr. Opin. Colloid Interface Sci.* **2006**, *11*, 164–170. [CrossRef]

69. Xu, L.; Crawford, K.; Gorman, C.B. Effects of temperature and pH on the degradation of poly(lactic acid) brushes. *Macromolecules* **2011**, *44*, 4777–4782. [CrossRef]

70. Li, S.; McCarthy, S. Further investigations on the hydrolytic degradation of poly (DL-lactide). *Biomaterials* **1999**, *20*, 35–44. [CrossRef]

71. Proikakis, C.S.; Mamouzelos, N.J.; Tarantili, P.A.; Andreopoulos, A.G. Swelling and hydrolytic degradation of poly(d,l-lactic acid) in aqueous solutions. *Polym. Degrad. Stab.* **2006**, *91*, 614–619. [CrossRef]

72. Loh, X.J. The effect of pH on the hydrolytic degradation of poly(ε-caprolactone) -block-poly(ethylene glycol) copolymers. *J. Appl. Polym. Sci.* **2013**, *127*, 2046–2056. [CrossRef]

73. Reich, G. Use of DSC to study the degradation behavior of PLA and PLGA microparticles. *Drug Dev. Ind. Pharm.* **1997**, *23*, 1177–1189. [CrossRef]

74. Paul, M.A.; Delcourt, C.; Alexandre, M.; Degée, P.; Monteverde, F.; Dubois, P. Polylactide/montmorillonite nanocomposites: Study of the hydrolytic degradation. *Polym. Degrad. Stab.* **2005**, *87*, 535–542. [CrossRef]

75. França, D.C.; Bezerra, E.B.; Morais, D.D.S.; Araújo, E.M.; Wellen, R.M.R. Effect of hydrolytic degradation on mechanical properties of PCL. *Mater. Sci. Forum* **2016**, *869*, 342–345. [CrossRef]

76. Wang, Z.; Wang, Y.; Ito, Y.; Zhang, P.; Chen, X. A comparative study on the *in vivo* degradation of poly(L-lactide) based composite implants for bone fracture fixation. *Sci. Rep.* **2016**, *6*, 20770. [CrossRef]

77. Sotres, J.; Jankovskaja, S.; Wannerberger, K.; Arnebrant, T. Ex-Vivo Force Spectroscopy of Intestinal Mucosa Reveals the Mechanical Properties of Mucus Blankets. *Sci. Rep.* **2017**, *7*, 7270. [CrossRef]

78. Cao, X.; Bansil, R.; Bhaskar, K.R.; Turner, B.S.; LaMont, J.T.; Niu, N.; Afdhal, N.H. pH-dependent conformational change of gastric mucin leads to sol-gel transition. *Biophys. J.* **1999**, *76*, 1250–1258. [CrossRef]

79. Bhaskar, K.R.; Gong, D.H.; Bansil, R.; Pajevic, S.; Hamilton, J.A.; Turner, B.S.; LaMont, J.T. Profound increase in viscosity and aggregation of pig gastric mucin at low pH. *Am. J. Physiol.* **1991**, *261*, G827–G832. [CrossRef]

80. Celli, J.P.; Turner, B.S.; Afdhal, N.H.; Ewoldt, R.H.; McKinley, G.H.; Bansil, R.; Erramilli, S. Rheology of gastric mucin exhibits a pH-dependent sol-gel transition. *Biomacromolecules* **2007**, *8*, 1580–1586. [CrossRef]

Article

Novel PEEK Copolymer Synthesis and Biosafety—I: Cytotoxicity Evaluation for Clinical Application

Joon Woo Chon [1,†], **Xin Yang** [2,†], **Seung Mook Lee** [2], **Young Jun Kim** [2,*], **In Sung Jeon** [3,†], **Jae Young Jho** [3,*] and **Dong June Chung** [1,*]

[1]   Department of Polymer Science & Engineering, Sungkyunkwan University, Suwon 16419, Korea; chonjoon@skku.edu
[2]   School of Chemical Engineering, Sungkyunkwan University, Suwon 16419, Korea; xin12.25@skku.edu (X.Y.); tmdanr1102@naver.com (S.M.L.)
[3]   School of Chemical and Biological Engineering, Seoul National University, Seoul 08826, Korea; ajeon88@gmail.com
*   Correspondence: youngkim@skku.edu (Y.J.K.); jyjho@snu.ac.kr (J.Y.J.); djchung@skku.edu (D.J.C.)
†   These persons contributed equally to this paper.

Received: 23 September 2019; Accepted: 1 November 2019; Published: 2 November 2019

Abstract: In this research, we synthesized novel polyetheretherketone (PEEK) copolymers and evaluated the biosafety and cytotoxicity of their composites for spinal cage applications in the orthopedic field. The PEEK copolymers and their composites were prepared through a solution polymerization method using diphenyl sulfone as a polymerization solvent. The composite of PEEK copolymer showed good mechanical properties similar to that of natural bone, and also showed good thermal characteristics for the processing of clinical use as spine cage. The results of an in vitro cytotoxicity test did not show any evidence of a toxic effect on the novel PEEK composite. On the basis of these cytotoxicity test results, the PEEK composite also proved its in vitro biosafety for application to an implantable spine cage.

Keywords: PEEK copolymer synthesis; PEEK composite; Spine cage application; In vitro biosafety

---

1. Introduction

The spine plays two important and distinct roles. It provides a strong central axis for the appendicular skeleton and protects the spinal cord and roots of delicate nerves connected to the brain. Therefore, the artificial bone graft materials used in spinal cage development must have proper strength and stiffness, as well as the capability to bond to vertebrae. Metallic materials, such as titanium alloy and stainless steel, have been commonly used in spinal cages with competent mechanical and biologically inert properties [1,2]. Despite the outstanding benefits of metallic materials for spinal cage applications, several concerns have been raised due to the stress shielding effect, low biocompatibility, release of ionic effluence, magnetic image interference, and additional second surgery required for removal. Typically, the Young's modulus of these alloys (titanium alloy: 80–125 GPa, magnesium alloy: 41–45 GPa) is shown to be higher than that of human bone (3–20 GPa). The mismatch of the Young's moduli between metallic materials and the human bone could induce a stress shielding effect on the bone [3], which can lead to implant loosening, bone thickening, and chronic inflammation [4].

To overcome the several concerns associated with the use of metallic materials, polyaryletherketones (PAEKs), including the polyetheretherketone (PEEK) group, have been employed as biomaterials for spinal cages [5]. PEEK polymers have specific thermal processing conditions due to their crystal structures [6]. Therefore, a copolymerization method between various PEEKs was applied to fit the mechanical properties of PEEK polymers [7]. PEEK polymers find applications as high performance super engineering plastics. However, due to its high glass transition temperature (T_g) and high melting point (T_m) [8,9] the processing temperature of PEEK is very high. Several modifications were reported in order to reduce the processing temperature, such as lowering the T_m and softening the backbone of the PEEK [10–12]. In the present paper, P(E2–E4)K was synthesized using HQ, PD, and 4,4′-DFBP. As the reactant, PD has the structure of "-ether-phenyl-ether-phenyl-ether-phenyl-ether" which contains four ethers (E4: EEEE). Therefore, the introduction of PD into P(E2–E4)K greatly affected the distribution of the flexible and rigid segments on their molecule chains.

PEEK polymers were growing interest for their clinical application as a plate in fracture fixation due to their stiffness and similar Young's moduli to that of the human bone [13]. According to recent reports, their Young's moduli can be fitted to closely match that of the human bone (3–20 GPa) by preparing a composite with a reinforcement treatment or modifying the chemical structure by increasing the molecular weight [14–16].

Regarding the mechanical strength of cortical bone, the flexural and compressive strength of human cortical bone have been reported to be in the range of 103–238 and 130–213 MPa, respectively [17,18]. Since the mechanical strength of PEEK is in the low range, PEEK should be further reinforced to fit in the high range in order to be applied for various patients. One of carbon fillers that is commonly used as a reinforcement for the spinal cage is carbon fiber (CF) because of its good mechanical properties, thermal resistance, wearability, and biocompatibility [19,20]. Another carbon filler that can be used is graphene oxide (GO), a promising filler for both mechanical and biological applications due to its high surface area and hydrophilic oxygen-rich functional groups on the surface [21]. Several studies show that the mechanical properties of various polymers were enhanced by adding GO but decreased after adding 1 wt % of GO or more due to the aggregation of GO nanosheets [22–24]. According to recent studies, a low concentration of GO has been confirmed to be non-toxic and biocompatible because of the oxygen-rich functional groups which enhance cell adhesion and spreading [25–27].

In this study, we aimed to verify the potential in vitro biocompatibility of various reinforced PEEK composites for applicability to spinal surgery. Further, to enhance mechanical properties, a PEEK composite which blended with novel PEEK copolymers and various reinforcements was developed. An in vitro analysis was performed using an MTT assay to confirm the cytotoxicity of the PEEK composites. A live/dead cell assay was used to visually determine the cell viability. The cell activity via the charge of phosphatase derived from the living cell membrane was detected by the alkaline phosphatase (ALP) assay.

2. Materials and Methods

2.1. Materials

Hydroquinone (HQ), sodium carbonate (Na_2CO_3), phosphate-buffered saline (PBS), penicillin-streptomycin, trypsin-EDTA, a Live/Dead Cell assay kit, and MTT-assay kit were purchased from Sigma-Aldrich Chem. Co. (St. Louis, MO, USA). Further, 4,4′-Difluorobenzophenone (4,4′-DFBP) and diphenyl sulfone (DPS) were purchased from TCI Co. Ltd. (Tokyo, Japan). Fetal bovine serum and Dulbecco's modified Eagle medium (DMEM) were purchased from GE Healthcare Life Sciences (Logan, UT, USA). 4,4′-(1,4Phenylenebis(oxy)) diphenol (PD) was kindly provided by the Organosynthesis Lab., Dept. of Chemistry of Incheon National University (Incheon, Korea). The ALP assay kit was purchased from AnaSpec. Co. Inc. (Fremont, CA, USA). L-929 mouse fibroblast cells were obtained from the Korean Cell Line Bank Co. (Seoul, Korea).

P(E2–E4)K powder, graphene oxide powder (purity: >99 wt %, lateral dimension: $\geq$7 μm, thickness: 1.1~1.3 nm, LS-Chem, Ochang, Korea), and milled carbon fiber (PX 35 Milled Carbon Fibers, average length: 150 μm, average diameter: 7.2 μm, specific gravity: 1.81 g/cm^3, Zoltek Co., St. Louis, MO, USA) were used for the preparation of composite samples using synthesized PEEK copolymers.

2.2. PEEK Polymerization and Composite Preparation

Novel PEEK (P(E2)K and P(E4)K) and its copolymer (P(E2–E4)K) were polymerized by a solution polymerization method with various formulation for controlling the number of ether groups in the repeating unit. (Figure 1) After preparing PEEK and its copolymer, two kinds of reinforcements (CF: carbon fiber and GO: graphene oxide) were blended with the PEEK copolymers using suspension blending method (Table 1). The contents of GO were fixed to 0.5 wt % (see Supplementary Materials).

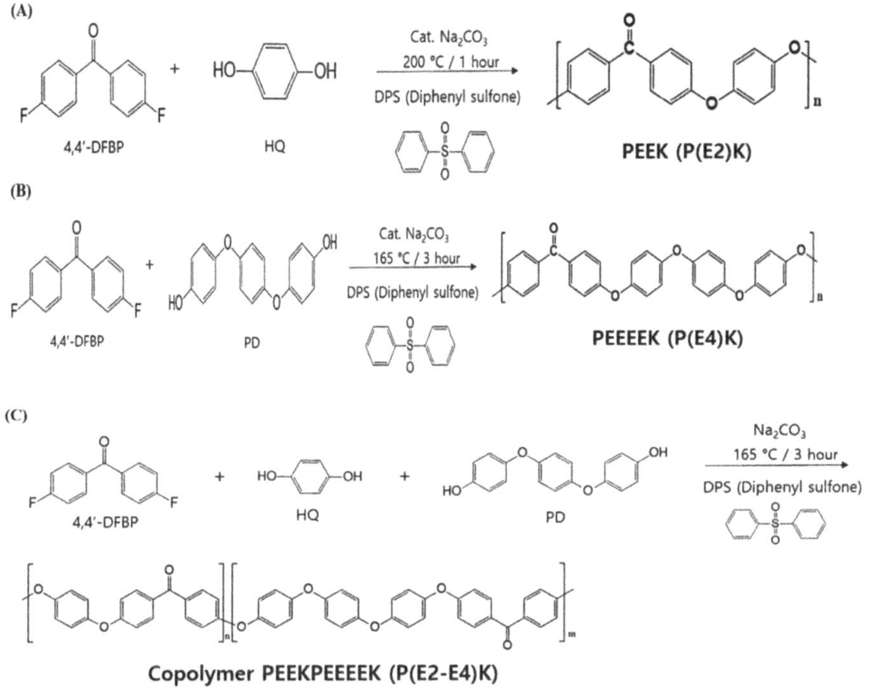

Figure 1. Chemical synthesis schemes of the PEEK polymers for controlling the number of ether groups in the repeating unit. (**A**) PEEK polymerization for P(E2)K, (**B**) PEEEEK polymerization for P(E4)K, and (**C**) the copolymerization of P(E2)K and P(E4)K to P(E2–E4)K.

Table 1. Formulation ratios of the PEEK composites.

Sample No.	Samples Composition (wt %)
No. 1	P(E2-E4)K only
No. 2	P(E2-E4)K/CF(30)
No. 3	P(E2-E4)K/GO(0.5)/CF(30)

2.2.1. Synthesis of Poly(ether ether ketone) P(E2)K

A mixture of hydroquinone (HQ, 10.00 mmol), 4,4′-difluorobenzophenone (4,4′-DFBP, 10.00 mmol), ground sodium carbonate (Na_2CO_3, 10.2 mmol), and diphenyl sulfone (DPS, 200 wt %) was purged by vacuum/nitrogen cycles using a Schlenk line. This mixture was stirred under nitrogen atmosphere and heated at 160 °C for 2 h, 250 °C for 2 h, at 320 °C for 1 h, successively. The obtained grey slurry was cooled to room temperature and grounded to a fine powder. The obtained product was washed in hot acetone (3 times) and hot DI water (3 times). Finally, the product was dried under high vacuum (ca. 0.05 mbar) at 140 °C for 12 h, to afford a grey solid.

2.2.2. Synthesis of Poly(ether ether ether ether ketone) P(E4)K

A mixture of 4,4′-(1,4phenylenebis(oxy)) diphenol (PD, 10.00 mmol), 4,4′-DFBP (10.00 mmol), Na_2CO_3 (10.2 mmol), and DPS (200 wt %) was purged by vacuum/nitrogen cycles using a Schlenk line. This mixture was stirred under nitrogen atmosphere and heated successively at 160 °C for 2 h, at 235 °C for 2 h, and finally at 290 °C for 5 h. The obtained brown slurry was cooled to room temperature and grounded to a fine powder. The obtained product was washed with hot acetone (3 times) and hot DI water (3 times). Finally, the product was dried under high vacuum (ca. 0.05 mbar) at 140 °C for 12 h, to produce a brown solid.

2.2.3. Synthesis of the Random Copolymer P(E2–E4)K

A mixture of HQ (9.00 mmol), PD (1.00 mmol), 4,4′-DFBP (10.00 mmol), Na_2CO_3 (10.2 mmol), and DPS (200 wt %) was purged by vacuum/nitrogen cycles using a Schlenk line. This mixture was stirred under nitrogen atmosphere and successively heated at 160 °C for 2 h, 240 °C for a further 2 h, and finally at 300 °C for 5 h. The obtained grey slurry was cooled to room temperature and ground to a fine powder. The obtained product was washed with hot acetone (3 times) and hot DI water (3 times). Finally, the product was dried under high vacuum (ca. 0.05 mbar) at 140 °C for 12 h, to produce a grey solid.

2.2.4. Fabrication of the PEEK Composite

To prepare the blend powders, P(E2–E4)K powder and fillers were separately dispersed in a beaker containing ethanol, followed by ultrasonication for 30 min. Subsequently, each dispersed powder sample was mixed with another using a magnetic stirrer for 12 h, according to the composition provided in Table 1. The composite suspension was filtered and vacuum-dried at 60 °C for 24 h.

For an evaluation of the mechanical properties and bio-safety, the blended powder samples were molded into $10 \times 10 \times 4$ mm^3 plates using a mini-injection molding machine (Bautek Co., Uijeongbu-si, Korea) at processing temperatures of 390 °C, with a preset molding temperature of 190 °C. A flexural test specimen of $80 \times 10 \times 4$ mm^3 and compression test specimen of $10 \times 10 \times 4$ mm^3 were also prepared using the above procedure. To ensure a similar degree of crystallinity, all the samples were annealed at 220 °C for 4 h.

2.3. Characterization and Mechanical Properties of the Synthesized PEEK Polymers and Composites

Using an Ubbelohde viscometer (Xylem Inc., Rye Brook, New York, USA), the inherent viscosities (η_{inh}*) of the synthesized PEEK polymers (including copolymer) solutions (dissolved in concentrated sulfuric acid (98%) with a concentration of 0.5 g/dL) were measured at 30 °C. To confirm the thermal properties, differential scanning calorimetric (DSC) was measured with polymer powder samples using a DSC Q20 (TA Instruments, New Castle, PA, USA) under nitrogen flow, with a heating rate of 10 °C/min (25 to 350 °C). Perkin Elmer TGA 7 thermal analyzer (Perkin Elmer, Boston, MA, USA) was used to characterize the thermogravimetric properties of synthesized PEEK polymers. In brief, 10 mg of sample was placed in a sample pan and test was performed under nitrogen flow, with a heating rate of 10 °C /min (50 to 800 °C). Fourier transform infrared spectroscopy (FT-IR) spectra were obtained using a Nicolet™ iSTM 50 (Thermo Fisher Scientific, Waltham, MA, USA). Spectra were recorded using a spectral width ranging from 500 to 2000 cm^{-1} and an accumulation of 32 scans.

The flexural strength and compressive strength were measured according to the ISO 178 and ISO 604 standards, respectively. A universal test machine (Lloyd LR10K, West Sussex, UK) with a load cell of 10 kN was used for the test. The cross-head speeds were 2 mm/min and 1 mm/min for flexural and compression testing, respectively. The average of five measurements was obtained from seven specimens for each test.

2.4. In vitro Cytotoxicity Test of the PEEK Composites

Every sample for in vitro cytotoxicity test was extracted for 72 h at 36.5 °C with 4 g of P(E2–E4)K composites in 20 ml PBS. The formulation ratios of the additives in the P(E2–E4)K composites samples are shown in Table 1. Every cell experiment (The MTT-assay and ALP-assay) showing the cell viability and cell activity were carried out according the ISO 10993–5 standard. All experiments were carried out 5 times. Briefly, L-929 mouse fibroblasts were seeded (1×10^5 cell/mL) into each well of a tissue culture polystyrene dish (TCPS). The cell was incubated at 36.5 °C in a 5% CO_2 environment with DMEM medium containing 10% FBS and 1% penicillin-streptomycin. For the assay, each extracted sample solution described above was added to the TCPS wells and incubated with the cells for 2 days.

For the MTT-assay, DMEM (100 µL) and MTT (25 µL) solution were added to each well after 1%~20% of diluted extracted solution (100 µL) and the mixture was incubated for 4 h under the same conditions. After additional incubation, the supernatant was removed from each well by aspiration. Next, DMSO (100 µL) and glycine buffer (12.5 µL) were sequentially added to each well. For the ALP-assay, the same cell cultivation process as the MTT-assay was performed. To determine the cell viability and activity, the ultraviolet-visible (UV-Vis) absorbance of the solutions in each well was measured at 570 nm for the MTT-assay and at 405 nm for the ALP-assay using a SpectraMax M5 microplate reader (Molecular Devices Korea LLC, Seoul, Korea). For the live/dead Cell assay, samples were prepared using the same extraction protocol as the MTT assay. The specific extracted concentration was harsher than MTT assay (100% extracts solution/diluted solution to 50% concentration of original extracts) [28]. Next, calcein-AM and propidium iodide (PI) were added to each well followed by incubation for 15 min at 36.5 °C in a 5 % CO_2 atmosphere. Fluorescence images were captured and analyzed using a Nikon Eclipse Ti microscope (Nikon Instruments Inc., Tokyo, Japan).

3. Results and Discussion

3.1. Synthesis and Characterizations of P(E2)K, P(E4)K, and P(E2–E4)K

Aromatic poly(ether ketone)s and copolymers were synthesized by typical nucleophilic displacement reaction. As shown in Figure 1, P(E2)K was prepared by HQ and 4,4′-DFBP with a molar ratio of 1:1.P(E4)K was prepared with PD and 4,4′-DFBP. P(E2)K showed a Tm value that was too high to process with extrusion or an injection molding, and P(E4)K had a relatively low molecular weight that diminished the mechanical properties. Therefore, its copolymers were synthesized to address polymer's disadvantages. P(E2–E4)K was synthesized using HQ, PD, and 4,4′-DFBP. These three polymers were prepared in diphenyl sulfone under similar conditions (Table 2) and their molecule chains all consisted of a flexible segment (-phenyl-ether-phenyl-) and rigid segment (-phenyl-ketone-phenyl-). The reactant, PD, has the structure of "-ether-phenyl-ether-phenyl-ether-phenyl-ether", which contains four ethers (E4: EEEE). Therefore, the introduction of PD into P(E2–E4)K greatly affected the distribution of the flexible and rigid segments on their molecule chains.

In both cases, the obtained polymers were insoluble in a wide range of organic solvents, except strong Brønsted acids (e.g., concentrated H_2SO_4). Therefore, the inherent viscosities were determined by measuring the flow time of their solutions in concentrated sulfuric acid (H_2SO_4) at 30°C. These polymers showed inherent viscosity values from 0.63 to 1.43 dL/g. P(E4)K showed the lowest inherent viscosity value among all prepared polymers. We assumed that the introduction of ether groups reduced the thermal stability of P(E4)K. However, a considerable reduction in viscosity was not observed for P(E2–E4)K copolymers where PD feed was 10%. Since we were unable to find the exact values of Mark–Houwink constants of K and α. We could not estimate the Mv of these polymers. Here, we present the inherent viscosity values instead. Comparing with previously reported inherent viscosity values of PEEK [10,29], we are certain that high molecular weight PEEK homo and copolymers are prepared.

The FTIR spectroscopy was further utilized to characterize the obtained polymers for estimating the copolymer composition. Figure 2 shows FTIR spectra of the PEEK homo and copolymers (absorbance vs. wavenumber). The bands at 1250, 1489 and 1650 cm^{-1} are due to C–O–C stretch, the aromatic C=C stretch and C=O stretch, respectively. The bands at 1489 and 1650 cm^{-1} are used as analytical bands since they are well separated from other bands. No specific differences between the three samples from the FTIR spectra because the functional groups of them are same. However, the phenyl group density of P(E4)K is higher than that of P(E2)K, which is used for estimating polymer composition. The ratio of peak area of C=O to aromatic C=C for P(E2)K, P(E4)K, and P(E2–E4)K were 2.43, 4.72, and 2.56. Considering the theoretical peak area ratio for P(E2–E4)K is 2.66, the observed peak area ratio of 2.56 for the copolymer can be accepted.

Table 2. Synthesis data for the preparation of aromatic PEEK homo and copolymers.

Polymer	Polymerization Solvent	Polymerization Temperature (°C)	Reaction Time (hr)	Yield (%)	η_{inh}* (dL/g)
P(E2)K	Ph$_2$SO$_2$ (DPS)	160/250/320	2/2/1	96	1.20
P(E4)K	Ph$_2$SO$_2$ (DPS)	160/235/290	2/2/5	94	0.63
P(E2–E4)K	Ph$_2$SO$_2$ (DPS)	160/240/300	2/4/5	97	1.43

* Inherent viscosity measured in a 'U'-tube viscometer for a 0.5 g/dL solutions in 96% H$_2$SO$_4$ at 30 °C.

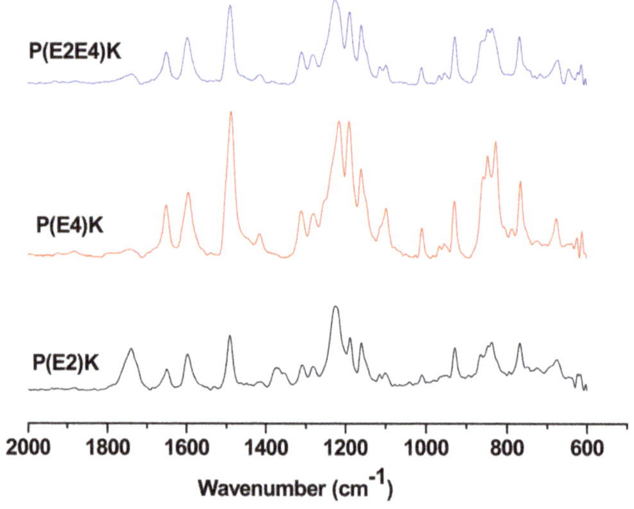

Figure 2. FT-IR spectra of the PEEK homo and copolymers. (Absorbance vs. Wavenumber).

3.2. Thermal Properties of P(E2)K, P(E4)K, and P(E2–E4)K

As shown in Figure 3, the T_m values of P(E2)K, P(E4)K, and P(E2–E4)K were investigated by DSC, and were 343, 315, and 333 °C respectively. As one of the most important parameters of high-performance plastics, the thermal property results indicate that introduction of a flexible segment decrease T_m. The structure and property (T_m and T_g) relationships for poly(ether ether ketone) are well known [30,31]. The incorporation of a flexible ether linkages (C–O–C) decreases T_m and T_g and the incorporation of ketone linkages increase T_m and T_g.

The TGA was performed under a nitrogen atmosphere. As shown in Figure 4, the three polymers exhibited excellent thermal stability. Their onset weight loss temperatures were all greater than 480 °C, particularly that of the P(E2–E4)K, probably owing to the relatively high molecular weight. These three polymers exhibited excellent properties based on the TGA and DSC test results which demonstrated that the sequence distributions did not affect their thermal properties.

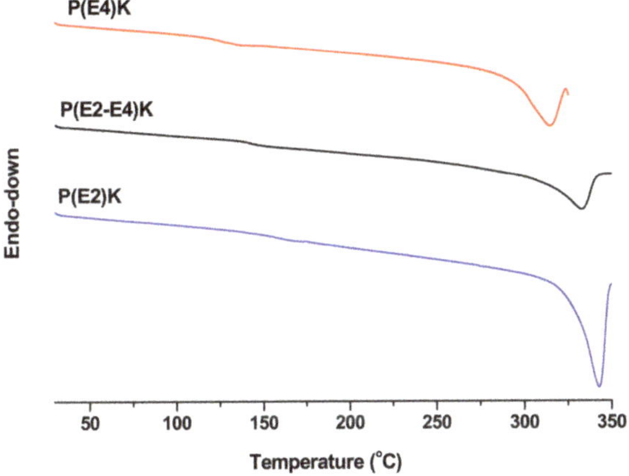

Figure 3. DSC thermograms of the PEEK homo and copolymers.

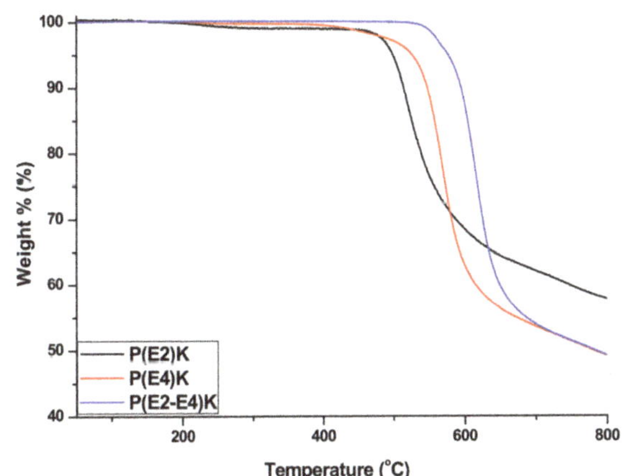

Figure 4. TGA curves of the PEEK homo and copolymers in a nitrogen atmosphere.

3.3. Mechanical Properties

The flexural and compressive strength test results are shown in Table 3. The flexural strengths for the samples No. 1, No. 2, and No. 3 are 155, 240, and 250 MPa, respectively. The relationship between the flexural stress and strain in the composite samples is demonstrated in Figure 5. No. 2 and No. 3 exhibited higher yield stress and modulus but lower break energy compared to those of No. 1. The flexural strength of No. 2 and No. 3 were enhanced by 54.8% and 61.3%, respectively compared to that of No. 1 due to the high stiffness and modulus characteristics of the CFs [32]. The addition of GO increased the flexural strength by 10 MPa from No. 2 sample since GO plays a role in load transfer through strong interaction with copolymer matrix (see Supplementary Materials).

The incorporation of GO improves the mechanical strength through two mechanisms: formation of π–π stacking interaction between the conjugated structure of GO and the benzene rings of copolymer, and hydrogen bonding between hydroxyl and carboxylic groups of GO and carbonyl group in the matrix [33,34]. The composite samples generally showed a sharp drop in strain after 2.8% due to the rigid characteristics of CF and GO fillers. This suggests a shift from ductile to brittle fracture behavior. As shown in Figure 6a, sample No. 1 showed smooth surface in conjunction with river patterns, indicating a typical ductile fracture behavior. When GO and CFs were added (Figure 6b–c), the fracture surface became much rougher with multilayer structure due to many interface debonding and pullout of CFs. Most of CFs in both No. 2 and No. 3 showed a smooth surface with little copolymer attached, indicating a weak interfacial adhesion between CF and copolymer matrix. This could be the main reason for the failure mode.

The compressive strengths for No. 1, No. 2, and No. 3 in MPa were 125, 161, and 164, respectively, as shown in Table 3. Compared to the compressive strength of No. 1, those of No. 2 and No. 3 exhibited a sharp increase by 28.9% and 31.2%, respectively. As shown in Figure 7, the carbon fillers not only increased both yield strength and modulus, but also reduced the break energy, similar to the flexural strength test results. Furthermore, the compressive strength was slightly increased by 1.8% with the addition of GO. However, the improved compressive strength was primarily governed by the mechanical properties of the CFs rather than the properties of GO.

Table 3. Flexural and compressive strength test results of the P(E2–E4)K composites.

Sample Name	Sample Composition (wt %)	Flexural Strength (MPa)	Compressive Strength (MPa)
No. 1	P(E2–E4)K	155 ± 3.6	125 ± 1.5
No. 2	P(E2–E4)K/CF(30)	240 ± 3.9	161 ± 2.9
No. 3	P(E2–E4)K/GO(0.5)/CF(30)	250 ± 4.0	164 ± 3.1

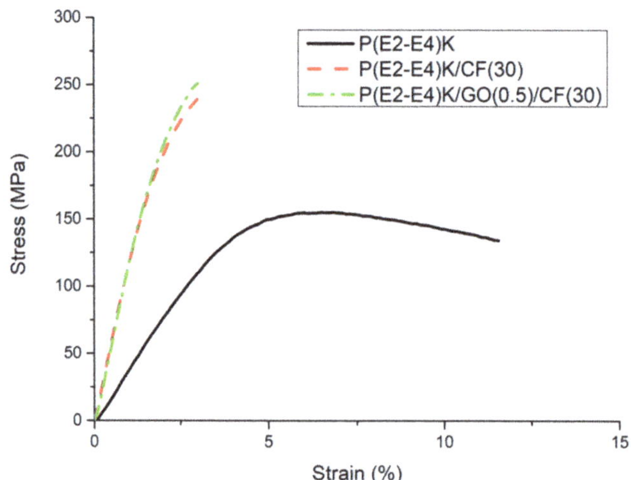

Figure 5. Flexural stress-strain curves of the P(E2–E4)K composite samples.

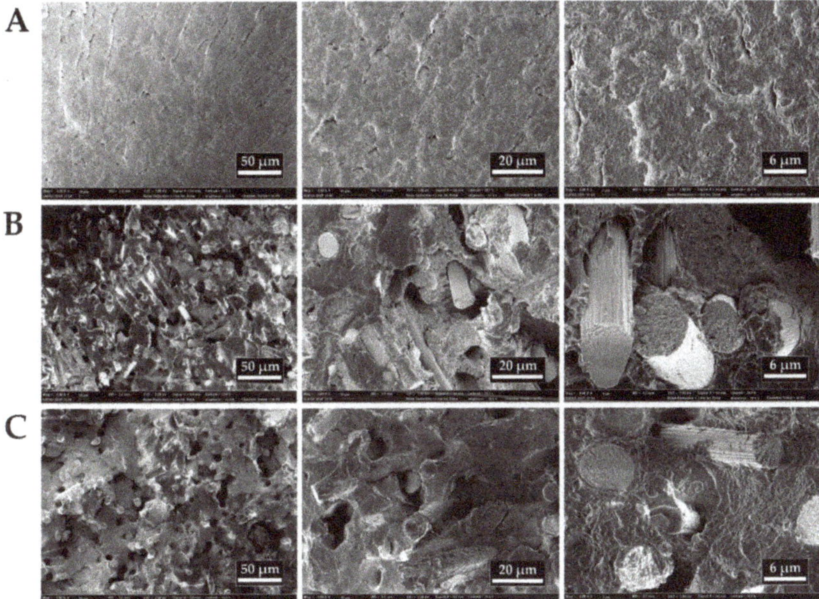

Figure 6. SEM images of flexural fracture surfaces of (**A**) No. 1, (**B**) No. 2, and (**C**) No. 3 samples.

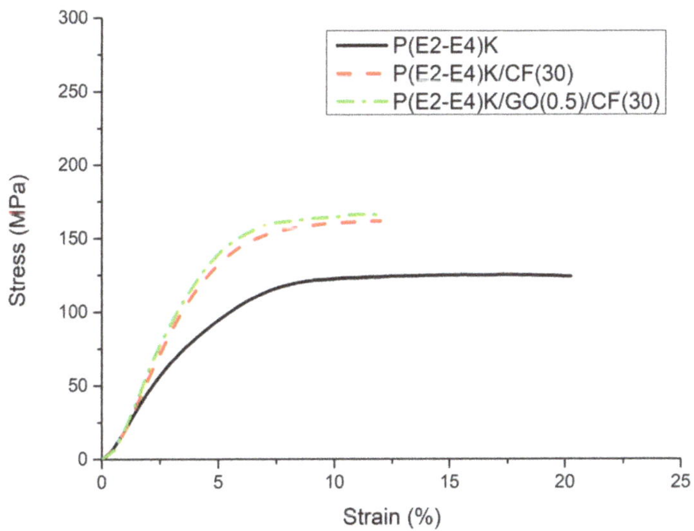

Figure 7. Compression stress-strain curves of the P(E2–E4)K composite samples.

3.4. In Vitro Cytotoxicity Test of the PEEK Composites

As an effective assessment tool for cytotoxicity, the reduction of mitochondria in living cells was evaluated to convert MTT tetrazolium (yellow) into MTT formazan (purple) to confirm the cell viability [35]. Variation of UV-Vis absorption forces at 570 nm varies in proportion to the reduction of MTT formazan. As shown in Figure 8, even if the concentration of the extract increased from 1% to 20% of the total volume of the medium, the cell survival capability at the equivalent advantage level (over 80%) was shown. As a result, all the samples showed non-toxic behavior in accordance with ISO 10993–5 standards. The ALP-assay kit was used to detect changes in the phosphate activity in living cell membranes to confirm the cell activity. As shown in Figure 9, P(E2–E4)K composite samples with various reinforcements showed an appropriate level of ALP activity compared to the neat sample (No. 1) case. Even when the concentration increased to 20% of the total media volume, the ALP activity was greater (or nearly identical) than that of the neat sample. As shown in the results of the live/dead cell assay (Figure 10), each sample had abundant live cells (green colored cells) without any dead cells (red colored cells), even when the concentration was greater than those in the other cytotoxicity tests. Furthermore, the equivalent cell population behavior was observed between the control group and each sample. These results mean that the PEEK composites containing various types of additives exerted no toxic effects and exhibited appropriate cell activity according to the in vitro cytotoxicity analysis data, per the ISO standard.

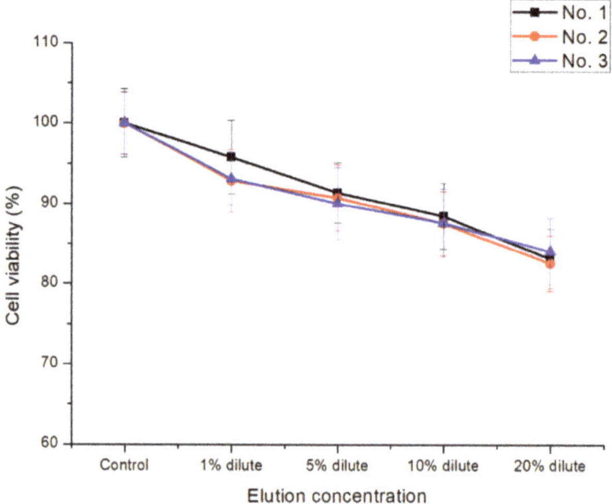

Figure 8. Cell viability results of the P(E2–E4)K composites via the MTT assay. Statistical analysis was conducted by one-way analysis of variance (ANOVA) and Turkey's test; $p < 0.05$.

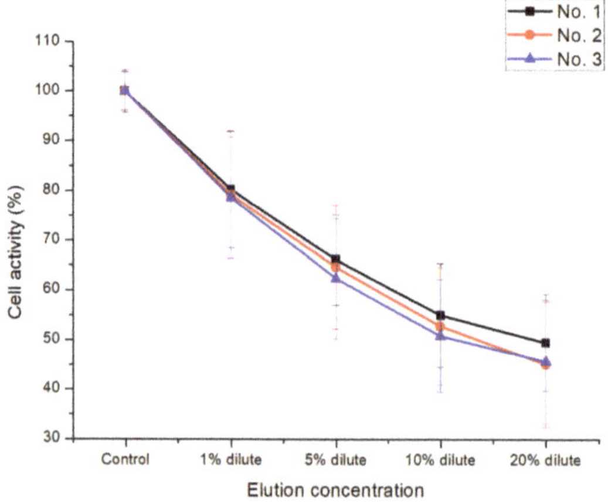

Figure 9. Cell activity results of the P(E2–E4)K composites via the ALP assay. Statistical analysis was conducted by one-way analysis of variance (ANOVA) and Turkey's test; $p < 0.05$.

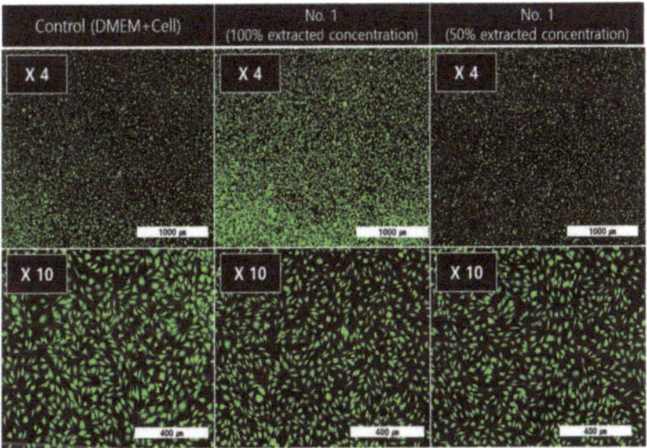

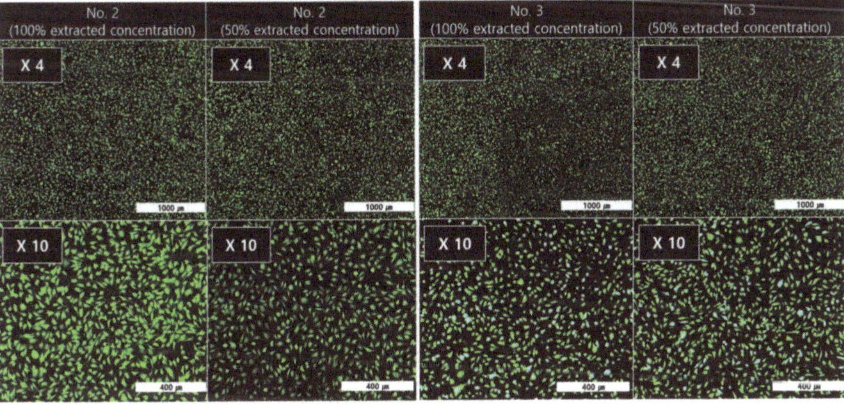

Figure 10. Images of cell viability via Live/Dead Cell-assay.

4. Conclusions

In this study, novel PEEK composites were synthesized for use in spinal cages as a prosthetic. (P(E2)K, P(E4)K), and their copolymer were synthesized via a typical nucleophilic displacement reaction, and a P(E2–E4)K composite with various reinforcements was prepared. The P(E2–E4)K composite showed enhanced thermal and mechanical properties compared to P(E2)K and P(E4)K for spinal cage applications. Based on the in vitro analysis, all the P(E2–E4)K composite samples were classified as non-toxic according to the ISO standard. Furthermore, each P(E2–E4)K composite sample had low cytotoxicity, even when its extracted volume increased to 20%. Based on the ALP assay, the P(E2–E4)K composite sample showed an appropriate positive value, indicating the capability of adjusting the PEEK composition for use in a spinal cage. These cytotoxicity data promote the applicability of the PEEK composites for orthopedic surgery.

Supplementary Materials: The following are available online at http://www.mdpi.com/2073-4360/11/11/1803/s1.

Author Contributions: Conceptualization, J.W.C. and D.J.C.; methodology, J.W.C. and D.J.C.; formal analysis, X.Y. and S.M.L.; investigation, J.W.C., X.Y., S.M.L., and I.S.J.; writing—original draft preparation, J.W.C., X.Y., and I.S.J.; writing—review and editing, J.W.C. and D.J.C.; supervision, Y.J.K.; project administration, J.Y.J.

Funding: This work was supported by the Technology Innovation Program (Project No.; 10062163, Project Title; Developing economical manufacturing techniques for aromatic polyketones and their reinforced composites (compressive strength > 180 MPa) used for medical applications), which is funded by the Ministry of Trade, industry & Energy (MOTIE, Korea).

Acknowledgments: We appreciate Jin Ho Kim (Department of Chemistry, Incheon National University, Incheon, 22012, Korea) for providing various types of novel monomers for this work.

Conflicts of Interest: The authors declare no conflict of interest.

References

1.	Park, J.B.; Lakes, R.S. *Biomaterials: An Introduction*, 3rd ed.; Springer: Berlin/Heidelberg, Germany, 2007; pp. 100–136.
2.	Blackwood, D.J. Biomaterials: Past Successes and Future Problems. *Corros. Rev.* **2003**, *21*, 97–124. [CrossRef]
3.	Seal, C.K.; Vince, K.; Hodgson, M.A. Biodegradable surgical implants based on magnesium alloys—A review of current research. *IOP Conf. Ser. Mater. Sci. Eng.* **2009**, *4*, 012011. [CrossRef]
4.	Salahshoor, M.; Guo, Y. Biodegradable Orthopedic Magnesium-Calcium (MgCa) Alloys, Processing, and Corrosion Performance. *Materials* **2012**, *5*, 135–155. [CrossRef] [PubMed]
5.	Kurtz, S.M.; Devine, J.N. PEEK biomaterials in trauma, orthopedic, and spinal implants. *Biomaterials* **2007**, *28*, 4845–4869. [CrossRef]
6.	James, C.S. Polyetheretherketone (PEEK): Processing-structure and properties studies for a matrix in high performance composites. *Polym. Compos.* **1986**, *7*, 159–169.
7.	Christian, B.; David, J.W.; Frank, E.K.; William, J.M. The sodium salts of sulphonated poly(aryl-ether-ether-ketone) (PEEK): Preparation and characterization. *Polymer* **1987**, *28*, 1009–1016.
8.	Dawson, P.C.; Blundell, D.J. X-ray data for poly(aryl ether ketones). *Polymer* **1980**, *21*, 577–578. [CrossRef]
9.	Mehmet-Alkan, A.A.; Hay, J.N. The crystallinity of PEEK composites. *Polymer* **1993**, *34*, 3529–3531. [CrossRef]
10.	Attwood, T.E.; Dawson, P.C.; Freeman, J.L.; Hoy, L.R.; Rose, J.B.; Staniland, P.A. Synthesis and properties of polyaryletherketones. *Polymer* **1981**, *22*, 1096–1103. [CrossRef]
11.	Clendlinning, R.A. Poly(aryl ether ketone) Block Copolymers. U.S. Patent 4,786,694, 22 November 1988.
12.	Clendlinning, R.A. Modified poly(aryl ether ketones) derived from biphenol. E.P. Patent 0,266,132, 4 May 1987.
13.	Skinner, H.B. Composite technology for total hip arthroplasty. *Clin. Orthop. Relat. Res.* **1988**, *235*, 224–236. [CrossRef]
14.	Mark, H.F.; Bikales, N.M.; Overberger, C.G.; Menges, G.; Kroschiwitz, J.I. *Encyclopedia of Polymer Science and Engineering*, 2nd ed.; Wiley: New York, NY, USA, 1988; pp. 313–320.
15.	Roel, F.M.R.K.; Steven, M.V.G.; Arthur, D.G.; Oner, F.C. Polyetheretherketone (PEEK) cages in cervical applications: A systematic review. *Spine J.* **2015**, *15*, 1446–1460.
16.	Hongli, P.; Hailong, T.; Shaobo, Q.; Ning, W.; Yu, Q.W. Progress of titanium strut for cervical reconstruction with nano-graphene oxide loaded hydroxyapatite/polyamide composite and interbody fusion after corpectomy with anterior plate fixation. *Artif. Cells Nanomed. Biotechnol.* **2019**, *47*, 3094–3100.
17.	Caeiro, J.R.; González, P.; Guede, D. Biomechanics and bone (& II): Trials in different hierarchical levels of bone and alternative tools for the determination of bone strength. *Rev. Osteoporos. Metab. Miner.* **2013**, *5*, 99–108.
18.	Zioupos, P.; Currey, J.D. Changes in the stiffness, strength, and toughness of human cortical bone with age. *Bone* **1998**, *22*, 57–66. [CrossRef]
19.	Saleem, A.; Frormann, L.; Iqbal, A. High performance thermoplastic composites: Study on the mechanical, thermal, and electrical resistivity properties of carbon fiber-reinforced polyetheretherketone and polyethersulphone. *Polym. Compos.* **2007**, *28*, 785–796. [CrossRef]
20.	Moore, R.; Beredjiklian, P.; Rhoad, R.; Theiss, S.; Cuckler, J.; Ducheyne, P.; Baker, D.G. A comparison of the inflammatory potential of particulates derived from two composite materials. *J. Biomed. Mater. Res.* **1997**, *34*, 137–147. [CrossRef]
21.	Shin, S.R.; Zihlmann, C.; Akbari, M.; Assawes, P.; Cheung, L.; Zhang, K.; Manoharan, V.; Zhang, Y.S.; Yüksekkaya, M.; Wan, K.-T.; et al. Reduced graphene oxide-GelMA hybrid hydrogels as scaffolds for cardiac tissue engineering. *Small* **2016**, *12*, 3677–3689. [CrossRef]
22.	Pang, W.; Ni, Z.; Chen, G.; Huang, G.; Huang, H.; Zhao, Y. Mechanical and thermal properties of graphene oxide/ultrahigh molecular weight polyethylene nanocomposites. *RSC Adv.* **2015**, *5*, 63063–63072. [CrossRef]
23.	Wang, Y.; Shi, Z.; Fang, J.; Xu, H.; Yin, J. Graphene oxide/polybenzimidazole composites fabricated by a solvent-exchange method. *Carbon* **2011**, *49*, 1199–1207. [CrossRef]

24. He, M.; Chen, X.; Guo, Z.; Qiu, X.; Yang, Y.; Su, C.; Jiang, N.; Li, Y.; Sun, D.; Zhang, L. Super tough graphene oxide reinforced polyetheretherketone for potential hard tissue repair applications. *Compos. Sci. Technol.* **2019**, *174*, 194–201. [CrossRef]

25. Wang, K.; Ruan, J.; Song, H.; Zhang, J.; Wo, Y.; Guo, S.; Cui, D. Biocompatibility of graphene oxide. *Nanoscale Res. Lett.* **2011**, *6*, 8. [CrossRef] [PubMed]

26. Girase, B.; Shah, J.S.; Misra, R.D.K. Cellular mechanics of modulated osteoblasts functions in graphene oxide reinforced elastomers. *Adv. Eng. Mater.* **2012**, *14*, B101. [CrossRef]

27. Chung, C.; Kim, Y.K.; Shin, D.; Ryoo, S.R.; Hong, B.H.; Min, D.H. Biomedical applications of graphene and graphene oxide. *Acc. Chem. Res.* **2013**, *46*, 2211–2224. [CrossRef] [PubMed]

28. ISO 10993-5. *International Standard*, 3rd ed.; ISO: Geneva, Switzerland, 2009.

29. Mitsuru, U.; Masaki, S. Synthesis of Aromatic Poly (ether ketones). *Macromolecules* **1987**, *20*, 2675–2678.

30. Hsieh, K.H.; Chern, Y.C.; Ho, K.S.; Wang, Y.Z.; Chan, B.W.; Chen, L.W. Effect of dihydroxydiphenyl ether on Poly (ether ether ketone). *J. Polym. Res.* **1996**, *3*, 83–88. [CrossRef]

31. Rao, V.L.; Sivadasan, P. Synthesis and properties of polyether ketones. *Eur. Polym. J.* **1994**, *30*, 1381–1388. [CrossRef]

32. Bhatt, P.; Goel, A. Carbon Fibres: Production, Properties and Potential Use. *Mater. Sci. Res. India* **2017**, *14*, 52–57. [CrossRef]

33. Peng, S.; Feng, P.; Wu, P.; Huang, W.; Yang, Y.; Guo, W.; Gao, C.; Shuai, C. Graphene oxide as an interface phase between polyetheretherketone and hydroxyapatite for tissue engineering scaffolds. *Sci. Rep.* **2017**, *7*, 46604. [CrossRef]

34. Heo, Y.; Im, H.; Kim, J. The effect of sulfonated graphene oxide on Sulfonated Poly (Ether Ether Ketone) membrane for direct methanol fuel cells. *J. Membr. Sci.* **2013**, *425*, 11–22. [CrossRef]

35. Johan, V.M.; Gertjan, J.L.K.; Jacqueline, C. *Cancer Cell Culture*; Humana Press: Totawa, NJ, USA, 2011; pp. 237–245.

 *polymers*

Article

One Year Evaluation of Material Properties Changes of Polylactide Parts in Various Hydrolytic Degradation Conditions

Angela Andrzejewska

Department of Biomedical Engineering, Faculty of Mechanical Engineering, UTP University of Science and Technology in Bydgoszcz, Prof. S. Kaliskiego 7 Avenue, 85-796 Bydgoszcz, Poland; angela.andrzejewska@utp.edu.pl

Received: 20 August 2019; Accepted: 11 September 2019; Published: 13 September 2019

Abstract: Biodegradable biocompatible materials are widely used in medical applications. Determining the possibility of using biodegradable materials depends on determining the changes in their parameters over time due to degradation. The current scientific research on biodegradable materials has presented results based on research methods characterized by the different geometry and cross-section size of the specimen, type of degradation medium, or different pH value of the medium or maximum degradation time. This paper presents the results of a one-year study on the influence of the type of degradation medium on the changes in mechanical behavior and the uptake of the degradation medium by biodegradable specimens with large cross-sections. In addition, a prototype of a test stand was created, which allowed for the specimens to be stored vertically to ensure regular medium exposure and eliminate the interaction of the surface of the tested specimens with the sides of the container. The obtained results allowed the statistical significance of differences in the mechanical parameters determined in the uniaxial tensile test after 2, 4, 6, 12, 26, 39, and 52 weeks of degradation to be indicated depending on the type of degradation medium. It was proven that the changes in mechanical behavior depend on the percentage change in the mass of the specimens during degradation. The percentage change in mass depends on the type of degradation medium. Based on the results of this research, it was noted that in long-term degradation above 12 weeks, buffered sodium chloride solution is the optimal choice for the degradation medium. However, distilled water or physiological saline solution can be used as an alternative during the degradation period for up to 12 weeks.

Keywords: polylactide; hydrolytic degradation; mechanical properties

1. Introduction

Polymeric materials, just like metals and ceramics, are widely used in medical applications. In the group of polymeric materials, synthetic, biological, and hybrid materials can be distinguished. In [1], the authors presented a detailed list of synthetic polymeric materials for biomedical applications. These materials are given in detail: polyolefins, poly(tetrafluoroethylene) (PTFE), poly(vinyl chloride) (PVC), silicone, methacrylates, polyesters, polyethers, polyamides, and polyurethanes. In addition, the materials were divided and described according to their intended use to produce temporary in vivo applications; general surgical implants; orthopedic implants; vascular and cardio-vascular intervention; plastic, reconstructive, and cosmetic surgery as well as ophthalmology, dentistry, or neurosurgery.

Synthetic polymers can also be classified according to their biodegradability under physiological conditions. The group of synthetic, biodegradable polymers deserves special attention. After fulfilling their function, they are cleared out of the body by natural methods. Synthetic, biodegradable polymers are used to produce surgical sutures [2,3], bone anastomosis plates [4], materials for tissue scaffolds [5–7], and scaffolds for repairing tendons and ligaments [8].

Due to the increasing interest and wide variety of application possibilities of biodegradable, biocompatible polymers, it is necessary to evaluate the changes in the behavior of these materials under the influence of the degradation of the aqueous conditions. The test procedure for biodegradable materials, especially for surgical applications, has been defined in ISO 15814, which describes in detail the type of degradation medium used, degradation times, and test methods. The standard recommends the use of buffered sodium chloride solution, pH = 7.4, heated to 37 °C. The pH of the medium should be controlled. If necessary, the medium should be buffered or replaced periodically. The standard also specifies the degradation intervals between the verification of selected parameters. The intervals of 0, 2, 4, 8, 16, and 26 weeks are recommended for short-term degradable polymers (e.g., for PLGA), and for long-term degradable polymers (e.g., pure PLA) of 0, 6, 12, 26, 39, and 52 weeks, respectively.

In the research described in scientific publications, authors have used different specimen types, types of degradation medium, and maximum degradation times as well as the intervals between the individual intervals of the verification of changes within the material. The adopted test methods apply to biocompatible materials such as polylactide and its enantiomers (PLLA, PDLLA) [9–11], polyglycolide [10], polycaprolactone [11,12], polylactide-co-glycolide [13,14], polyurethanes [15], trimethylene polycarbonate [16], polylactide with polycaprolactone [17], polydoxanone [17], and polylactide based copolymers, for example, polylactide-co-glycolide composites with calcium dihydrogen phosphate, calcium hydrophosphate, calcium phosphate and calcium carbonate [18], polylactide-co-δ-valerolactone [19], polylactide with polycaprolactone [17,20], and the poly-DL-lactide-co-polycarbonate of trimethylene [16]. Composites with biodegradable materials (e.g., polylactide with carbon fiber [21]) are also subjected to degradation tests.

In the study by Barbeck et al. [9], specimens were placed in a simulated physiological fluid (SBF) heated to 37 °C and stored for eight weeks. In the paper by Bartkowiak-Jowsa et al. [10], the process of hydrolytic degradation was carried out in distilled water at a pH 7.0 and temperature 37 °C. The total time of the storage of specimens in the medium was 24 months, and the studies were performed at monthly intervals in the first year and at bimonthly intervals in the second year. In the studies Scaffaro et al. presented in [11], specimens were stored in a buffer solution with a variable pH of 4.0, 7.0, and 10.0 and heated to 37 °C. The incubation time was 50 days and the tests were performed after 4, 21, and 50 days. The authors of subsequent papers [12–16,18,20] used buffered sodium chloride (PBS) at pH = 7.4, which was heated to 37 °C, as a degradation medium. These studies were characterized by a variable incubation period of 90 days [12], 12 months [13], 30 days [14], 38 weeks [15], 53 weeks [16], and 8 weeks [18,20].

In the study by Vieira et al. [17], three different degradation media (i.e., water, PBS, and sodium chloride (NaCl) solution) were investigated and heated to 37 °C. The total time of specimen degradation was 12 or 28 weeks. In the case of the studies Fernández et al. presented in [19], the experiments were also conducted in PBS, but in contrast to the previously quoted studies, the pH of the solution was 7.2. The incubation time of the specimens was 98 days. However, in Liu et al. [21], the degradation medium was the HANKs solution, which, as above, was heated to 37 °C. The total conditioning time was 49 or 73 weeks.

Summarizing the above analyses of the current state of knowledge, the diversity of the adopted research methodology for determining the influence of degradation on selected parameters of tested biodegradable materials can be observed. Moreover, most of the tests were conducted for specimens with small cross-sections (e.g., surgical sutures).

The aim of this study was to determine the influence of the most often used types of degradation media on the rate of mechanical behavior changes and the absorption of the degradation medium by a biocompatible polymeric material with a long degradation time. The results of this research will be used to determine the limits of applicability of selected media in the degradation of the model of biodegradable material.

2. Materials and Methods

The aim of the work was achieved by using appropriate materials and research methods. In order to verify the influence of the medium type on the changes in the behavior of the polymeric material as a model material, granules made of biodegradable polylactide with the commercial name of Ingeo™ Biopolymer 3100HP (NatureWorks LLC, Minnetonka, MN, USA) were used. The material was characterized by a density of 1.24 g/cm^3 and a melt mass flow rate (MFR) of 24 g/10 min was selected. Specimens from biodegradable material were formed by injection molding on a hybrid Engel E-Victory 310/110 (Engel Austria GmbH, Schwertberg, Austria) injection molding machine equipped with an Engel Viper 6 robot. The geometry of the molds used in the test was in accordance with ISO 527-1, type 1A.

The polymer specimens were exposed to hydrolytic degradation in three degradation media (i.e., buffered saline solution – PBS (Biocorp Polska Sp. z o.o, Warsaw, Poland), 0.9% saline solution (0.9% NaCl, self-made), and distilled water (H$_2$O$_{dest.}$, self-made)). The degradation time was 0, 2, 4, 6, 12, 26, 39, and 52 weeks. During the degradation phase, the specimens were stored vertically in relation to their longer axis in a prepared in-house research stand. The test stand was registered for patent protection in the Patent Office of the Republic of Poland under number (W.127335, 15.05.2018). The prototype of the stand allowed for a constant temperature of 37 ± 1 °C to be maintained. The pH value of the degradation medium was controlled using an Elmetron CP-411 pH-meter (ELMETRON, Zabrze, Poland). The medium was changed at least every four weeks.

Tests of the mechanical properties were performed on an INSTRON ElectroPuls E3000 (INSTRON, Norwood, MA, USA) equipped with an electromagnetic actuator with a force range of ±3 kN. The speed of moving the crosshead of the testing machine was 1 mm/min. The INSTRON static clip-on extensometer was used to determine the material deformation and modulus of elasticity. The distance between the measuring arms was 12.5 mm. Based on the measured values of stresses and deformations, the following values were determined: tensile strength (σ_M), breaking strength (σ_B), Young modulus (F), and toughness (Q). In each group, n = 8 specimens were tested. Specimens were tested at room temperature (T = 23 ± 2 °C, humidity 50 ± 10%) after removal from the medium and dried with a paper towel.

The change in the relative mass of the specimens during degradation (Δm) was determined from the data collected during the weighing process. The initial mass (m_0) of the specimen was determined before it was placed in the degradation medium and after a specified time of degradation, the mass was measured in time (m_t). The mass change was estimated based on the following equation: $\Delta m = (m_t - m_0)/m_0 \times 100\%$. AS 220.X2, RADWAG (Radom, Poland) laboratory scales were used to determine the mass of specimens.

The obtained results were analyzed using software such as GraphPad and Excel, and descriptive statistics and statistical significance tests were used. The normality of the distribution of quantitative variables was determined using the Shapiro–Wilk test. Verification was carried out at the level of significance of α = 0.05. The statistical significance of the differences in the achieved results was determined with the use of one-way ANOVA variance analysis. The aim of these analyses was to verify the hypothesis of the equality of mean values of the tested variable in several populations ($k \geq 2$). Fisher's LSD test was used to determine the differences of the studied results.

3. Results and Discussion

Four parameters (tensile strength (σ_M), breaking strength (σ_B), Young modulus (E), and absorbed energy (Q)) were used in the analysis of the influence of the type of degradation medium on the changes in the mechanical behavior of the material. Changes in the mentioned mechanical parameters obtained from the uniaxial tensile test were observed with relation to the time of degradation (in weeks) and to the mass change (Δm). Figure 1 shows an example of the stress–strain curves of the specimens during degradation at selected intervals in various mediums in relation to the unconditioned specimens. However, the tendencies of changes are presented in Figures 2 and 3.

The stress–strain curves presented in Figure 1 show that in the case where the specimens were immersed in solution, the relative elongation of the specimens increased and the tensile strength decreased. Due to the degradation processes over six months of immersion time, the stress–strain curves changed. In this case, the relative elongation does not increase with the decrease in tensile strength, and the specimens cracked when the maximum tensile force was reached.

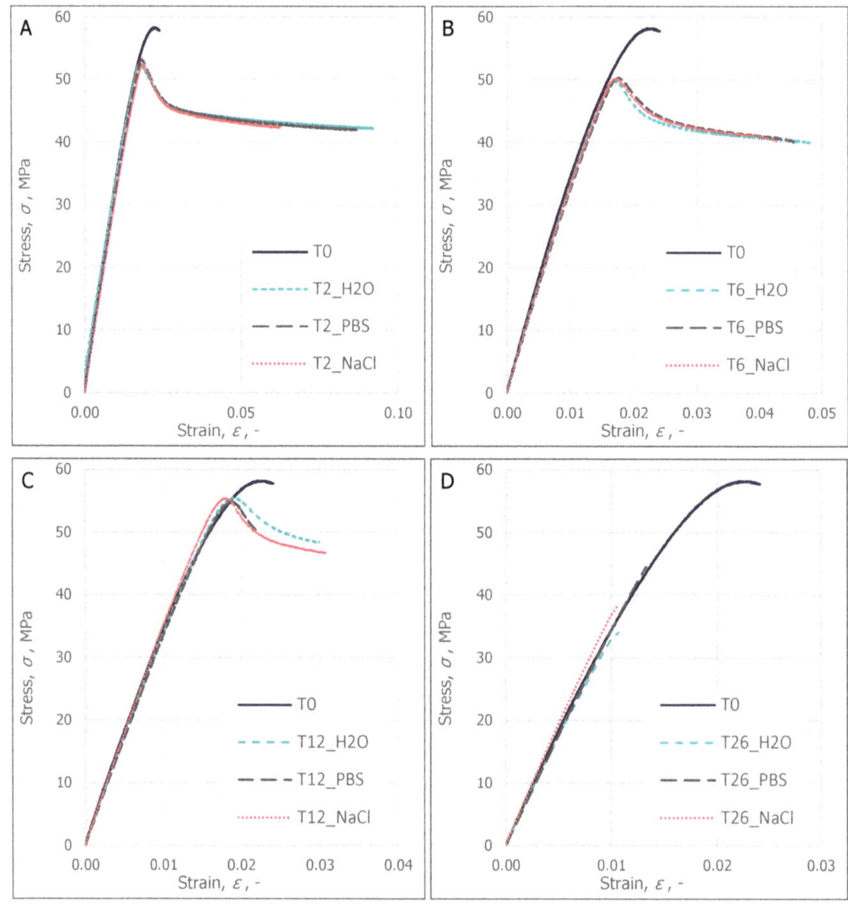

Figure 1. The stress–strain curves of the specimens during degradation in various mediums: (**A**) Comparison between specimens before degradation (T0) and after two weeks of degradation; (**B**) Comparison between specimens before degradation and after six weeks of degradation; (**C**) Comparison between specimens before degradation and after twelve weeks of degradation; (**D**) Comparison between specimens before degradation and after twenty-six weeks of degradation.

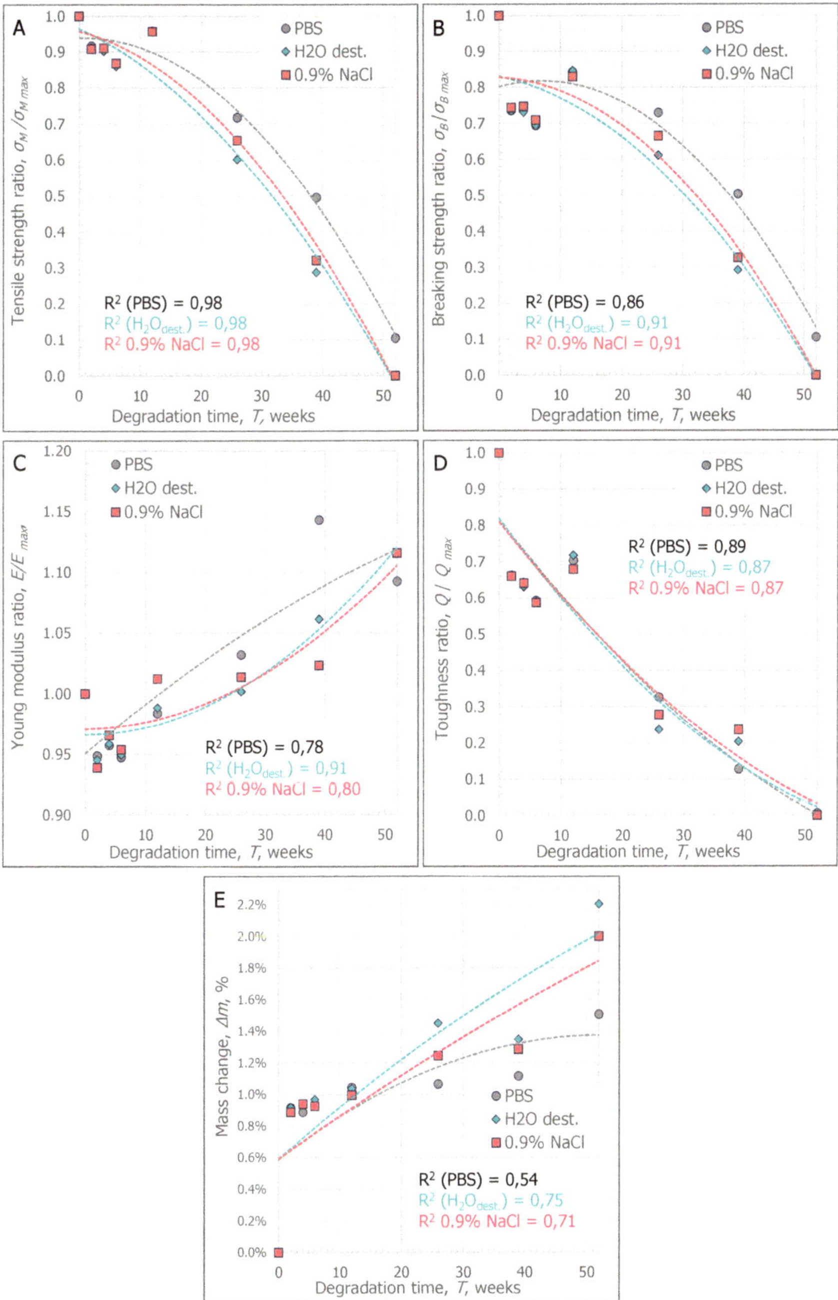

Figure 2. Tendencies of the changes in material parameters during degradation: (**A**) Tensile strength ratio; (**B**) Breaking strength ratio; (**C**) Young modulus ratio; (**D**) Toughness ratio; (**E**) Mass change.

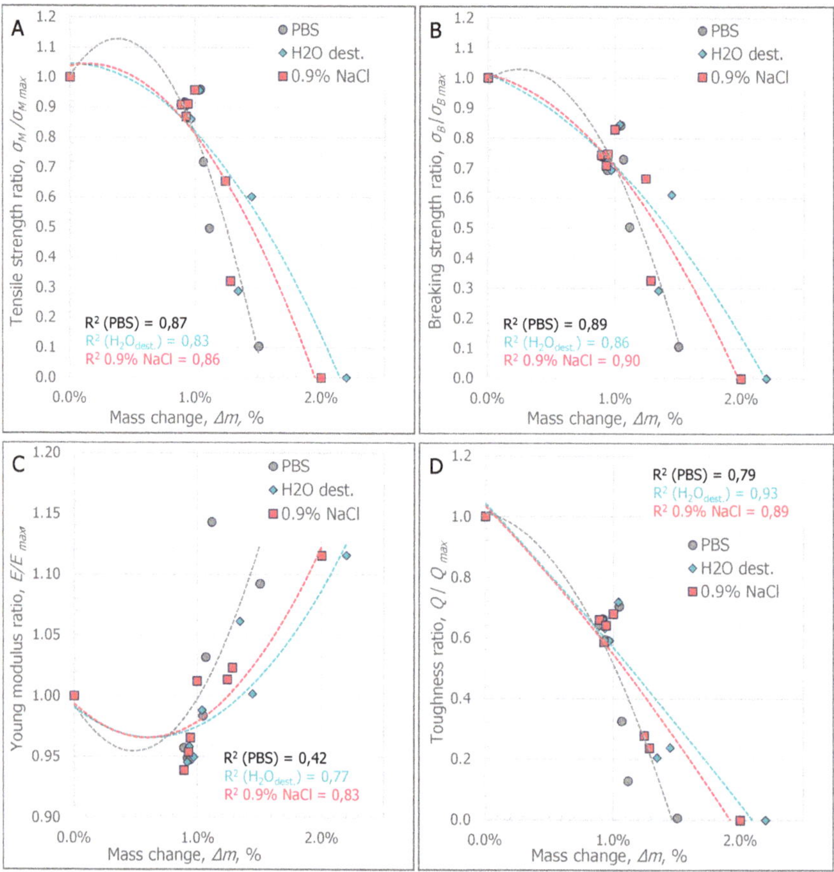

Figure 3. Tendencies of changes in the material parameters in relation to mass change: (**A**) Tensile strength ratio; (**B**) Breaking strength ratio; (**C**) Young modulus ratio; (**D**) Toughness ratio.

The tendencies of changes in the mechanical parameters were described with a model of square function. Comparisons of the statistical significance of the differences in the results were made in two ways. In the first analysis, the differences between the values obtained in each week of degradation for each type of medium were compared. In the second stage, the significance of the differences between three types of degradation media in specific weeks of degradation were compared.

The value of the coefficients of determination R^2 shows that the method applied to the description of degradation tendency with a square function allows for a value of the regression line fitting above $R^2 = 0.8$ to be obtained, thus, giving a good curve fitting for most of the analyzed parameters (i.e., the comparison of the relationship between the change in tensile strength, braking strength, and toughness during degradation). The adopted model of the Young modulus changed and the relative mass changed dependent on the time of degradation by means of a square function showed on average fit.

In Figure 2A, the average tendencies of decrease in tensile strength during degradation depending on the type of degradation medium, are shown. Table 1 shows the mean values with the standard deviation and the median of the tensile strength parameter. In the case of specimens degraded in 0.9% sodium chloride solution and distilled water, the tensile strength in T52 was not determined because the specimens broke in the grips of the testing machine. In the table below, the "x" symbol is used. This statement also refers to the other parameters. Additionally, it was assumed that the specimens marked with the symbol "x" reached values close to 0 MPa (σ_M, σ_B), 0 MJ/m^3 (Q), and $\geq$4000 MPa (E; according to the PBS test).

Table 1. Mean and median values of the tensile strength parameter (σ_M, MPa) during degradation.

Value of σ_M	Medium 1:PBS	Medium 1:PBS	Medium 2:0.9% NaCl	Medium 2:0.9% NaCl	Medium 3:H$_2$O$_{dest.}$	Medium 3:H$_2$O$_{dest.}$
Time, T, Weeks	Mean ± STD	Median	Mean ± STD	Median	Mean ± STD	Median
0	57.92 ± 0.66	58.22	57.92 ± 0.66	58.22	57.92 ± 0.66	58.22
2	53.02 ± 0.13	53.01	52.62 ± 0.48	52.39	52.78 ± 0.34	52.65
4	52.62 ± 0.22	52.47	52.76 ± 0.20	52.81	52.28 ± 0.29	52.18
6	50.08 ± 0.31	50.25	50.33 ± 0.22	50.23	49.83 ± 0.48	49.91
12	55.68 ± 0.42	55.71	55.44 ± 0.26	55.39	55.37 ± 0.30	55.50
26	42.12 ± 1.88	41.43	37.94 ± 3.38	38.74	34.86 ± 1.28	34.80
39	29.59 ± 3.26	27.73	18.62 ± 3.20	17.39	16.67 ± 1.95	16.65
52	6.03 ± 2.47	5.09	x	x	x	x

Based on the presented curves, it was observed that the specimens stored in PBS showed a slower tendency toward a decrease in tensile strength than the specimens stored in 0.9% sodium chloride solution and distilled water. The rate of degradation in this case was stated as a 50% decrease in tensile strength in relation to the value determined at T0. Specimens stored in PBS peaked at T39. In contrast to the specimens stored in PBS, the specimens stored in 0.9% NaCl or H$_2$O$_{dest.}$ were characterized by a decrease in tensile strength of almost 70%. In order to compare the differences between the strength values determined in different weeks of degradation depending on the type of medium used, they were significantly different for each analyzed medium. Comparisons between three media types in a given week of degradation demonstrated that in the weeks between T2 and T12, the medium type did not significantly affect the differences between the values obtained for PBS, 0.9% NaCl, and H$_2$O$_{dest.}$. Only in the period from T26 to T52 did the type of medium influence the strength values reached by the specimens for the specimens degraded in PBS, which confirmed the observed tendency of a slower decrease in the strength values than in the other two media. No statistically significant differences were found in the period from T26 to T52 for the comparison of 0.9% NaCl and H$_2$O$_{dest.}$.

In Figure 2B, the average tendencies of the decrease in breaking strength during the degradation depending on the type of the applied degradation medium are presented. Table 2 shows the mean values with their standard deviation and the median of the breaking strength parameter.

Table 2. Mean and median values of the breaking strength parameter (σ_B, MPa) during degradation.

Value of σ_B	Medium 1:PBS	Medium 1:PBS	Medium 2:0.9% NaCl	Medium 2:0.9% NaCl	Medium 3:H$_2$O$_{dest.}$	Medium 3:H$_2$O$_{dest.}$
Time, T, Weeks	Mean ± STD	Median	Mean ± STD	Median	Mean ± STD	Median
0	57.00 ± 0.42	56.96	57.00 ± 0.42	56.96	57.00 ± 0.42	56.96
2	41.98 ± 0.33	41.92	42.42 ± 0.99	42.26	41.92 ± 0.41	42.14
4	42.20 ± 0.43	42.01	42.61 ± 0.32	42.54	41.68 ± 0.26	41.73
6	39.61 ± 0.44	39.44	40.47 ± 0.87	40.30	39.58 ± 0.34	39.58
12	48.04 ± 1.39	47.84	47.30 ± 1.24	46.64	48.22 ± 0.71	47.91
26	41.61 ± 1.88	41.43	37.94 ± 3.38	38.74	34.86 ± 1.28	34.80
39	28.72 ± 3.26	27.73	18.62 ± 3.20	17.39	16.67 ± 1.95	16.65
52	6.03 ± 2.47	5.09	x	x	x	x

It was observed that in the period from T26 to T52, the value of the breaking strength of the specimens was equal to their tensile strength value. Similarly, as in the case of the tensile strength parameter, the statistical significance of the differences in breaking strength between the weeks of degradation in each of the three media and between the three media in a selected week of degradation is shown.

In Figure 2C, the average growth tendencies of the Young modulus during degradation depending on the type of the medium are shown. Table 3 shows the mean values with their standard deviation and the median value of the Young modulus parameter.

Table 3. Mean and median values of the Young modulus parameter (E, MPa) during degradation.

Value of E	Medium 1:PBS	Medium 1:PBS	Medium 2:0.9% NaCl	Medium 2:0.9% NaCl	Medium 3:$H_2O_{dest.}$	Medium 3:$H_2O_{dest.}$
Time, T, Weeks	Mean ± STD	Median	Mean ± STD	Median	Mean ± STD	Median
0	3585.45 ± 123.32	3598.36	3585.45 ± 123.32	3598.36	3585.45 ± 123.32	3598.36
2	3419.68 ± 60.50	3382.19	3366,91 ± 49.81	3368.43	3389.44 ± 92.53	3382.61
4	3426.34 ± 27.08	3432.22	3462.73 ± 98.92	3462.27	3436.71 ± 54.45	3436.74
6	3399.84 ± 29.20	3386.44	3419.93 ± 72.91	3401.63	3404.59 ± 26.06	3410.28
12	3540.73 ± 53.45	3517.70	3629.45 ± 54.69	3662.88	3543.33 ± 39.08	3545.98
26	3703.38 ± 291.30	3599.78	3634.50 ± 139.96	3582.97	3591.46 ± 116.25	3635.94
39	3972.72 ± 296.05	3999.60	3668.81 ± 170.59	3701.04	3804.92 ± 350.35	3690.25
52	3916.82 ± 829.14	3914.08	x	x	x	x

At the beginning of T2-to-T6 degradation, a decrease in the Young modulus by about 5% was observed. After T12, this value slightly increased by 5–15% relative to T0. However, no statistically significant differences were found between the influence of the degradation medium on the changes in the Young modulus. This trend was confirmed throughout the degradation period and in comparisons between the media in a given week of degradation.

In Figure 2D, the average tendencies of the toughness decrease during the degradation depending on the type of the medium are presented. Table 4 presents the mean values with their standard deviation and the median of the toughness parameter.

Table 4. Mean and median values of the toughness parameter (Q, MJ/m^3) during degradation.

Value of Q	Medium 1: PBS	Medium 1: PBS	Medium 2: 0.9% NaCl	Medium 2: 0.9% NaCl	Medium 3: $H_2O_{dest.}$	Medium 3: $H_2O_{dest.}$
Time, T, weeks	Mean ± STD	Median	Mean ± STD	Median	Mean ± STD	Median
0	0.812 ± 0.032	0.821	0.812 ± 0.032	0.821	0.812 ± 0.032	0.821
2	0.538 ± 0.005	0.536	0.536 ± 0.007	0.538	0.538 ± 0.017	0.536
4	0.522 ± 0.008	0.525	0.521 ± 0.007	0.522	0.513 ± 0.012	0.508
6	0.482 ± 0.013	0.487	0.477 ± 0.003	0.476	0.480 ± 0.004	0.479
12	0.571 ± 0.004	0.570	0.552 ± 0.005	0.553	0.583 ± 0.006	0.584
26	0.265 ± 0.033	0.276	0.226 ± 0.038	0.236	0.193 ± 0.015	0.192
39	0.104 ± 0.032	0.096	0.050 ± 0.014	0.045	0.038 ± 0.006	0.040
52	0.005 ± 0.004	0.004	x	x	x	x

Based on the determined tendencies of decrease in the toughness parameter, it can be observed that in the period T2-to-T12, its value decreased by nearly 40%, and in the following weeks of degradation, it decreased to about 80–90% of the initial value. For T52 specimens, the toughness parameter, which was close to 0 MJ/m^3, was determined only for PBS specimens.

In Figure 1E, the average tendencies of increased percentage change of mass during degradation depending on the type of the chosen medium are presented. Table 5 presents the values of mean, standard deviation, and the median of the percentage mass change parameter.

Table 5. Mean and median values of the mass change parameter (Δm, %) during degradation.

Value of Δm	Medium 1: PBS	Medium 1: PBS	Medium 2: 0.9% NaCl	Medium 2: 0.9% NaCl	Medium 3: $H_2O_{dest.}$	Medium 3: $H_2O_{dest.}$
Time, T, Weeks	Mean ± STD	Median	Mean ± STD	Median	Mean ± STD	Median
0	x	x	x	x	x	x
2	0.90 ± 0.03	0.91	0.91 ± 0.03	0.91	0.90 ± 0.04	0.92
4	0.91 ± 0.04	0.92	0.96 ± 0.08	0.98	0.95 ± 0.02	0.96
6	0.95 ± 0.01	0.95	0.89 ± 0.13	0.95	0.92 ± 0.07	0.95
12	1.04 ± 0.03	1.03	1.04 ± 0.02	1.05	1.00 ± 0.03	1.01
26	1.07 ± 0.01	1.07	1.45 ± 0.33	1.25	1.24 ± 0.12	1.24
39	1.11 ± 0.03	1.11	1.35 ± 0.01	1.35	1.28 ± 0.02	1.28
52	1.51 ± 0.11	1.48	2.20 ± 0.18	2.16	2.00 ± 0.02	2.01

On the grounds of the determined tendencies of increase of the percentage parameter of mass change, it can be observed that in the period from T2 to T12, its value increased by nearly 1%, and in subsequent weeks of degradation, it increased to over 2% of the initial value. In the case of the specimens tested in T52, the specimens conditioned in distilled water and 0.9% saline solution achieved a mass increase of more than 2%, while in the case of the specimens stored in PBS, the value did not exceed 1.5% of the initial value.

The analyses proved the statistical significance of the differences in toughness between the specimens studied in the three media during the degradation period. Moreover, from T12, the type of degradation medium had an influence on the results achieved. Specimens stored in PBS needed more energy to destroy them than specimens stored in 0.9% NaCl or $H_2O_{dest.}$. No statistically significant differences between the use of 0.9% NaCl or $H_2O_{dest.}$ were found in studies over 12 weeks.

In this study, the mechanical tests showed a higher value of strength and toughness at 12 weeks than the specimens at a short degradation time. A similar phenomenon was described in Tokiwa et al. [22] where the authors indicated that the phenomenon of crystallization during degradation was the justification for the changes affecting the improvement of material strength. As a result of crystallization, new bonds developing in the polymer chain reduce their susceptibility to creep [10].

Figure 3A–D show the curves of the average tendencies of changes in specific mechanical parameters in relation to the percentage change in the mass of specimens.

Initially, the change in the mass of the specimens is related to the absorption of the medium by the polymer specimen. The percentage change in mass is then stabilized. This phenomenon has been observed by other researchers in relation to polymer and elastomer tests, which is described in the paper by Younes et al. [23]. During the stabilization period, the relative mass change value is between approximately 0.89% and approximately 0.95% of the initial mass. In addition, it has been pointed out that during this period, changes in mechanical parameters are followed by no more than a 30% decrease in the tensile strength, breaking strength, or Young modulus, and a nearly 50% decrease in toughness. This period occurs approximately to the twelfth week of degradation, where based on the statistical analyses, there were no significant differences in the type of degradation medium used.

A progressive increase in the percentage value of mass change was found over 12 weeks of degradation and the mass of the specimens increased above 1% relative to the initial mass. Increasing the percentage change in the specimen mass above 1% is associated with a change in material properties. In each analyzed case, the cracking method of the specimen changed from ductile failure to brittle failure ($\sigma_M = \sigma_B$). In addition, the increase in medium absorption, and thus, the percentage change in specimen mass, of more than 1.5%, significantly hindered the test process. The specimens stored in $H_2O_{dest.}$ and 0.9% NaCl during the T52 week were characterized by a mass change of more than 2%, which made it impossible to test them. In both cases, the test specimens broke in the grips of the testing machine.

4. Summary and Conclusions

The results of the research presented in this paper compared the influence of the type of medium on the speed of changes in mechanical parameters during 52 weeks of degradation as well as the absorption level of the medium as defined by the percentage change in the mass of the specimens. The obtained results showed, in general, a lower strength of the specimens stored in 0.9% NaCl and $H_2O_{dest.}$. The phenomenon of a faster decrease of strength parameters was observed for these specimens from T26. Moreover, based on statistical analyses, it was shown that if the degradation is carried out for no longer than 12 weeks, the type of medium does not significantly influence the changes of the strength parameters. In the time frame of up to 12 weeks, no significantly higher degradation medium uptake by the tested specimens was found. Only after 12 weeks was it observed that specimens conditioned in 0.9% NaCl and $H_2O_{dest.}$ had significantly higher medium consumption than specimens stored in PBS.

The presented model of changes in material properties due to medium absorption shows how the different chemical composition of the medium can significantly influence the rate of polymer material degradation. Therefore, a quantitative approach to degradation changes should lead to a more rational choice of degradation medium in future research.

The observed tendencies of changes in strength parameters allow for the optimal medium to be chosen (i.e., buffered saline solution), which, due to its properties, opposes significant changes in pH due to the hydrolysis of a biocompatible biodegradable material. Moreover, linking the properties of the buffer solution with the properties of physiological fluids in a living human or animal body allowed us to conclude that it corresponded much better to the internal environment than to distilled water or a 0.9% sodium saline solution. Therefore, a slower degradation rate in PBS than in other degradation media is a phenomenon of high demand.

The results presented in the paper contribute to the current state of knowledge by analyzing the influence of the type of degradation medium on changes in the mechanical behavior of biodegradable biocompatible materials for specimens with larger cross-sections than in the case of surgical sutures or plates for small bone anastomoses. Nevertheless, against the background of the obtained results, the geometric dimensions of biodegradable biomedical solutions used in bone anastomoses should be studied in more detailed studies.

The authors in [24] showed that the surface of a matrix made of polylactide was less susceptible to the rate of hydrolysis than its internal part and noted that if the polymeric matrix was initially crystalline or crystallizes during degradation, the internal part of a large-sized product would degrade faster than its surface without the formation of hollow residual structures.

Moreover, due to the noted higher uptake of the degradation medium by specimens stored in 0.9% NaCl and $H_2O_{dest.}$, future degradation tests should be extended with an analysis of the composition of the medium after degradation. In the studies described in the paper, only the changes in the pH of the medium were verified.

5. Patents

It was possible to maintain the conditions of hydrolytic degradation by developing a prototype of a test stand. The improved design of the test stand was submitted for protection to the Patent Office of the Republic of Poland under no. W.127335 (15.05.2018).

Author Contributions: It is single-authored paper.

Funding: This research received no external funding.

Conflicts of Interest: The authors declare no conflicts of interest.

References

1. Maitz, M.F. Applications of synthetic polymers in clinical medicine. *Biosurf. Biotribol.* **2015**, *1*, 161–176. [CrossRef]
2. Lee, H.S.; Park, S.H.; Lee, J.H.; Jeong, B.Y.; Ahn, S.K.; Choi, Y.M.; Choi, D.J.; Chang, J.H. Antimicrobial and biodegradable PLGA medical sutures with natural grapefruit seed extracts. *Mater. Lett.* **2013**, *95*, 40–43. [CrossRef]
3. Joseph, B.; George, A.; Gopi, S.; Kalarikkal, N.; Thomas, S. Polymer sutures for simultaneous wound healing and drug delivery—A review. *Int. J. Pharm.* **2017**, *524*, 454–466. [CrossRef] [PubMed]
4. dos Santos, T.M.B.K.; Merlini, C.; Aragones, Á.; Fredel, M.C. Manufacturing and characterization of plates for fracture fixation of bone with biocomposites of poly (lactic acid-co-glycolic acid) (PLGA) with calcium phosphates biocermics. *Mater. Sci. Eng. C* **2019**, *103*, 109728. [CrossRef] [PubMed]
5. Xue, Y.; Sant, V.; Phillippi, J.; Sant, S. Biodegradable and biomimetic elastomeric scaffolds for tissue-engineered heart valves. *Acta Biomater.* **2017**, *48*, 2–19. [CrossRef] [PubMed]
6. Du, L.; Yang, S.; Li, W.; Li, H.; Feng, S.; Zeng, R.; Yu, B.; Xiao, L.; Nie, H.-Y.; Tu, M. Scaffold composed of porous vancomycin-loaded poly(lactide-co-glycolide) microspheres: A controlled-release drug delivery system with shape-memory effect. *Mater. Sci. Eng. C* **2017**, *78*, 1172–1178. [CrossRef] [PubMed]
7. Shi, Y.Z.; Liu, J.; Yu, L.; Zhong, L.Z.; Jiang, H.B. β-TCP scaffold coated with PCL as biodegradable materials for dental applications. *Ceram. Int.* **2018**, *44*, 15086–15091. [CrossRef]
8. Buschmann, J.; Bürgisser, G.M. 8-Synthetic polymer scaffolds for tendon and ligament repair: Materials and biomechanics. In *Biomechanics of Tendons and Ligaments*; Buschmann, J., Bürgisser, G.M., Red, W., Eds.; Woodhead Publishing: Sawston, UK, 2017; pp. 225–250. ISBN 978-0-08-100489-0.
9. Barbeck, M.; Serra, T.; Booms, P.; Stojanovic, S.; Najman, S.; Engel, E.; Sader, R.; Kirkpatrick, C.J.; Navarro, M.; Ghanaati, S. Analysis of the in vitro degradation and the in vivo tissue response to bi-layered 3D-printed scaffolds combining PLA and biphasic PLA/bioglass components—Guidance of the inflammatory response as basis for osteochondral regeneration. *Bioact. Mater.* **2017**, *2*, 208–223. [CrossRef]
10. Bartkowiak-Jowsa, M.; Bedzinski, R.; Kozlowska, A.; Filipiak, J.; Pezowicz, C. Mechanical, rheological, fatigue, and degradation behavior of PLLA, PGLA and PDGLA as materials for vascular implants. *Meccanica* **2013**, *48*, 721–731. [CrossRef]
11. Scaffaro, R.; Lopresti, F.; Botta, L. Preparation, characterization and hydrolytic degradation of PLA/PCL co-mingled nanofibrous mats prepared via dual-jet electrospinning. *Eur. Polym. J.* **2017**, *96*, 266–277. [CrossRef]
12. Bosworth, L.A.; Downes, S. Physicochemical characterisation of degrading polycaprolactone scaffolds. *Polym. Degrad. Stab.* **2010**, *95*, 2269–2276. [CrossRef]
13. Engineer, C.; Parikh, J.; Raval, A. Effect of copolymer ratio on hydrolytic degradation of poly(lactide-co-glycolide) from drug eluting coronary stents. *Chem. Eng. Res. Des.* **2011**, *89*, 328–334. [CrossRef]
14. Deng, M.; Zhou, J.; Chen, G.; Burkley, D.; Xu, Y.; Jamiolkowski, D.; Barbolt, T. Effect of load and temperature on in vitro degradation of poly(glycolide-co-L-lactide) multifilament braids. *Biomaterials* **2005**, *26*, 4327–4336. [CrossRef] [PubMed]
15. da Silva, G.R.; da Silva-Cunha, A.; Behar-Cohen, F.; Ayres, E.; Oréfice, R.L. Biodegradation of polyurethanes and nanocomposites to non-cytotoxic degradation products. *Polym. Degrad. Stab.* **2010**, *95*, 491–499. [CrossRef]
16. Yang, J.; Liu, F.; Yang, L.; Li, S. Hydrolytic and enzymatic degradation of poly(trimethylene carbonate-co-d,l-lactide) random copolymers with shape memory behavior. *Eur. Polym. J.* **2010**, *46*, 783–791. [CrossRef]
17. Vieira, A.C.; Vieira, J.C.; Guedes, R.M.; Marques, A.T. Degradation and Viscoelastic Properties of PLA-PCL, PGA-PCL, PDO and PGA Fibres. *Mater. Sci. Forum* **2010**, *636*, 825–832. [CrossRef]
18. Ara, M.; Watanabe, M.; Imai, Y. Effect of blending calcium compounds on hydrolytic degradation of poly(DL-lactic acid-co-glycolic acid). *Biomaterials* **2002**, *23*, 2479–2483. [CrossRef]
19. Fernández, J.; Etxeberria, A.; Sarasua, J.R. In vitro degradation of poly(lactide/d-valerolactone) copolymers. *Polym. Degrad. Stab.* **2015**, *112*, 104–116. [CrossRef]

20. Vieira, A.C.; Guedes, R.M.; Tita, V. Constitutive modeling of biodegradable polymers: Hydrolytic degradation and time-dependent behavior. *Int. J. Solids Struct.* **2014**, *51*, 1164–1174. [CrossRef]
21. Liu, S.; Wu, G.; Chen, X.; Zhang, X.; Yu, J.; Liu, M.; Zhang, Y.; Wang, P. Degradation Behavior In Vitro of Carbon Nanotubes (CNTs)/Poly(lactic acid) (PLA) Composite Suture. *Polymers* **2019**, *11*, 1015. [CrossRef]
22. Tokiwa, Y.; Jarerat, A. Biodegradation of poly (L-lactide). *Biotechnol. Lett.* **2004**, *26*, 771–772. [CrossRef] [PubMed]
23. Younes, H.M.; Bravo-Grimaldo, E.; Amsden, B.G. Synthesis, characterization and in vitro degradation of a biodegradable elastomer. *Biomaterials* **2004**, *25*, 5261–5269. [CrossRef] [PubMed]
24. Beslikas, T.; Gigis, I.; Goulios, V.; Christoforides, J. Crystallization Study and Comparative in Vitro – in Vivo Hydrolysis of PLA Reinforcement Ligament. *Int. J. Mol. Sci.* **2011**, *12*, 6597–6618. [CrossRef] [PubMed]

Article

Pluronic F127-Folate Coated Super Paramagenic Iron Oxide Nanoparticles as Contrast Agent for Cancer Diagnosis in Magnetic Resonance Imaging

Hieu Vu-Quang [1,2,3,*], Mads Sloth Vinding [3], Thomas Nielsen [3], Marcus Görge Ullisch [3], Niels Chr. Nielsen [3], Dinh-Truong Nguyen [2] and Jørgen Kjems [3,4]

[1] NTT High-Tech Institute, Nguyen Tat Thanh University, Ho Chi Minh City 70000, Vietnam
[2] School of Biotechnology, Tan Tao University, Long An 82000, Vietnam; truong.nguyen@ttu.edu.vn
[3] Interdisciplinary Nanoscience Center (iNANO), Aarhus University, DK-8000 Aarhus, Denmark; msv@cfin.au.dk (M.S.V.); tn@ase.au.dk (T.N.); ullisch@inano.au.dk (M.G.U.); ncn@chem.au.dk (N.C.N.); jk@mbg.au.dk (J.K.)
[4] Department of Molecular Biology, Aarhus University, DK-8000 Aarhus, Denmark
* Correspondence: vqhieu@ntt.edu.vn

Received: 2 March 2019; Accepted: 13 April 2019; Published: 25 April 2019

Abstract: Contrast agents have been widely used in medicine to enhance contrast in magnetic resonance imaging (MRI). Among them, super paramagnetic iron oxide nanoparticles (SPION) have been reported to have low risk in clinical use. In our study, F127-Folate coated SPION was fabricated in order to efficiently target tumors and provide imaging contrast in MRI. SPION alone have an average core size of 15 nm. After stabilizing with Pluronic F127, the nanoparticles reached a hydrodynamic size of 180 nm and dispersed well in various kinds of media. The F127-Folate coated SPION were shown to specifically target folate receptor expressing cancer cells by flow cytometry analysis, confocal laser scanning microscope, as well as in vitro MRI. Furthermore, in vivo MRI images have shown the enhanced negative contrast from the F127-Folate coated SPION in tumor-bearing mice. In conclusion, our F127-Folate coated SPION have shown great potential as a contrast agent in MRI, as well as in the combination with drug delivery for cancer therapy.

Keywords: SPION; contrast agent; MRI; cancer diagnosis; folate receptor; pluronic F127

---

1. Introduction

Since the first discovery of the difference between cancer and normal tissue in the 1970's by Raymond Vahan Damadian, a number of cancer diagnostic MRI methods have been developed [1]. Many inventions and ideas have been introduced to improve the imaging contrast, in order to achieve high diagnostic accuracy. One idea in particular, is the use of MRI contrast agents. There are two contrast agent types have been approved for use in medicine, including gadolinium based T1 contrast agents and super paramagnetic iron oxide nanoparticles (SPION) based T2 contrast agents [2–4]. Among them, SPION are more desirable due to their low toxicity when used [5]. Resovist and Feridex are two types of SPION that have recently been made available on the drug market. They have not only been approved for use in MRI of the liver, but have also entered clinical trials (phase II) for lymph node metastatic diagnosis [6]. However, these diagnoses are "indirect" as they rely on the position and migration of macrophages in the organs. Thus, it leads to difficulty in diagnosis when working with organs and tissues that have a small macrophage population, such as cancers. On the other hand, the 'direct' diagnosis, that is based on the accumulation of SPION in the tumor region, is preferable because it can provide the specific contrast in the images.

Immunological barriers play a crucial role in the distribution of nanoparticles. After parenteral injection, nanoparticles circulate in the system and are recognized by circulating monocytes and macrophages.

They are then phagocytized and degraded in the reticuloendothelial system (RES) before reaching the target site [7]. Thus, in order to achieve the long blood circulated half-life, and overcome immune barriers, nanoparticles must be camouflaged with stealth materials [8,9]. Among various types of such materials, Poly(Ethylene Glycol) (PEG) is the most frequently used material to produce stealth nanoparticles [9].

Pluronic F127 belongs to poloxamer group, which is biocompatible and is widely used in clinics for various purposes [10,11]. F127 is an amphiphilic polymer which consists of two PEG chains and one Poly(Propylene Oxide) PPO block. Therefore, F127 has been used as nonionic emulsion surfactant, as well as stealth material in many studies [12–14]. In emulsion, the hydrophobic PPO block can anchor into the organic phase while the hydrophilic PEG blocks are exposed to the water phase. After the evaporation of the organic phase, the PPO blocks absorb the nanoparticles hydrophobic surface, whereas the PEG blocks create the new nanoparticle stealth surface [15].

Folic acid is the most frequently employed targeting ligand since its receptor is significantly overexpressed on many types of cancer, whilst it presents in low and non-detectable levels in normal cells [16–18]. Thus, folic acid has been conjugated to nanoparticles in order to achieve the active targeting accumulation into the target tumor [18].

In our study, we aim to develop F127-Folate coated SPION as an MRI contrast agent. F127-Folate coated SPION specifically target the folate expressing cancer cells, which can provide the contrast in MRI.

2. Materials and Methods

All reagents were purchased from Sigma-Aldrich, St Louis, MO, USA unless otherwise specified.

2.1. Conjugation of Pluronic F127 and Folic Acid

The conjugation of folic acid to Pluronic F127 was done as described by Jia-Jyun et al., 2009. [19] (Figure 1). Briefly, folic acid (0.4 mmol) was activated by 1,1' Carbonyldiimidazole (CDI) (0.44 mmol) in 6 mL of dry dimethyl sulfoxide (DMSO) and stirred for 24 h in a dark place. Dry F127 (0.1 mmol) was then added to the mixture and stirred for another 24 h at room temperature. Following this, the reaction was diluted with 50% distilled water and dialyzed (tube: MWCO 3500) against deionized water for 3 days (water was changed twice a day). Next, the solution was freeze-dried for 3 days. The lyophilized outcome was further purified by dissolving in acetone and filtration. The final product was again freeze-dried before being analyzed by NMR and stored at −20 °C until use.

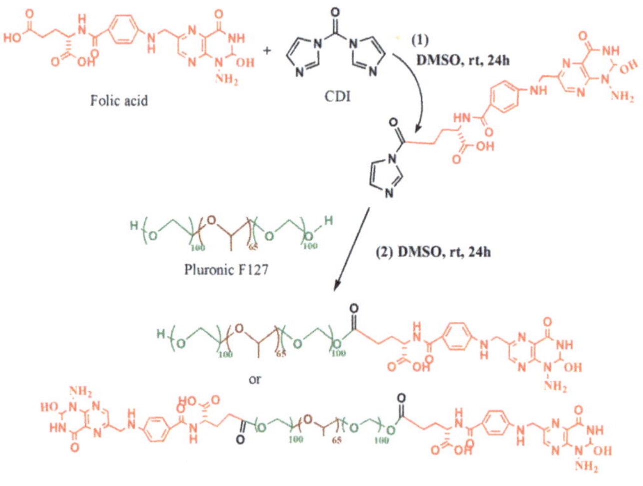

Figure 1. Schematic illustration of chemical reaction. (1) Activation of folic acid by 1,1' Carbonyldiimidazole (CDI) in dry dimethyl sulfoxide (DMSO) and darkness for 24 h, (2) conjugation of CDI-Folate to Pluronic F127, 24 h darkness.

2.2. Synthesis Magnetic Nanoparticles

Magnetic nanoparticles SPION were prepared by the co-precipitation of Fe(III) and Fe(II) in an alkaline solution [15]. Briefly, a mixture of 0.003 mole $FeCl_3 \cdot 6H_2O$, 0.015 mole $FeCl_2 \cdot 4H_2O$ and 500 µL of oleic acid (90%) in 45 mL deionized water was vigorously stirred under nitrogen pressure for 30 min. While being stirred for 30 min, 3 mL of 5 M NH_4OH was slowly dropped to the mixture. The black slurry product was extensively washed three times with deionized water, re-dispersed in 15 mL n-hexane and sonicated for 10 min. The solution was centrifuged at 12,000 rpm for 20 min, and the supernatant was discarded. The pellet was re-dispersed in 15 mL of *n*-hexane and vigorously vortexed. The solution was again centrifuged at 2000 rpm for 10 min to eliminate the large aggregation.

2.3. Iron Determination

The SPION (10 µL) were reduced in 10 µL of HCl 37% before introducing 180 µL of 0.2 X of FerroZine agent (Hatch, Germany). The mixture was incubated for 10 min in a 96-well plate. A standard curve was built in every measurement with three replicate wells. The absorbance of samples was read at 562 nm wavelength by Fluostar Otima (BMG Labtech, Otenberg, Germany).

2.4. F127-Folate Coated SPION and F127 Coated SPION Preparation and Particles Characterization

The SPION (5 mg) in 1 mL n-hexan was mixed with 13 mg F127 and 2 mg F127-Folate in 10 mL deionized water. The mixture was emulsified by vortexing for 2 min and further sonicated in water bath for 10 min. The n-hexan was allowed to evaporate overnight by magnetic stirring at 200 rpm/min. The nanoparticles were further washed with deionized water by centrifugation at 30,000 g for 30 min. This process was conducted in triplicate.

For fluorescent labeling, Nile Red (0.05 mg)—a hydrophobic fluorescent dye- was added to the n-hexane before emulsion. Nile Red absorbed to the hydrophobic layer of particles, thus making F127-Folate coated SPION/Nile Red [19].

The particle core size, morphology and diffraction (Figure A1) were also visualized by TEM instrument (TEMEDAX-20, Hillsboro, OR, USA) on a carbon coated copper grid.

The particle hydrodynamic size, polydispersity index (PDI), and zeta potential were characterized using Malvern Zetasizer Nano ZS instrument, Wocestershire, UK. The F127 coated SPION were synthesized in the same way and used as the control nanoparticles.

2.5. Cell Viability and Iron Uptake Concentration

KB cells were seeded at the density of 10,000 cells per well in a 96-well plate and incubated overnight in folic acid free RPMI 1640, 10% Fetal Bovine Serum (FBS), 1% antibiotic mixture (100 u/mL Penicillin, and 100 µg/mL Streptomycin) (Gibco). On the following day, cells were incubated with various concentrations of F127 coated SPION and F127-Folate coated SPION from 100 µg/mL to 3.125 µg/mL for 24 h (four replicate wells). Cell viability was done following the MTT assay (Sigma, St. Louis, MO, USA) while the iron concentration was determined as previously described.

2.6. Prussian Blue Staining

The F127–Folate coated SPION and F127 coated SPION (50 µg Fe/mL) were inoculated into 8 well tissue culture chambers (Sarstedt, Germany) and incubated for 3 h. Following this, the cells were washed 3 times with PBS and fixed in HCHO (4%) for 30 min. Iron nanoparticles were stained with a freshly mixed equal volume of HCl (4%) and potassium ferrocyanide (4%) for 15 min. After that, cells were rinsed twice with water and incubated with 250 µL Nuclear Fast Red for 5 min. In the final step, cell chambers were rinsed with water, dried, and mounted with cover slips.

2.7. Confocal Laser Scanning Imaging

For confocal laser scanning microscope (CLSM), cells were incubated with F127-Folate coated SPION/Nile Red and F127 coated SPION/Nile Red for 3 h. For the last 15 min of incubation, cell membranes were stained with 50 µg/mL wheat germ agglutinin- Alexa 488 (Life Technology, Carlsbad, CA, USA) for 15 min. Chambers were washed 3 times with PBS and fixed with HCHO 4% in 30 min. Cell chambers were rinsed with water, dried, and mounted with prolong gold antifade mountant with DAPI (Life Tecnology, Carlsbad, CA, USA). Fluorescent imaging was taken by confocal laser scanning microscope (Carl Zeiss, Okberkochen, Germany).

2.8. Flow Cytometer Assay (FACs)

F127-Folate coated SPION/Nile Red and F127 coated SPION/Nile Red (50 µg/mL) were inoculated into cell cultured wells and incubated for 3 h in the absence and (50 ng/mL) presence of free folic acid. Cells were then washed and harvested for FACs analysis (Beckman Coutler, Pasadena, CA, USA). Four replicated wells were made in total.

2.9. Magnetic Resonance Imaging

All of the MRI experiments were performed in a 16.4 Tesla vertical bore Bruker Avance II spectrometer running Paravision 6.0 (Bruker, Billeria, MA, USA) software. Data were analyzed with MATLAB (Mathworks, Natick, MA, USA).

2.10. In Vitro Scanning

The amounts of 4×10^5 cells in the FACs experiment were transferred to glass tubes and fixed in 200 µL of 5% agarose. T2 and T2* maps were generated with the multi slice multi echo (MSME) and multi gradient echo (MGE) sequences, respectively. Echo times were varied to estimate the T2 and T2* parameters (MSME: TE = 5 ms to 80 ms; MGE TE = 2 ms to 36 ms) and the repetition time (TR) was 5 s in each case. The field of view (FOV) was 22×22 mm^2, the matrix size 128×128, slice thickness (TH) 0.5 mm, and number of acquisitions (NA) 2.

The R2 relaxation of nanoparticles in the Appendix A was estimated by the MSME sequence.

2.11. In Vivo MRI Scanning

Tumors were implanted by subcutaneous injection of 10^6 KB cells to the left flank of Nude Balb/c female mice (Charles River — 8 weeks old). Tumors were allowed to grow for 2 weeks. All of the mice were fed at the Institute of Biomedicine, Aarhus University following institute standard protocol.

Mice were anesthetized with the mixture of Ketamine/Xylazine/PBS during the scan. Tumor bearing mice were scanned with MRI before and 24 h post nanoparticles administration (100 µg Fe/20 g mouse). T2 weighted images were obtained with MSME: FOV 26×26 mm^2, TR/TE 4000/10.13 ms, NA 2, matrix size 128×128, TH 1 mm, number of slices: 5. The signal to noise ratio (SNR) of tumor rim, tumor core, whole tumor, and back muscle were estimated using MATLAB. Since the nanoparticles do not target the back muscle, the SNR of back muscle was chosen for signal intensities comparison. To compare the signal-intensity change in the tumor post injection and pre-injection, the signal ratio between tumor and back muscles was calculated. Then, the reduced signal was calculated as shown in the equation below.

$$\% \ reduced \ signal = \frac{post - injection \frac{SNR \ (tumor)}{SNR \ (back \ muscle)}}{pre - injection \frac{SNR \ (tumor)}{SNR \ (back \ muscle)}} \times 100\% \tag{1}$$

3. Results and Discussion

Our study focused on developing folate receptor targeting SPION that could target cancer cells and enhance the contrast in MRI. In order to achieve successful accumulation in the tumor, the SPION must have steric surface to prevent them from systemic clearance and a targeting ligand is needed for tumor penetration. In this study, we prepared F127-Folate coated SPION to target the folate receptor expressing cancer cell. Our in vitro experiments, including FACs, CLSM, Prussian blue staining, and MRI have confirmed the specific targeting of F127-Folate coated SPION. The pilot in vivo MRI showed an enhanced contrast in the tumor of targeted nanoparticles.

3.1. Synthesis of Pluronic F127- Folate coated and F127 coated SPION

The conjugation of folic acid to Pluronic F127 was produced in accordance with the previous report (Figure 1) [19]. Firstly, folic acid was activated by the reaction of its carboxylic group to one of a carbonyldiimidazone group of CDI (reaction 1). Secondly, the remaining carbonyldiimidazone group reacted with the hydroxyl group of F127 (reaction 2). The molar ratio between folic acid: CDI: F127 was 1:1.1:5, so that at least one of the hydroxyl groups of F127 was conjugated to the folic acid. The NMR spectrums of folic acid and F127-Folate (Figure 2) showed the overlay of folic acid signal at peak 1 (8.5 ppm) (pteridine proton), peak 2 (7.6 ppm), and peak 3 (6.6 ppm) (aromatic proton), confirming the success of conjugation.

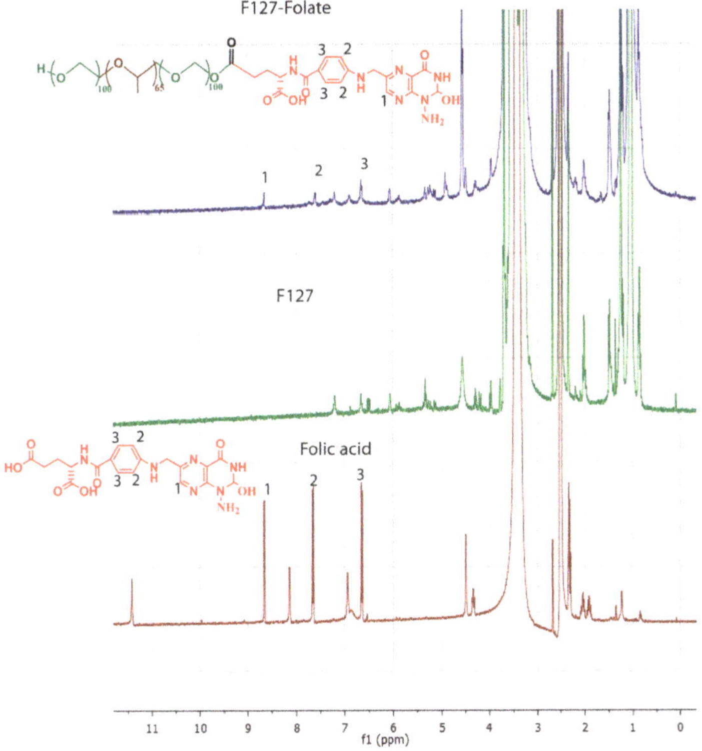

Figure 2. NMR spectrum of (1) folic acid, (2) Pluronic F127, (3) Pluronic F127 – Folate.

The F127 coated SPION were synthesized as shown in Figure 3. In contrast to oleic acid coated SPION, the nanoparticles, after being coated with F127, were well dispersed in water (Figure 3.3). It was hypothesized that the hydrophobic block poly (propylene oxide) of F127 is bound to the hydrophobic oleic layer of SPION, while the hydrophilic block poly (ethylene oxide) exposed itself to the water phase [13,15,20,21]. Besides F127, the same coating procedure was applied to poly (vinyl alcohol) PVA and Pluronic F68 (Figure 3.2). The results showed that SPION were aggregated after the n-hexane evaporation and centrifugation. This result could be due to the non-interaction between hydroxyl group of PVA and oleic acid layers, while the hydrophobic block of F68 was not long enough to anchor onto oleic-SPION surface. Jain et al., (2009) suggest that the ratio between hydrophobic and hydrophilic block of the coating polymer plays an important role during the coating process [13].

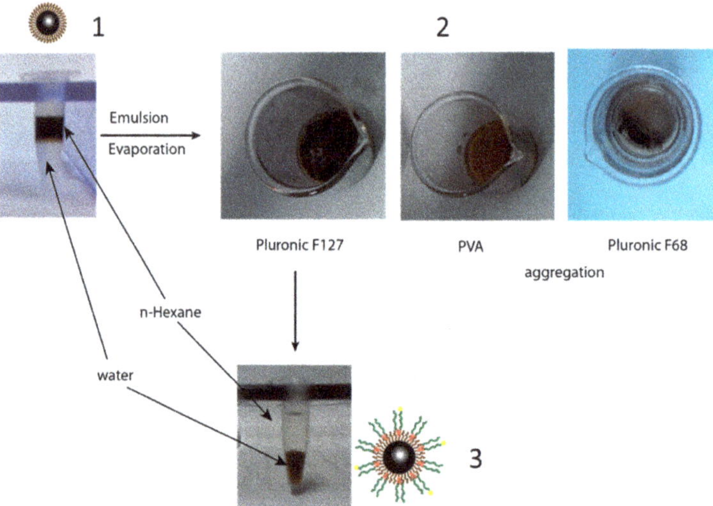

Figure 3. Coating procedure of F127 onto the oleic-super paramagnetic iron oxide nanoparticles (SPION). (1) Oleic SPION dispersed in n-Hexane, (2) polymer coated SPION after the evaporation of *n*-Hexane, (3) the dispersion of F127 coated SPION in water.

3.2. Particles Size and Zeta Potential

3.2.1. Size of Particles

The core size of oleic coated SPION was 12 ± 5 nm (Figure 4a), which was similar to that of the core of nanoparticles after coating with F127 or F127-Folate (Figure 4b). The means that the hydrodynamic size of the nanoparticles was 180 to 190 nm in water, PBS, and cell culture medium, while the PDI of the nanoparticles was mainly below 0.15 (Table 1). Moreover, the nanoparticle sizes in three different environments (water, PBS, culture medium) were stable and well dispersed. Our F127 coated SPION were stable in distilled water for over six months, however nanoparticle size slightly reduced as PDI increased (size/PDI: 185 nm/0.135 versus 156 nm/0.182). It was reported that the stabilization of SPION with F127 at the surface provides the nanoparticles with a steric surface composed of a hydrophilic PEG block. The hydrophilic surface prevents the nanoparticles from aggregating and protein binding [15]. Furthermore, in the systemic condition, the stealth nanoparticles in the size range of 10 to 200 nm could escape from renal filtration and RES [20,22].

Table 1. Hydrodynamic size of nanoparticles (DLS) in water, PBS, cell culture medium, and after six months.

Sample	Water–PDI (nm)		PBS–PDI (nm)		Cell Medium–PDI (nm)		Particles after 6 Months, Water (nm)	
F127 Coated SPION	185.2 ± 4.0	0.135 ± 0.01	192.3 ± 9.6	0.138 ± 0.03	182.3 ± 4.2	0.141 ± 0.1	156.7 ± 4.3	0.182 ± 0.02
F127-Folate coated SPION	183.6 ± 2.8	0.122 ± 0.01	194.4 ± 7.8	0.139 ± 0.01	174.4 ± 7.8	0.109 ± 0.02	158.6 ± 4.7	0.175 ± 0.02

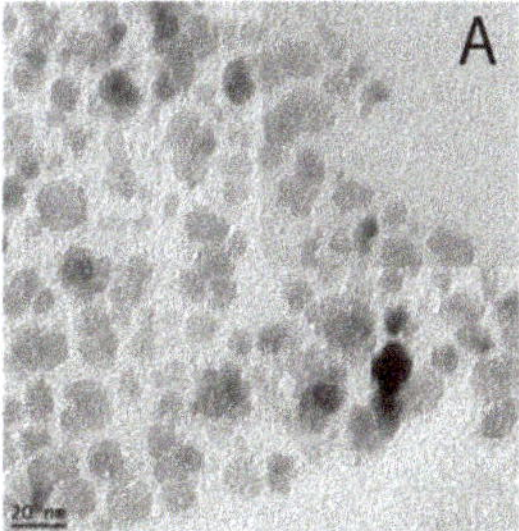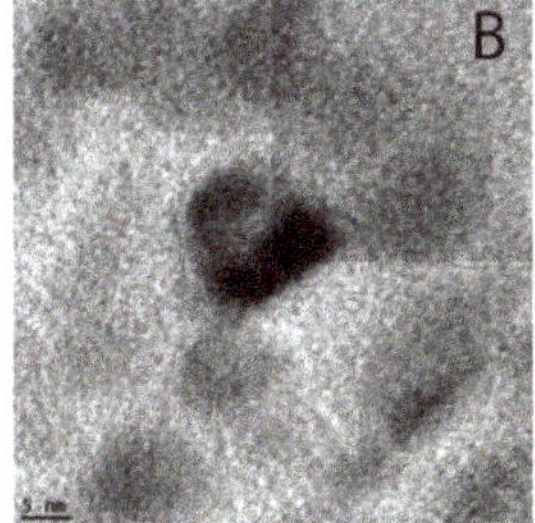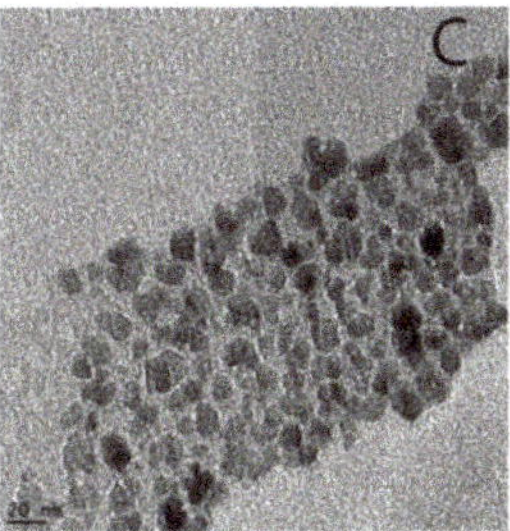

Figure 4. TEM images of SPION. (**a**) Oleic coated SPION, (**b**) oleic coated SPION at magnification 350k ×, (**c**) F127 coated SPION. The scale bars are 20 nm, 5 nm, and 20 nm.

3.2.2. Zeta Potential

Nanoparticle charge also has an influence on their blood half-life. Negative charge would reduce the interaction between nanoparticles and blood plasma, thereby enhancing their circulation time [13]. As shown in Table 2, zeta potential of F127 coated SPION and F127-Folate coated SPION were −20.2 mV and −17.15 mV, respectively, which was entirely consistent with previous reports on F127 coated SPION [19,21,23]. These results also supported the explanation about the non-aggregation of nanoparticles in the cell culture medium (Table 1). Thus, F127-Folate coated SPION would have a long circulation time without leading to any aggregation in the bloodstream.

Table 2. Zeta potential of F127 coated SPION and F127-Folate coated SPION in water. (n = 4).

Sample	Zeta Potential
F127 coated SPION	−20.2 ± 0.5 mV
F127-Folate coated SPION	−17.15 ± 1.8 mV

3.3. Cell Viability

Cell viability determines the influence of nanoparticles on cell growth and survival. The results showed that the toxicity slightly rose with the increased concentration of SPION. In comparison between lowest and highest iron concentration used to determine cell viability, the cell viability was 92% and 87% at 3.125 µg/mL and 100 µg/mL, respectively (Figure 5). However, the reduction of cell viability was not statistically significant. Thus, the F127 coated SPION and F127-Folate coated SPION were nontoxic to cells with a long incubation period (24 h). Furthermore, Jia-Jyun et al., (2009) showed that these types of magnetic nanoparticles are safe for the cells, even though a higher iron concentration was used.

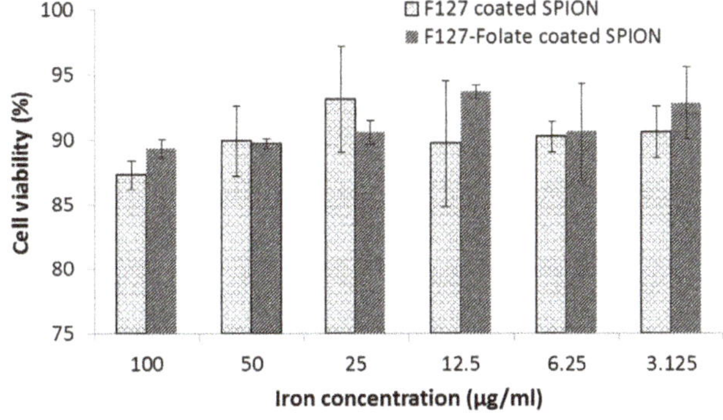

Figure 5. Cell viability of F127 coated SPIO and F127-Folate coated SPION after 24 h of incubation. MTT assay, n = 4, vertical axis: % percentage of viability compares to untreated cells.

3.4. In Vitro Uptake of Nanoparticles

Various experiments were carried out in order to prove the enhanced uptake of F127-Folate coated SPION to folate receptor expression KB cells.

3.4.1. Prussian Blue Staining and Confocal Laser Scanning (CLSM)

In order to confirm the uptake of nanoparticles in the cells, two staining methods, including Prussian blue and CLSM, were performed. Prussian blue staining is a method to display the accumulation of SPION or iron in cells by forming a complex with potassium ferricyanide. As shown in Figure 6, SPION appeared in blue while cell bodies that were pink. In particular, Figure 6c shows that F127-Folate coated SPION were present inside the cells, while there were only a few of them with the incubation of F127 coated SPION, and none in the negative control. On the other hand, CLSM could determine the uptake of nanoparticles using labeled fluorophores. As shown in Figure 7, the specific targeting of F127-Folate coated SPION/Nile Red to KB cells were visible (Figure 7c), while the control particles had unclear fluorescence signals. Both of the Prussian staining and CLSM results were entirely consistent with each other, indicating the enhanced uptake of F127-Folate coated SPION to KB cells, whereas only an insignificant amount of F127 coated SPION was found in the incubated cells. The results suggest that nanoparticles could accumulate in the cells in both a passive and active way. In the passive way, Pluronic F127 could enable the penetration of nanoparticles through the cell membrane by inserting their hydrophobic chains to the phospholipid bilayers, thereby intensifying an endocytic process [24–26]. In the active way, F127-Folate could specifically target and internalize into the cell via folate receptor mediated endocytosis [16,19]. Furthermore, the appearance of iron oxide nanoparticles (Figure 6) and fluorescent dye (Figure 7) in the cells would draw a potential application of multifunctional nanoparticles in molecular imaging and drug delivery [15].

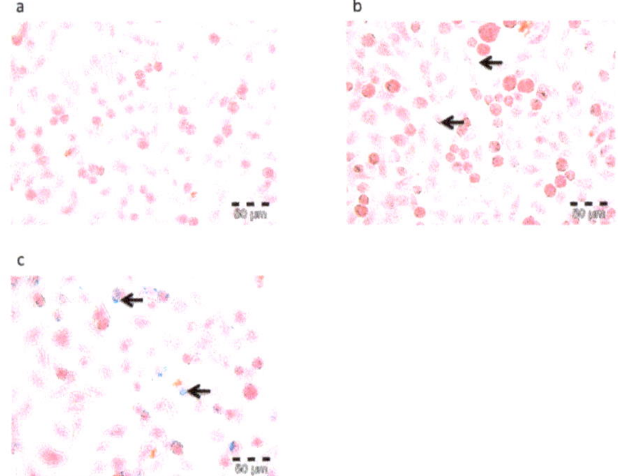

Figure 6. Prussian blue staining of KB cells. Cells were incubated with (**a**) cells only – negative control, F127 coated SPION (**b**), and F127-Folate coated SPION (**c**) for 3 h. Blue color indicates SPION, pink colors indicate cell body, and arrows show the position of iron, scale bar: 50 µL.

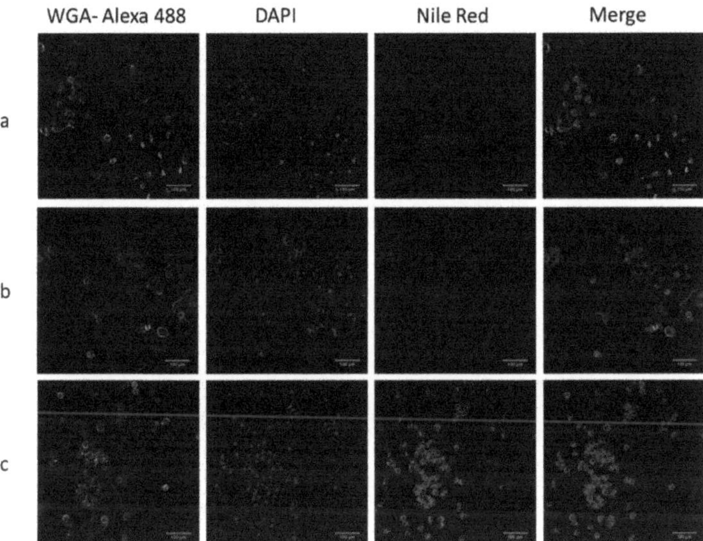

Figure 7. Confocal laser scanning microscope images of particles and KB cells. Cells were incubated with particles for 3 h. (**a**) Negative control, (**b**) F127 coated SPION and Nile Red, (**c**) F127-Folate coated SPION and Nile Red. Green: Wheat germ agglutinin-Alexa fluoro 488, Red: Nile Red, scale bar: 100 µm.

3.4.2. Iron Concentration Measurement and Flow Cytometry Assays (FACs)

In order to quantify the number of SPION that were taken up by cancer cells, the iron concentration of cells and FACs were confirmed. In the iron concentration measurement, the absorbance of sample after reaction between SPION and Ferrozine were measured. As shown in Figure 8, the iron concentration in cells tended to increase proportionally with incubated nanoparticles concentration. In comparison between two nanoparticle types, there was more iron in F127-Folate coated SPION treated cells, than to the ones treated with F127 coated SPION.

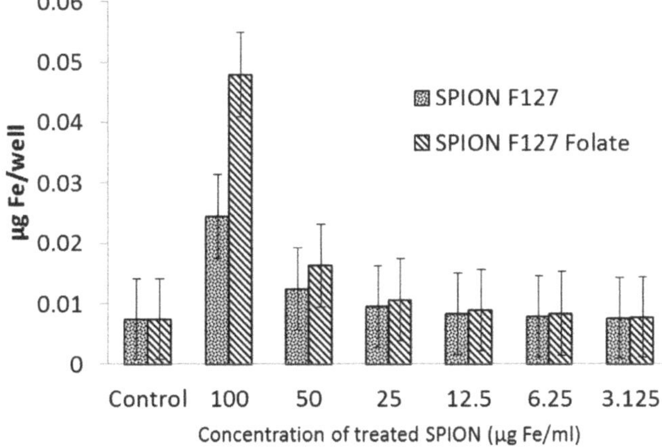

Figure 8. Iron concentration in KB cells, 3 h after incubation. From the left: Cells, 100, 50, 25, 12.5. 6.25, 3.125 µg/mL. ($n = 4$), Y axis: µg/well.

On the other hand, FACs were used to determine the uptake efficiency of nanoparticles via fluorescent intensities. In the experiment, KB cells were incubated with the nanoparticles in the medium with and without free folic acid (50 ng/mL). As shown in Figure 9, there were more signals in the nanoparticles treated sample than in negative control. More specifically, cells that were incubated with F127-Folate coated SPION showed an uptake enhancement in comparison with cells treated with F127 coated SPION. The results were entirely consistent with the previous iron concentration experiment, where the amount of accumulated F127-Folate coated SPION in the cells was 1.7 times higher than for those were incubated with F127 coated SPION. Furthermore, the F127-Folate coated SPION treated cells in the presence of folic acid, showed lower uptake than that of cells incubated in folic acid free media (65% versus 80%). With the existence of competitive subtracts, folic acid molecules initially target the folate receptor [27] and inhibit the binding of F127-Folate coated SPION; meaning the uptake of nanoparticles was reduced. The result determined the specific folate receptor targeting of F127-Folate coated SPION.

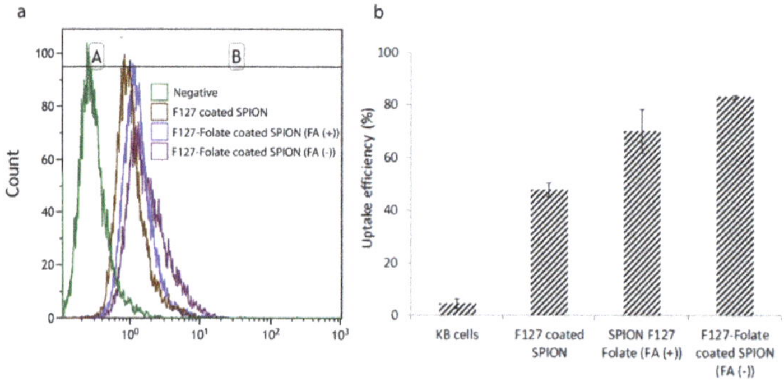

Figure 9. Flow cytometry analysis (FACs). KB cells were incubated with 50 µg/mL iron F127 coated SPION; F127-Folate coated SPION in the media with and without folic acid (5 ng/mL). Histogram (**a**) shows the cell fluorescent signal, graph (**b**) presents uptake efficiency. Gate region A was chosen for non-uptake particles, while gate region B indicates the uptake (n = 4, cell counted number: 10,000, iron concentration for incubation 50 µL/mL).

These results are consistent with previous experiments (CLSM and Prussian blue) and other studies. Thereby, it is suggested that our F127-Folate coated SPION could specifically target folate receptor expression cancer cells without causing any harmful consequences [13,19].

3.5. Magnetic Resonance Imaging

3.5.1. T2 Relaxation of Nanoparticles

Figure A2 (Appendix A) shows the exponential curve of R2 relaxation in various iron concentrations. The relaxation between iron concentration and R2 relaxation is a linear curve. In our study, it was a $y = 0.5046x$.

3.5.2. In Vitro T2 Relaxation

In order to examine the targeting efficiency and evaluate the MRI contrast agent to the cells, in vitro T2 and T2* relaxation maps were made. As shown in Figure 10, cells that were incubated with SPION showed a reduction in T2 and T2* time compared to negative control (cells and agarose). In particular, F127-Folate coated SPION treated cells showed a shorter T2 and T2* relaxation time compared to non-target F127 coated SPION. Besides, T2 and T2* relaxation times were reduced more considerably in cells that were incubated in the absence of folic acid than cells treated in the presence of folic acid (T2: 53 ms versus 60 ms, T2*: 3.32 ms versus 4.01 ms). This result repeatedly proved the specific targeting and uptake enhancement of F127-Folate coated SPION to folate receptor expressing cells. It also showed great potential for the use of particles for in vivo MRI.

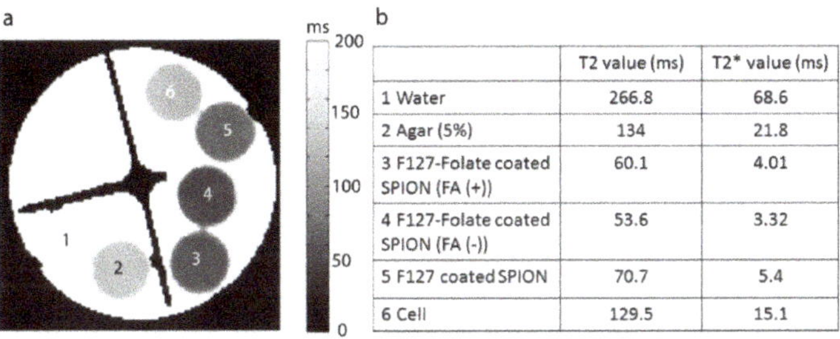

	T2 value (ms)	T2* value (ms)
1 Water	266.8	68.6
2 Agar (5%)	134	21.8
3 F127-Folate coated SPION (FA (+))	60.1	4.01
4 F127-Folate coated SPION (FA (-))	53.6	3.32
5 F127 coated SPION	70.7	5.4
6 Cell	129.5	15.1

Figure 10. (**a**) T2 map of cells in NMR tubes, (1) water, (2) agarose 0.5%, (3) F127- Folate coated SPION in folic acid media,(4) F127-Folate coated SPION in free Folic acid media, (5) F127 coated SPION, (6) cells in 0.5% agarose. (**b**) T2 and T2* value from T2 map (a) and T2* map (images not show). Number of cells: 4.5×10^5 cells/ 200 µL in 0.5 % agarose.

3.5.3. In Vivo T2 Weighted Images

In order to prove the success in cancer targeting in vivo, the nanoparticles should be able to circulate in the system, target cancer via folate receptors, and provide enhanced contrast in MRI. In our study, tumor-bearing mice were intravenously administrated with F127 coated SPION and F127-Folate coated SPION (Figure 11). The signal intensities were measured from different regions including whole tumor, tumor rim, and tumor core. These were compared to back muscle in order to estimate the influence of SPION on T2 weighted MR images after 24 h of injection. The reduced signal from whole tumor, tumor rim, and tumor core relative to back muscle (Table 3) was decreased from about 15% to 20% after F127-Folate coated SPION administration (24 h); meanwhile, no change in the reduced signal in these regions was observed in the mice that were administered with F127 coated SPION. Thus, the reduced signal in the tumor suggests the successful accumulation of F127-Folate coated SPION in the tumor 24 h after injection, while F127 coated SPION were washed out. However, the presence of F127-Folate coated SPION in tumors was random and possibly depended on the distribution of blood vessels [28,29].

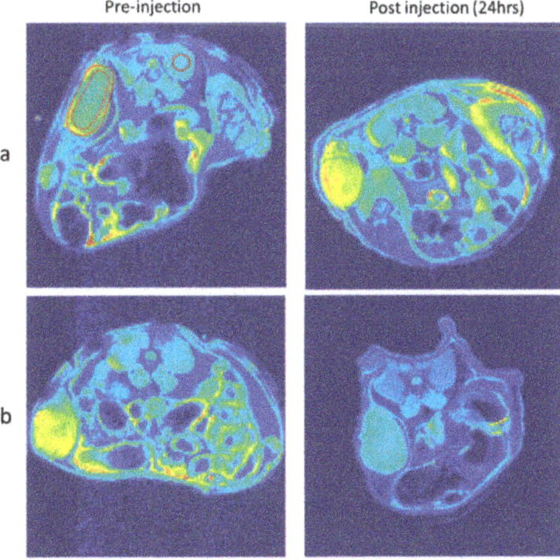

Figure 11. T2 weighted images of tumor bearing mice pre-injection and post injection. F127 coated SPION (**a**) and F127-Folate coated SPION (**b**) nanoparticles. TR/TE 4000/10.16 ms, TH 1 mm, matrix size 128 × 128, number of average: two.

Table 3. The regional signal to noise ratio (SNR) from whole tumor, tumor core, and tumor rim and relatively to SNR from the back muscle.

Position	F127 Coated SPION			F127-Folate Coated SPION		
	Pre-Injection	Post-Injection	Reduced Signal (%)	Pre-Injection	Post-Injection	Reduced Signal (%)
Tumor rim/back muscle	1.5	1.5	99.8	1.6	1.4	85.6
Tumor core/back muscle	1.3	1.5	117.7	1.8	1.4	80.3
Whole tumor/back muscle	1.3	1.4	108.1	1.7	1.4	83.2

This study suggests the potential use of F127-Folate coated SPION as a theranostic agent in the future research. The nanoparticles could specifically provide the enhanced contrast in MRI, and, at the same time, carry chemotherapeutic drugs such as Doxorubicin for chemotherapy.

4. Conclusions

In this study, F127-folate coated SPION was synthesized to target folate receptor expressing cancer. After coating, the nanoparticles were stable for a long time and in a number of different environments. In vitro results proved the specific targeting of the nanoparticles to KB cells via folate receptors, as well as the enhanced negative contrast in MRI. The pilot in vivo MRI study showed about 20% CNR reduction in tumor bearing mice that were administrated with F127-folate coated SPION.

Author Contributions: H.V.-Q. conducted the majority of the research and drafted the manuscript; M.S.V., T.N., M.G.U., N.C.N. supervised the MRI study and participated in the preparation of the manuscript; J.K. supervised the experiments; D.-T.N. prepared the manuscript.

Funding: This research received no external funding.

Conflicts of Interest: The authors declare no conflict of interest.

Appendix A

Figure A1. Diffraction of SPION. Magnetite Fe_3O_4: (1) 3; (2) 2.56; (3) 2.11; (4) 1.73; (5) 1.63; (6) 1.49; (7) 1.29. Diffraction pattern confirms the main chemical structure of nanoparticle is Fe_3O_4.

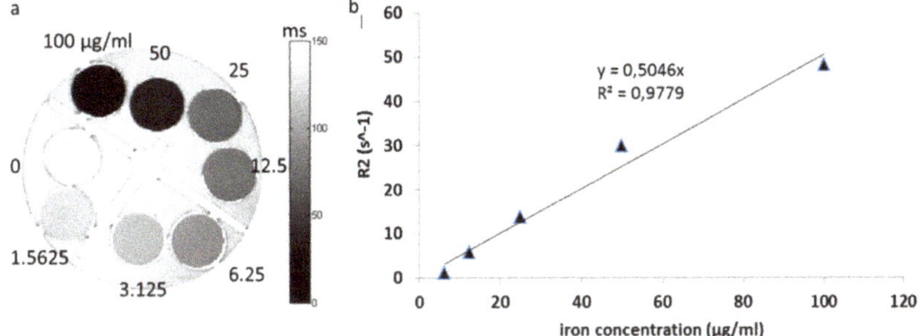

Figure A2. T2 relaxation of F127 coated SPION. F127 coated SPION was diluted in various concentrations, ranging from 100 µg to 1.5625 µg/mL and aliquot to NMR tubes. Next, T2 map was taken by choosing different TE (from 7.76 ms to 62 ms in steps of 10), TR of 5000 ms, number, or average (NA) of 20, matrix size of 256 x 256, field of view (FOV) of 22×22 mm^2, slice thickness (TH) of 0.5 mm. (**a**) T2 map of F127 coated SPION, (**b**) R2 relaxation time graph represents the correlation between iron concentration and R2 relaxation time. Numbers in (**a**) indicate iron concentration (µg/mL).

References

1. Macchia, R.J.; Termine, J.E.; Buchen, C.D.; Raymond, V.; Damadian, M.D. Magnetic resonance imaging and the controversy of the 2003 Nobel Prize in Physiology or Medicine. *J. Urol.* **2007**, *178*, 783–785. [CrossRef]

2. Caravan, P.; Ellison, J.J.; McMurry, T.J.; Lauffer, R.B. Gadolinium(III) Chelates as MRI Contrast Agents: Structure, Dynamics, and Applications. *Chem. Rev.* **1999**, *99*, 2293–2352. [CrossRef]

3. Kim, D.K.; Zhang, Y.; Voit, W.; Rao, K.V.; Muhammed, M. Synthesis and characterization of surfactant-coated superparamagnetic monodispersed iron oxide nanoparticles. *J. Magn. Magn. Mater.* **2001**, *225*, 30–36. [CrossRef]

4. Laurent, S.; Forge, D.; Port, M.; Roch, A.; Robic, C.; Vander Elst, L.; Muller, R.N. Magnetic iron oxide nanoparticles: Synthesis, stabilization, vectorization, physicochemical characterizations, and biological applications. *Chem. Rev.* **2008**, *108*, 2064–2110. [CrossRef]

5. Singh, N.; Jenkins, G.J.S.; Asadi, R.; Doak, S.H. Potential toxicity of superparamagnetic iron oxide nanoparticles (SPION). *Nano Rev.* **2010**, *1*, 5358. [CrossRef]

6. Harisinghani, M.G.; Barentsz, J.; Hahn, P.F.; Deserno, W.M.; Tabatabaei, S.; van de Kaa, C.H.; de la Rosette, J.; Weissleder, R. Noninvasive Detection of Clinically Occult Lymph-Node Metastases in Prostate Cancer. *N. Engl. J. Med.* **2003**, *348*, 2491–2499. [CrossRef]

7. Frangioni, J.V. New Technologies for Human Cancer Imaging. *J. Clin. Oncol.* **2008**, *26*, 4012–4021. [CrossRef]

8. Sosnovik, D.; Nahrendorf, M.; Weissleder, R. Magnetic nanoparticles for MR imaging: Agents, techniques and cardiovascular applications. *Basic Res. Cardiol.* **2008**, *103*, 122–130. [CrossRef]

9. Jokerst, J.V.; Lobovkina, T.; Zare, R.N.; Gambhir, S.S. Nanoparticle PEGylation for imaging and therapy. *Nanomed. Nanotechnol. Biol. Med.* **2011**, *6*, 715–728. [CrossRef] [PubMed]

10. Orringer, E.P.; Casella, J.F.; Ataga, K.I.; Koshy, M.; Adams-Graves, P.; Luchtman-Jones, L.; Wun, T.; Watanabe, M.; Shafer, F.; Kutlar, A.; et al. Purified poloxamer 188 for treatment of acute vaso-occlusive crisis of sickle cell disease: A randomized controlled trial. *JAMA* **2001**, *286*, 2099–2106. [CrossRef] [PubMed]

11. Serbest, G.; Horwitz, J.; Jost, M.; Barbee, K. Mechanisms of cell death and neuroprotection by poloxamer 188 after mechanical trauma. *FASEB J.* **2006**, *20*, 308–310. [CrossRef]

12. Scherlund, M.; Malmsten, M.; Brodin, A. Stabilization of a thermosetting emulsion system using ionic and nonionic surfactants. *Int. J. Pharm.* **1998**, *173*, 103–116. [CrossRef]

13. Jain, T.K.; Foy, S.P.; Erokwu, B.; Dimitrijevic, S.; Flask, C.A.; Labhasetwar, V. Magnetic resonance imaging of multifunctional pluronic stabilized iron-oxide nanoparticles in tumor-bearing mice. *Biomaterials* **2009**, *30*, 6748–6756. [CrossRef] [PubMed]

14. Menon, J.U.; Kona, S.; Wadajkar, A.S.; Desai, F.; Vadla, A.; Nguyen, K.T. Effects of surfactants on the properties of PLGA nanoparticles. *J. Biomed. Mater. Res. Part A* **2012**, *100A*, 1998–2005. [CrossRef]

15. Jain, T.K.; Morales, M.A.; Sahoo, S.K.; Leslie-Pelecky, D.L.; Labhasetwar, V. Iron Oxide Nanoparticles for Sustained Delivery of Anticancer Agents. *Mol. Pharm.* **2005**, *2*, 194–205. [CrossRef]

16. Low, P.S.; Antony, A.C. Folate receptor-targeted drugs for cancer and inflammatory diseases. *Adv. Drug Deliv. Rev.* **2004**, *56*, 1055–1058. [CrossRef]

17. Ross, J.F.; Chaudhuri, P.K.; Ratnam, M. Differential regulation of folate receptor isoforms in normal and malignant tissues in vivo and in established cell lines. Physiologic and clinical implications. *Cancer* **1994**, *73*, 2432–2443. [CrossRef]

18. Kularatne, S.; Low, P. Targeting of Nanoparticles: Folate Receptor. In *Cancer Nanotechnology*; Grobmyer, S.R., Moudgil, B.M., Eds.; Humana Press: New York, NY, USA, 2010; Volume 624, pp. 249–265.

19. Lin, J.-J.; Chen, J.-S.; Huang, S.-J.; Ko, J.-H.; Wang, Y.-M.; Chen, T.-L.; Wang, L.-F. Folic acid–Pluronic F127 magnetic nanoparticle clusters for combined targeting, diagnosis, and therapy applications. *Biomaterials* **2009**, *30*, 5114–5124. [CrossRef] [PubMed]

20. Jain, T.K.; Reddy, M.K.; Morales, M.A.; Leslie-Pelecky, D.L.; Labhasetwar, V. Biodistribution, Clearance, and Biocompatibility of Iron Oxide Magnetic Nanoparticles in Rats. *Mol. Pharm.* **2008**, *5*, 316–327. [CrossRef] [PubMed]

21. Jain, T.K.; Richey, J.; Strand, M.; Leslie-Pelecky, D.L.; Flask, C.A.; Labhasetwar, V. Magnetic nanoparticles with dual functional properties: Drug delivery and magnetic resonance imaging. *Biomaterials* **2008**, *29*, 4012–4021. [CrossRef]

22. Choi, H.S.; Liu, W.; Liu, F.; Nasr, K.; Misra, P.; Bawendi, M.G.; Frangioni, J.V. Design considerations for tumour-targeted nanoparticles. *Nat. Nano* **2010**, *5*, 42–47. [CrossRef]

23. Gao, Q.; Liang, Q.; Yu, F.; Xu, J.; Zhao, Q.; Sun, B. Synthesis and characterization of novel amphiphilic copolymer stearic acid-coupled F127 nanoparticles for nano-technology based drug delivery system. *Colloids Surf. B Biointerfaces* **2011**, *88*, 741–748. [CrossRef]

24. Astafieva, I.; Maksimova, I.; Lukanidin, E.; Alakhov, V.; Kabanov, A. Enhancement of the polycation-mediated DNA uptake and cell transfection with Pluronic P85 block copolymer. *FEBS Lett.* **1996**, *389*, 278–280. [CrossRef]

25. Kabanov, A.V.; Alakhov, V.Y. Pluronic® Block Copolymers in Drug Delivery: From Micellar Nanocontainers to Biological Response Modifiers. *Drug Carrier Syst.* **2002**, *19*, 72. [CrossRef]

26. Melik-Nubarov, N.S.; Pomaz, O.O.; Dorodnych, T.; Badun, G.A.; Ksenofontov, A.L.; Schemchukova, O.B.; Arzhakov, S.A. Interaction of tumor and normal blood cells with ethylene oxide and propylene oxide block copolymers. *FEBS Lett.* **1999**, *446*, 194–198. [CrossRef]

27. Lee, R.J.; Low, P.S. Delivery of liposomes into cultured KB cells via folate receptor-mediated endocytosis. *J. Biol. Chem.* **1994**, *269*, 3198–3204.

28. Choi, H.; Choi, S.R.; Zhou, R.; Kung, H.F.; Chen, I.W. Iron oxide nanoparticles as magnetic resonance contrast agent for tumor imaging via folate receptor-targeted delivery1. *Acad. Radiol.* **2004**, *11*, 996–1004. [CrossRef]

29. Bettio, A.; Honer, M.; Müller, C.; Brühlmeier, M.; Müller, U.; Schibli, R.; Groehn, V.; Schubiger, A.P.; Ametamey, S.M. Synthesis and Preclinical Evaluation of a Folic Acid Derivative Labeled with 18F for PET Imaging of Folate Receptor–Positive Tumors. *J. Nuclear Med.* **2006**, *47*, 1153–1160.

 *polymers*

Article

Synthesis, Characterization and Cytotoxicity of Novel Thermoresponsive Star Copolymers of *N,N′*-Dimethylaminoethyl Methacrylate and Hydroxyl-Bearing Oligo(Ethylene Glycol) Methacrylate

Barbara Mendrek [1,*], Agnieszka Fus [2], Katarzyna Klarzyńska [2], Aleksander L. Sieroń [2], Mario Smet [3], Agnieszka Kowalczuk [1] and Andrzej Dworak [1]

[1]   Centre of Polymer and Carbon Materials, Polish Academy of Sciences, M. Curie-Sklodowskiej 34, 41-819 Zabrze, Poland; akowalczuk@cmpw-pan.edu.pl (A.K.); adworak@cmpw-pan.edu.pl (A.D.)

[2]   Department of Molecular Biology and Genetics, Medical University of Silesia, Medykow 18, 40-752 Katowice, Poland; afus@sum.edu.pl (A.F.); katarzyna.klarzynska@gmail.com (K.K.); alsieron@sum.edu.pl (A.L.S.)

[3]   Department of Chemistry, University of Leuven, Celestijnenlaan, 200F, B-3001 Leuven (Heverlee), Belgium; mario.smet@kuleuven.be

*   Correspondence: bmendrek@cmpw-pan.edu.pl; Tel.: +48-32-271-6077

Received: 19 October 2018; Accepted: 7 November 2018; Published: 12 November 2018

Abstract: Novel, nontoxic star copolymers of *N,N*-dimethylaminoethyl methacrylate (DMAEMA) and hydroxyl-bearing oligo(ethylene glycol) methacrylate (OEGMA-OH) were synthesized via atom transfer radical polymerization (ATRP) using hyperbranched poly(arylene oxindole) as the macroinitiator. Stars with molar masses from 100,000 g/mol to 257,000 g/mol and with various amounts of OEGMA-OH in the arms were prepared. As these polymers can find applications, e.g., as carriers of nucleic acids, drugs or antibacterial or antifouling agents, in this work, much attention has been devoted to exploring their solution behavior and their stimuli-responsive properties. The behavior of the stars was studied in aqueous solutions under various pH and temperature conditions, as well as in PBS buffer, in Dulbecco's modified Eagle's medium (DMEM) and in organic solvents for comparison. The results indicated that increasing the content of hydrophilic OEGMA-OH units in the arms up to 10 mol% increased the cloud point temperature. For the stars with an OEGMA-OH content of 10 mol%, the thermo- and pH-responsivity was switched off. Since cytotoxicity experiments have shown that the obtained stars are less toxic than homopolymer DMAEMA stars, the presented studies confirmed that the prepared polymers are great candidates for the design of various nanosystems for biomedical applications.

Keywords: star polymers; solution behavior; ATRP

---

1. Introduction

Although star polymers are often studied as structures for the transport of various bioactive substances [1,2], not enough attention has been devoted to their characteristics in the solutions used in biological experiments. Star macromolecules should ensure good solubility of active substances in aqueous media and protect the substances from undesirable inactivation and decomposition. These requirements are generally satisfied by star structures with a hydrophobic core and a hydrophilic shell. The function of the core is to accept a hydrophobic substance, while the shell ensures the solubility of the nanoparticles in body fluids and protects the transported substance from interactions with other active compounds. The protection of an active biocompound by the star polymer may also be achieved by its complexation with appropriate functional groups of the shell. The coating must also ensure the nontoxicity of the entire system and prevent aggregation with the components of body fluids. Furthermore, the star polymers should be stable in aqueous solutions, which is often enabled by the covalent binding of the core to the arms of the star. To ensure adequate biodistribution, the size of the star polymers is also important. The sizes of such nanoparticles described in the literature for biomedical applications are mostly in the range of 5–200 nm [3,4].

Unfortunately, it is not usually determined if the stars are self-organizing in the presence of salts or biological compounds present in the media or whether a stimulus present in the same temperature or pH range as found in pure water causes the polymers to respond. Moreover, any post-polymerization modification, such as the incorporation of various types of compounds into the structure of stars, e.g., fluorescent [5,6] or targeting moieties [7–9], may strongly influence the final solution properties. This fact is often not taken into account in studies. For the above reasons, in our work, we synthesized novel stars with reactive hydroxyl groups that can be used in future applications in biomedicine, and we thoroughly evaluated their properties in aqueous solutions and different media used for biological tests. Herein, we provide the knowledge essential for their intended applications, e.g., for the transport of nucleic acids in gene therapy or for the introduction of appropriate fluorophores to visualize the pathways of stars in the cells via fluorescence microscopy.

Star polymers with arms consisting of *N,N*-dimethylaminoethyl methacrylate and oligo(ethylene glycol) methacrylate (OEGMA) with different numbers of ethylene glycol units could be synthesized using different polymerization techniques, e.g., ATRP [9–12], reversible addition-fragmentation chain transfer (RAFT) [13,14] or group transfer polymerization (GTP) [15]. Both the "core first" and "arm first" approaches have been used to synthesize these structures by the aforementioned polymerizations. In the "core first" method, the multifunctional compound initiates the polymerization of the monomers [9,11,12]. This strategy facilitates the formation of well-defined polymers with precisely controlled numbers of arms and chain lengths. In the "arm first" method, the linear chains have first been synthesized and then crosslinked using a bifunctional coupling agent [10,14,15]. In this way, stars with a very large number of arms have been obtained; however, during core formation, gelation processes may occur and unreacted linear polymer chains may remain in the reaction mixture.

Of the synthesized DMAEMA-*co*-OEGMA star polymers, mainly OEGMA derivatives with methyl or ethyl end groups have been prepared [10–12,14,15]. Such stars have been used as nucleic acid carriers in gene therapy, and the cationic amino groups of these polymers condense DNA or RNA into so-called polyplexes, while poly[oligo(ethylene glycol) methacrylate] (POEGMA) segments are responsible for the increase in the biocompatibility [10,12,15]. There are also reports of the use of P(DMAEMA-*co*-OEGMA) stars as drug delivery systems for the transport of doxorubicin [16–18] and fluorescein [19]. In most cases, the solution behavior of such stars has not been discussed or the authors focus on the properties of the resulting star complexes with nucleic acids or drugs. In our previous papers, we emphasized the importance of investigating the solution behavior of the prepared stars for the design of effective nanocarriers [11,20–22].

The post-polymerization modifications of the abovementioned P(DMAEMA-*co*-OEGMA) stars are limited due to the sterically hindered access to the end groups of the star arms. The use of hydroxyl-bearing OEGMA may be a solution to this problem. To the best of our knowledge, OEGMA-OH has only been used for building stars in the work of Wang [9] who described the synthesis of an eight-armed star copolymer of DMAEMA and OEGMA-OH (M_n = 360 g/mol) onto a silsesquioxane core using ATRP. The hydroxyl groups of the OEGMA were used to facilitate the further functionalization of the stars with targeting peptides. This report clearly indicated that facile star functionalization is possible with this type of monomer. The authors claimed that those stars self-assemble into micelles with sizes ranging from 128 nm to 242 nm in aqueous solutions, but further studies of the behavior of the stars in solution were not carried out.

The studies performed in this paper are designed to show how the chemical structures of the stars and the ability to control their behavior in various solutions (DMEM, PBS, and organic solvents) as well as in water under various pH and temperature conditions, can guide their future use in biomedical applications. Using the "core-first" method, we synthesized new functional stars with 28 arms made of poly(*N,N*-dimethylaminoethyl methacrylate) and hydroxyl-bearing poly[oligo(ethylene glycol) methacrylate]. Hyperbranched poly(arylene oxindole) was used as the star core because of its large number of initiating bromoester groups, leading to a multiarmed star topology. The approach proposed here allowed the relatively simple introduction of reactive groups to the star structure by using functional monomers polymerizable via ATRP. Due to the thermo and pH responsivity of PDMAEMA, for our P(DMAEMA-*co*-OEGMA-OH) stars, we decided to determine how the addition of the OEGMA-OH comonomer would affect their behavior in solution in response to selected stimuli. Thus, the obtained polymers were thoroughly characterized in different solutions under various temperature and pH conditions. For all biomedical applications, the cytotoxicity of the synthesized polymers should be investigated. Hence, the viability of the HT-1080 model cell line in the presence of stars was evaluated. The stars were found to be nontoxic in the concentration ranges required for biological experiments, which allows us to use them in the near future for the preparation of star polymer/nucleic acid complexes and incorporate labeling moieties into their structure for molecular imaging applications.

2. Materials and Methods

2.1. Materials

N,N-Dimethylaminoethyl methacrylate (DMAEMA, 99%) was purchased from Merck and purified by distillation prior to use. Hydroxyl-bearing oligo(ethylene glycol) methacrylate with an average molar mass of M_n = 360 g/mol (OEGMA-OH) was purchased from Sigma Aldrich and purified according to the procedure described in [23]. Branched polyethylenimine (PEI) with M_n = 25,000 g/mol, 1,2-dichlorobenzene (99%), 1,1,4,7,10,10-hexamethyltriethylenetetramine (HMTETA, 97%), copper (I) bromide (CuBr, 99.999%), copper (II) bromide (CuBr$_2$, 99%), p-xylene (99%) and phosphate-buffered saline (PBS) were purchased from Sigma Aldrich and used as received. Dulbecco's modified Eagle's medium (DMEM) was also purchased from Sigma Aldrich and was dissolved in sterile distilled water in accordance with the manufacturer's instructions. DOWEX MARATHON MSC ion exchange resin was purchased from Sigma Aldrich and protonated using 1.6 M HNO$_3$. Methanol (99.8%) and tetrahydrofuran (THF, 99.8%) were purchased from POCh and used as received.

Alamar blue reagent (DAL1100) was obtained from Invitrogen and used as a 10% solution. Penicillin/streptomycin/amphotericin B mix, fetal bovine serum (FBS), L-glutamine and trypsin/EDTA (10×) were purchased from PAN Biotech. All reagents were added to the culture medium to proper final concentrations prior to use. L-glutamine was added to the culture medium every two weeks.

2.2. Synthesis of Star Polymers with Poly[N,N-Dimethylaminoethyl Methacrylate-co-Hydroxyl-Bearing Oligo(Ethylene Glycol) Methacrylate] Arms (P(DMAEMA-co-OEGMA-OH))

Example star polymer synthesis was as follows: hyperbranched poly(arylene oxindole) (PArOx, 80×10^{-3} g, 3.8×10^{-6} mol) (synthesis and characterization published elsewhere [24]), was dissolved with CuBr (1.53×10^{-2} g, 1.07×10^{-4} mol) and CuBr$_2$ (1.2×10^{-3} g, 5.35×10^{-6} mol) in 8.6 mL of 1,2-dichlorobenzene in a Schlenk flask under nitrogen with a magnetic stirrer and degassed using freeze-pump-thaw cycles. Then, HMTETA (2.46×10^{-5} g, 1.07×10^{-4} mol, and 2.9×10^{-5} mL) was added to the solution, and the mixture was degassed. The ratio of [PArOx]:[CuBr]:[CuBr$_2$]:[HMTETA] was 1:1:0.05:1 in all cases. Next, DMAEMA (1.7 g, 1.08×10^{-2} mol, 1.8 mL) in a 1:4 *v/v* solution with 1,2-dichlorobenzene was added, and the reaction mixture was degassed twice. The flask was placed in an oil bath at 40 °C. Next, after the desired conversion of DMAEMA, OEGMA-OH (0.39 g,

1.07×10^{-3} mol, 0.35 mL, degassed separately) was added to the reaction mixture, which was again thermostated at 40 °C in an oil bath. The monomer-to-initiator ratio was different for each synthesis to obtain stars with arms of different lengths and different OEGMA-OH contents. For details, see Table 1.

The monomers conversion was followed by high-performance liquid chromatography (HPLC) with p-xylene as an internal standard. Samples were taken during the polymerization and analyzed without purification. After the desired molar mass was obtained, THF (5 mL) was added, and the solution was passed through a DOWEX-MSC-1 ion exchange resin column to remove the copper. The resulting solution without copper was dialyzed against methanol and then against water (SpectraPor membrane with MWCO 1000 g/mol) and freeze dried to prevent polymer aggregation.

^{1}H NMR (600 MHz, CDCl$_3$) δ_{ppm}: 0.8–1.4 (CH_3C–), 1.7–2.2 (–CH_2C–) in the methacrylate backbone, 2.6–2.8 (–OCH_2CH_2N–), 4.0–4.2 (–OCH_2CH$_2$N–) and (–OCH_2CH$_2$O–), 2.2–2.5 (–N-CH_3), 3.6–3.8 (–OCH_2CH$_2$O–).

Table 1. Polymerization conditions for poly(*N,N*-dimethylaminoethyl methacrylate (DMAEMA)-*co*-hydroxyl-bearing oligo(ethylene glycol) methacrylate (OEGMA-OH)) stars.

Star	[DMAEMA]:[OEGMA]:[PArOx] [a]	Conversion DMAEMA:OEGMA [b] [%]	OEGMA-OH Content [c] [mol%]
P1	100:10:1	44:3	2.6
P2	50:5:1	62:29	2.8
P3	30:3:1	60:56	4.2
P4	40:10:1	36:30	10.0

[a] theoretical ratio per mol of initiating sites; [b] measured by HPLC; [c] calculated from ^{1}H NMR.

2.3. Cell Culture

Human HT-1080 fibrosarcoma cells (ATCC#CCL-121) were purchased from American Type Culture Collection (ATCC). Cells were cultured in cell culture medium (CCM) consisting of DMEM supplemented with 10% FBS, 1% L-glutamine, penicillin, streptomycin and amphotericin B, and the cells were cultured at 37 °C under 5% CO$_2$.

2.4. Cytotoxicity of Star Polymers

One day before the cytotoxicity assays, HT-1080 cells were seeded in 0.5 mL of CCM in 24-well plates at a density of 8×10^4 cells per well. The following day, the CCM was supplemented with tested star polymers and PEI to the desired final concentrations: 0 (control), 5, 10, 20, 30, 40, 50, 60, 70, 80, 90 and 100 µg/mL. The cells were incubated for an additional 24 h, then the medium was removed, and the plates were rinsed with prewarmed PBS buffer. The cell viability was evaluated with the Alamar blue assay. After adding 200 µL of prewarmed Alamar blue reagent in CCM to a final concentration of 10% in each well, the cells were incubated under standard conditions at 37 °C and 5% CO$_2$ for 1 h. Subsequently, 100 µL of the mixture from each well was transferred to a new well of a 96-well TPP™ plate (PerkinElmer, Waltham, MA, USA), and the fluorescence emission was monitored at 590 nm using a VICTOR™ Multilabel Plate Reader (PerkinElmer, USA) with a 560 nm excitation source. The cell viability was assessed based on the percent of live cells compared to the control cells not treated with the polymer.

2.5. Methods

High-performance liquid chromatography (HPLC) was used to determine the monomer conversion. An Agilent system (1260 Infinity) equipped with an Eclipse XDB-C18 column (84.6 × 150 mm, Agilent) and a UV/VIS diode array detector (Agilent, 1260 DAD VL, Santa Clara, CA, USA) was used. A linear gradient from 10% to 95% B within 40 min was applied (A: 0.1% TFA in water (Gradient Grade, POCh, Poland) and B: 0.1% TFA in acetonitrile (Gradient Grade, 99.9%, POCh,

Gliwice, Poland)). The flow rate was maintained at 0.5 mL/min. The chromatograms were recorded at 220 nm using Agilent ChemStation software.

Gel permeation chromatography with multiangle laser light scattering (GPC-MALLS) detection was used to determine the molar masses and molar mass dispersities of the star polymers. The system contained a differential refractive index detector (Δn-2010 RI WGE Dr. Bures) and a multiangle laser light scattering detector (DAWN EOS from Wyatt Technologies, Santa Barbara, CA, USA) and the following columns: GRAM gel guard, GRAM 100 Å, GRAM 1000 Å and GRAM 3000 Å (Polymer Standard Service, Mainz, Germany). GPC was performed in DMF at 45 °C with a nominal flow rate of 1 mL/min. The results were evaluated using ASTRA 5 software from Wyatt Technologies. The refractive index increments (dn/dc) of the stars were independently measured in DMF using a SEC-3010 dn/dc WGE Dr. Bures differential refractive index detector.

Dynamic light scattering (DLS) measurements were performed on a Brookhaven BI-200 goniometer (Brookhaven Instruments Corporation, New York, NY, USA) with vertically polarized incident light (λ = 632.8 nm) supplied by a He-Ne laser operating at 35 mW and a Brookhaven BI-9000 AT digital autocorrelator. The autocorrelation functions were analyzed using the constrained regularized algorithm CONTIN. The measurements were made at a 90° angle at 25 °C. The dispersity of particle sizes was given as $\frac{\mu_2}{\bar{\Gamma}^2}$, where $\bar{\Gamma}$ is the average value of the relaxation rates Γ, and μ_2 is its second moment. These values were obtained from cumulant analysis.

Before DLS analysis, the star polymer solutions (c = 1 mg/mL) were passed through membrane filters with nominal pore sizes of 0.2 µm (ANATOP 25 PLUS, Whatman, Maidstone, UK).

The cloud point transition temperatures of the prepared stars were determined using a Specord200 Plus (Analytik Jena, Jena, Germany) spectrophotometer equipped with a thermostated cuvette (heating rate 2 °C/min). The transmittance in DMEM culture medium and PBS (for all samples c = 1 mg/mL) was monitored at λ = 700 nm as a function of temperature. The cloud points were determined as the temperature at which the transmittance of the polymer solution reached 50% of its initial value.

In order to investigate the responsivity of the P(DMAEMA-*co*-OEGMA-OH) stars as a function of pH, the star polymers were directly dissolved in deionized water, and the pH of each solution was adjusted by adding appropriate amounts of 1 M KOH or 1 M HCl. The pH adjustments were performed using a pH meter (Elmetron, Zabrze, Poland), and the tested pH values ranged from 2 to 12.

Zeta potential measurements were performed in triplicate on a Zetasizer Nano ZS 90 (Malvern Panalytical, Malvern, UK) in disposable folded capillary cells. The adjusted solutions were equilibrated for one day at each pH value before measurement. The zeta potential (ζ) used in this work was calculated from the electrophoretic mobility, u, employing the Helmholtz–Smoluchowski Equation (1):

$$u = \varepsilon\zeta/\eta \tag{1}$$

where ε is the dielectric constant of the solvent, and η is the viscosity of the solvent.

3. Results and Discussion

3.1. Synthesis of Star Polymers with Poly[N,N-Dimethylaminoethyl Methacrylate-co-Hydroxyl-Bearing Oligo(Ethylene Glycol) Methacrylate] Arms (P(DMAEMA-co-OEGMA-OH))

Star polymers were obtained via the "core first" method using atom transfer radical polymerization (ATRP) (Scheme 1). The core of each star was a hyperbranched poly(arylene oxindole) (PArOx). The synthesis and characterization of PArOx was described previously [24]. The core provided 28 bromoester groups capable of initiating ATRP of methacrylate monomers. The absolute molar mass of PArOx, as measured by GPC-MALLS, was M_n = 21,000 g/mol and M_w/M_n = 2.2 [20], while the number of initiating groups was determined from the Frey equation [25], and that value related the number of dendritic and terminal units with the degree of polymerization.

In this work, to obtain stars with copolymer arms of a desired structure, we used a different approach than we used in our earlier work on stars with random copolymer DMAEMA and di(ethylene glycol) methyl ether methacrylate arms [11]. We assumed that DMAEMA and OEGMA-OH will form arm chains with a random distribution of monomer units, as was studied by Lang et al. [26] who reported that the values of the reactivity ratios of DMAEMA and OEGMA (M_n = 475 g/mol), a monomer with a similar structure to OEGMA-OH, estimated in ATRP were close to unity. Taking this fact into consideration, after the addition of the first monomer (DMAEMA), when its conversion was respectively high, a second monomer (OEGMA-OH) was added. This "one-pot" approach ensured that the OEGMA-OH units were built in mainly at the ends of the star arms. Stars designed in this way should easily undergo post-polymerization modifications due to expected facile access to OH groups in the multiarmed and sterically crowded star structure.

The conditions for the polymerization processes are shown in Table 1.

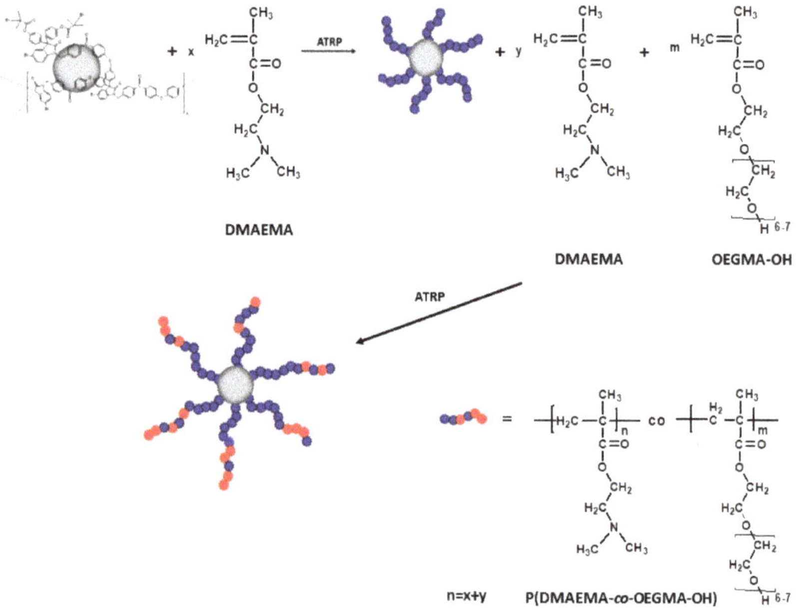

Scheme 1. Synthesis of P(DMAEMA-*co*-OEGMA-OH) star copolymers.

After the optimization of ATRP conditions, the solvent, temperature, time and the initiator-to-catalyst complex ratio were the same for all polymerizations (for details, see the Experimental Section). The only variable was the initial ratio of the monomers per mole of initiating sites (Table 1). In this way, star polymers with various lengths of the arms and different contents of OEGMA-OH in the arm were obtained (Table 1).

The content of OEGMA-OH in the star arms was calculated based on the ^{1}H NMR spectra taken in chloroform; the ratio of the signal of the methyl protons of the amino groups (c) in the DMAEMA to the signal of the methylene protons in the pendant chains of OEGMA-OH (f). Four stars with different molar contents of OEGMA-OH in the range from 2.6 to 10 mol% were obtained (Table 1). A representative proton NMR spectrum of a star polymer is shown in Figure 1.

The theoretical molar masses of all synthesized star polymers were calculated from the monomer conversion and compared with those obtained from GPC with multiangle laser light scattering (GPC-MALLS) detection. For GPC measurements, the refractive index increments (dn/dc) of the stars were independently measured in DMF to obtain absolute values of the molar masses (Table 2).

The theoretical molar masses of the star were calculated using Equation (2):

$$M_{theor.} = M_{PArOx} + ([C_{DMAEMA}]_0/[C_{PArOx}]_0) \times \Delta C_{DMAEMA} \times M_{DMAEMA}$$
$$+([C_{OEGMA-OH}]_0/[C_{PArOx}]_0) \times \Delta C_{OEGMA-OH} \times M_{OEGMA-OH}) \tag{2}$$

where:

$M_{theor.}$—the theoretical molar mass of the star, M_{PArOx}—the molar mass of the macroinitiator, $[C_{DMAEMA}]_0$—the initial molar concentration of DMAEMA, $[C_{PArOx}]_0$—the initial molar concentration of the initiator, ΔC_{DMAEMA}—the conversion of DMAEMA, M_{DMAEMA}—the molar mass of DMAEMA, $[C_{OEGMA-OH}]_0$—the initial molar concentration of OEGMA-OH, $\Delta C_{OEGMA-OH}$—the conversion of OEGMA-OH, and $M_{OEGMA-OH}$—the molar mass of OEGMA-OH.

The good correlation between the theoretical and measured molar masses (Table 2) of the obtained star polymers confirmed that the polymerization was controlled. The absolute values of the molar masses of the stars were in the range of 100,000–257,000 g/mol (Table 2).

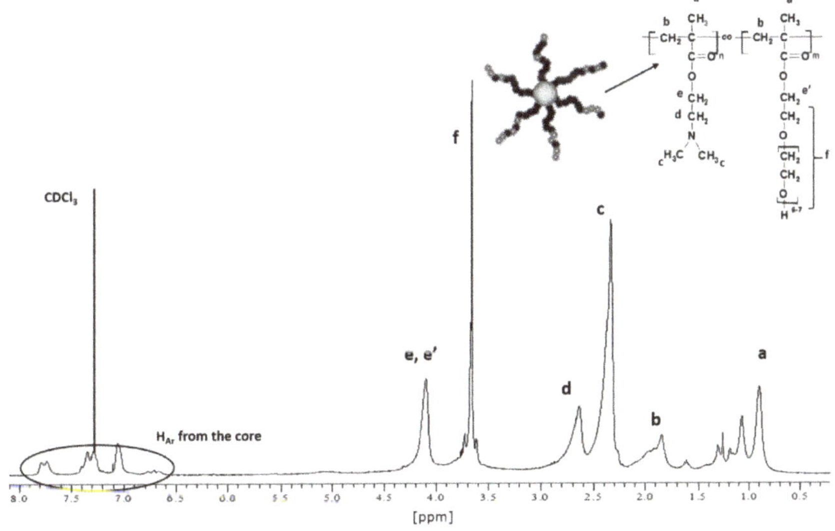

Figure 1. ^{1}H NMR spectrum (CDCl3, 600 MHz) of P(DMAEMA-*co*-OEGMA-OH) star (sample P4, Table 1).

Table 2. Molar masses of the P(DMAEMA-*co*-OEGMA-OH) star copolymers.

Star	M_n [a] [g/mol]	dn/dc [a] [mL/g]	M_w/M_n [a]	M_{theor} [b] [g/mol]
P1	257 000	0.067	2.4	240 000
P2	159 000	0.074	2.5	172 000
P3	132 000	0.081	2.3	117 000
P4	100 000	0.126	2.2	114 000

[a] from GPC-MALLS, [b] calculated from Equation (2).

Very often, the synthesis of star polymers via controlled radical polymerization is accompanied by the recombination of active radical species at the ends of the star arms. This reaction often occurs when the conversion of monomers incorporated into the star arms is high; this process is called "star-star coupling" [27,28]. If this occurs, the GPC chromatograms will show a shoulder in the high molar mass region. The chromatograms of all obtained star polymers are shown in Figure 2. The peaks are monomodal and uniform. This proves that despite the relatively high DMAEMA conversions, "star-star coupling" did not occur.

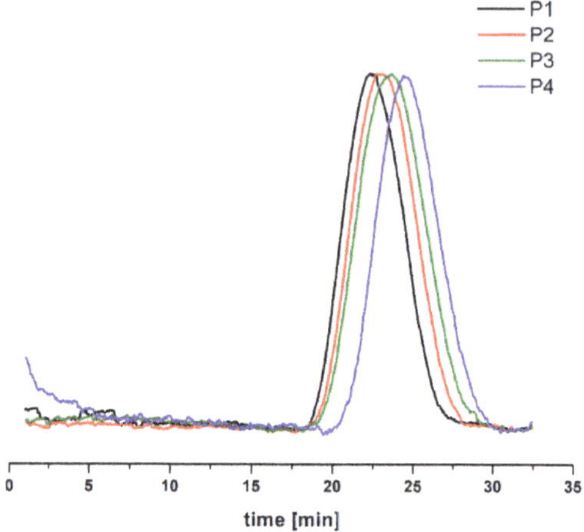

Figure 2. Chromatograms (RI traces) of P(DMAEMA-*co*-OEGMA-OH) star copolymers (DMF, 1 mL/min).

3.2. Solution Behavior of the (P(DMAEMA-co-OEGMA-OH)) Star Copolymers

Because the size of the stars is the most important factor for biomedical applications, the relationships between the structure and solution behavior of star polymers should be examined in detail. It should be noted that in the solution, the structures of star-like macromolecules resemble the micelles formed in the self-organization of linear, amphiphilic block copolymers, and star molecules are thus often called unimolecular micelles [29]. The nomenclature of the isolated star molecules in solution and their aggregates is not uniform; in this work, the most common nomenclature found in the literature was applied.

The sizes of the obtained stars in various solvents were measured using dynamic light scattering. Five solvents were used in these experiments, two organic solvents (acetone and ethanol), pure water and two media that are often used in biomedical experiments (PBS and the most commonly used cell culture medium, DMEM). Due to the presence of various types of inorganic salts and biocompounds in culture media, such studies are necessary to properly assess the structures formed in solution.

Hydrodynamic radius ($R_h^{90°}$) measurements were made in triplicate at 25 °C, and the average values are summarized in Table 3. The concentration of polymers in each solvent was equal to 1 mg/mL.

Table 3. Hydrodynamic radii of the P(DMAEMA-*co*-OEGMA-OH) stars in different solvents.

Star	M_n [g/mol]	Contour Length of Arm [nm]	$R_h^{90°}$ [nm]				
			Acetone	EtOH	Water (pH Dissolution)	DMEM	PBS
P1	257,000	11.3	10.0 ± 2	22.4 ± 2	17.0 ± 2 (6.8)	20.0 ± 2	14.4 ± 1
P2	159,000	8.2	11.0 ± 2	20.0 ± 3	18.0 ± 1 (6.9)	19.3 ± 2	13.3 ± 2
P3	132,000	5.0	13.6 ± 1	20.5 ± 2	18.0 ± 2 (6.8)	11.7 ± 1	10.7 ± 3
P4	100,000	4.3	9.0 ± 2	26.6 ± 1	41.0 ± 2 (6.6)	27.3 ± 2	14.0 ± 1

Additionally, Table 3 compares the contour length of the star arms calculated for the fully extended arms by multiplying the DP of the arm and the length of one repeating unit containing two carbon atoms in the backbone (0.252 nm) [30].

The choice of solvent was mainly guided by their possible interactions with the polymer. Acetone was chosen because it is a good solvent both for the arms and the PArOx core of the stars. The remaining solvents are more selective; they do not penetrate the dense hydrophobic structure of the PArOx core, and instead, they interact with the shell formed by the arms of the star.

The obtained values of the hydrodynamic radii in acetone and PBS (Table 3) are the lowest, which suggests that in these solvents, stars exist as single macromolecules. In the others, the average values of $R_h^{90°}$ are slightly higher (Table 3), indicating the aggregation of a few macromolecules. We are aware, however, that comparisons between the measured values of $R_h^{90°}$ and the calculated contour length do not allow us to directly conclude whether the stars are isolated or if they self-organize into several interconnected structures.

The representative size distributions of the star polymers (samples P1–P4, Table 1) in PBS are shown in Figure 3. The obtained results show that in the case of P3 and P4 stars, aggregates could be formed. The same trends were observed in acetone and ethanol, but in DMEM and water at the pH of dissolution, one very broad distribution was observed for all stars. However, based on the calculated contour lengths of the arms (Table 3) and the distribution of sizes (Figure 3), the formed aggregates are rather small and not the major species in the solution. Similar relationships between the size of stars in various solvents were observed for homopolymer DMAEMA stars [22] as well as for copolymer P(DMAEMA-*co*-DEGMA) stars with the PArOx core [11].

As the largest hydrodynamic diameter of the stars was in the range of fifty nanometers, we can conclude that the introduction of a new monomer, namely OEGMA-OH, only leads to limited aggregation in the solution, and that the obtained nanoparticles are still small enough for biomedical applications.

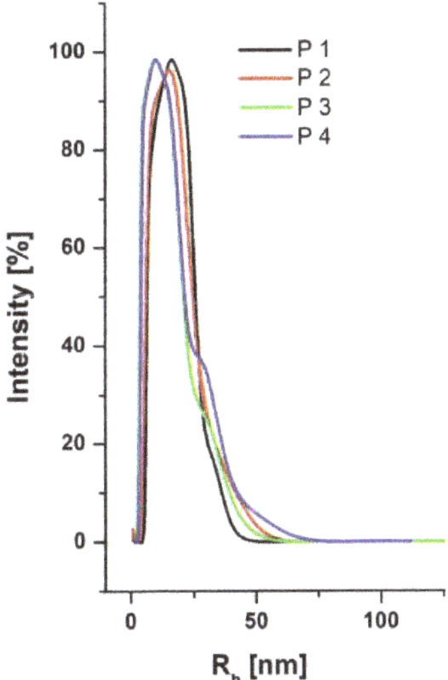

Figure 3. Distribution of the hydrodynamic radii of P(DMAEMA-*co*-OEGMA-OH) stars in PBS.

3.3. Temperature-Responsive Properties of the P(DMAEMA-co-OEGMA-OH) Stars

Since PDMAEMA itself is a pH- and thermoresponsive polymer [31,32], stars made of this monomer behave differently in solutions under different pH and temperature conditions [22,33,34]. The cloud point temperatures (T_{CP} values) of DMAEMA polymers depend on their molar mass and concentration as well as the charge on the amine groups in the chains [31,35]. Generally, at high pH values, weak charges on the polymer chains lead to a decrease of the observed T_{CP} values, while at low pH values, the electrostatic repulsions between the charged groups counteract the chain collapse, which increases the phase-transition temperature.

In the case of the star topology of the DMAEMA polymers, the thermoresponsive properties are much more complex because the T_{CP} may be influenced by the hydrophobicity of the core as well as the number of arms and the molar mass [31].

As the phase transition temperature is governed by the ratio of hydrophobic to hydrophilic units [36], the stars prepared in this paper are an interesting case. They possess a hydrophobic core that should decrease the T_{CP} values, but the arms of the stars consist of hydrophilic OEGMA-OH units, which should increase the transition temperature.

The responses of the obtained stars to temperature changes were investigated using UV–Vis spectroscopy (Figure 4). The T_{CP} values were determined as the temperatures at which the transmittance of the polymer solution decreased to 50%. The values of the cloud point temperature for a given star differ depending on the solvent used, as shown in Table 4. It should be noted that the stars (P1–P4, Table 1) were not thermoresponsive when the pH was high enough to deprotonate the arms. The presence of the salts in aqueous solutions of PBS and DMEM enabled measurements of the phase transition temperature due to the "salting-out" effect [37]. The measurements were performed in PBS (pH = 7.4) and culture medium (DMEM), and the same sample concentration of 1 mg/mL was used in all experiments.

The phase transition temperature range of the star copolymers was between 64.5–68.5 °C in PBS and 42.4–52.8 °C in DMEM. The lower T_{CP} in DMEM (approximately 20 °C lower than in PBS) is likely caused by the larger number of different salts in DMEM than in PBS.

As mentioned above, the amount of the hydrophilic OEGMA-OH units in the star arms also influenced the T_{CP}. P4 (Table 1), with the smallest molar mass and the highest content of OEGMA-OH (10 mol%), was not thermoresponsive in either PBS or DMEM. Such amount of hydrophilic monomer in the relatively short arms (Table 2) switched off the phase transition from coil to globule of the star arms, suppressing the "salting out" effect.

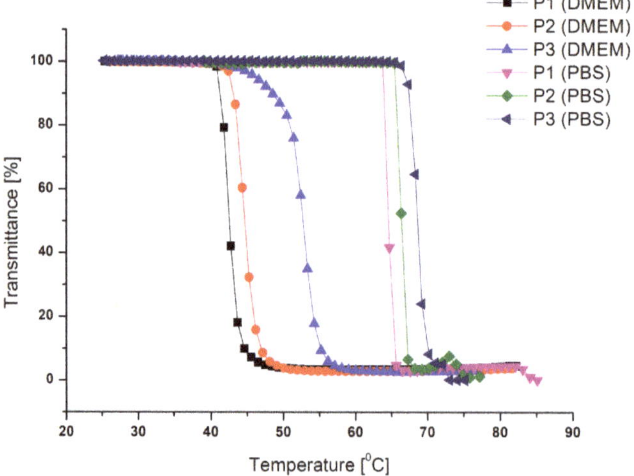

Figure 4. Transmittance as a function of temperature for P(DMAEMA-*co*-OEGMA-OH) stars in PBS buffer and Dulbecco's modified Eagle's medium (DMEM) (λ = 700 nm, c = 1 g/L).

Table 4. Phase transition temperature data of the P(DMAEMA-*co*-OEGMA-OH) star copolymers.

Star	M_n [a] [g/mol]	OEGMA-OH Content [mol%]	T_{CP} [°C]	
			PBS	DMEM
P1	257 000	2.6	64.5	42.4
P2	159 000	2.8	66.3	44.5
P3	132 000	4.2	68.5	52.8

In the case of polymers having a lower content of OEGMA-OH, the values of T_{CP} increase with increasing OEGMA-OH content regardless of their molar mass. Similar results were obtained by Trzebicka et al. [38] for POEGMA copolymers, and a linear increase in T_{CP} was observed with increasing content of the more hydrophilic comonomer.

The differences between the T_{CP} values of stars P1 and P2 are insignificant despite the fact that they have considerably different molar masses. The P3 sample has the highest T_{CP} because its OEGMA-OH content (4.2 mol%) (hydrophilic content) is the largest.

In our previous work, we studied the sensitivity of DMAEMA homopolymer stars with a PArOX core to temperature [22]. The thermoresponsivity of these stars was dependent on the pH of the solution, and a T_{CP} was observed only at pH 13, when the chains of the arms were unionized. Additionally, in DMEM at a neutral pH, no phase transition occurred. Here, we show that the introduction of a small fraction of OEGMA-OH into the PDMAEMA stars is sufficient to weaken the repulsive electrostatic interactions between the partially ionized arm chains and induce their collapse upon a change in temperature in DMEM.

The most suitable star polymer for further studies involving the thermoresponsive properties seemed to be star P1, as it showed with the highest molar mass and lowest OEGMA-OH content with a T_{CP} of approximately 42.4 °C, which is the closest value to physiological conditions.

3.4. pH Responsivity and Zeta Potential of the P(DMAEMA-co-OEGMA-OH) Stars

DMAEMA polymers are weak bases; they are fully protonated under acidic conditions and completely deprotonated under basic conditions [31,39].

We previously reported the effect of pH on the sizes of stars with PDMAEMA and P(DMAEMA-co-DEGMA) arms [11,22]. We observed that for PDMAEMA stars, with decreasing pH, the electrostatic repulsion from the charged amino groups stretched the arms, which resulted in an increase in the size of the nanoparticles present in the solution. At alkaline pH values, the hydrogen bonds formed between the amino groups of the PDMAEMA reduced the size of the nanostructures. The addition of DEGMA units to the star arms complicated their solution behavior as either one or two populations of nanoparticles in light scattering measurements corresponding to star and/or star aggregates were observed [11].

All P(DMAEMA-co-OEGMA-OH) stars are soluble in water throughout the whole pH range. For dynamic light scattering measurements, stars (samples P1–P4, Table 1) were directly dissolved in deionized water, and the pH of each solution was adjusted by adding the appropriate amount of 1 M KOH or 1 M HCl.

The change in the size of the PDMAEMA star as a function of the pH of the solution is shown in Figure 5. Throughout the tested pH range, one broad distribution of particles was observed, similar to what was seen with homopolymer DMAEMA stars [22]. Nevertheless, their behavior in solution is different from that of the stars in our previous research.

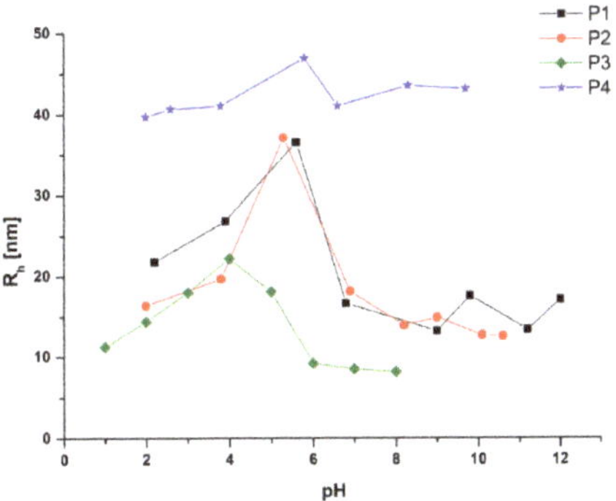

Figure 5. The average hydrodynamic radius as a function of pH for P(DMAEMA-co-OEGMA-OH) stars (the lines are guides for the eye) (c = 1 mg/mL, 90°, 25 °C).

Most importantly, the hydrodynamic radii of the nanoparticles do not exceed 50 nm at any of the tested pH levels, which allows their further use in biomedical applications. For all star polymers, the maximum $R_h^{90°}$ values were observed in the range of pH 4 to 6.9, which is probably caused by the formation of larger structures in that particular pH range.

Under low pH conditions and under high pH conditions, the sizes of P1–P3 stars remain small and most likely correspond to isolated macromolecules or aggregates of several stars. This conclusion is also supported by the lack of a definite dependency between the size of nanoparticles and the molar mass of the stars. The stars with 10 mol% OEGMA-OH reached the highest hydrodynamic radius of all tested stars (Figure 5). The maximum sizes are observed for the P4 sample, but the differences in the $R_h^{90°}$ values are not as great over the entire pH range as the differences seen for stars P1–P3. Based on these results, we can conclude that when the OEGMA-OH fraction is equal to or higher than 10%, it not only switches off the pH-responsive properties of the polymer but also promotes stronger aggregation compared to stars with lower contents of hydrophilic units. Based on these results, the assumption seems justified that in the case of the polymer P4 (Table 1), the switching off the temperature and pH responsivity is a complex result of the relatively high OEGMA content and the relatively small PDMAEMA fraction in the arms together with the low molar mass of the star.

For biomedical applications, the positive charge on the surface of nanoparticles facilitates their cellular uptake. This phenomenon is based upon the electrostatic interactions of the positively charged particle surfaces with the negatively charged cell membrane [12]. The zeta potential is an indicator of the surface charge of nanoparticles, and its value depends upon the pH of the solution.

For stars P1–P4, their zeta potential values decreased with increasing pH due to the loss of positively charged amino groups in the PDMAEMA (Figure 6). The change in the charge from positive to negative for all stars occurred at a pH value close to 10.

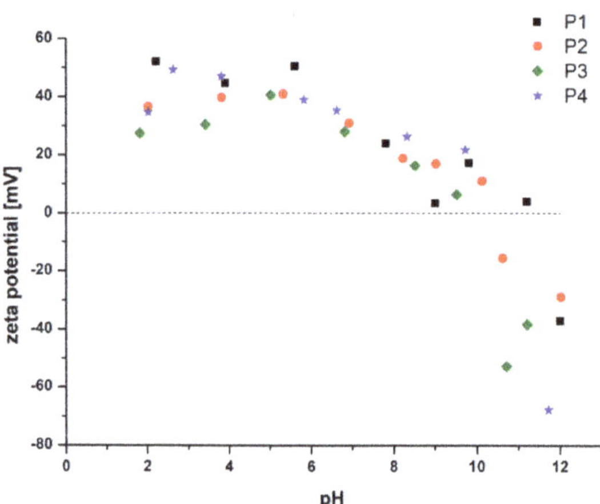

Figure 6. The zeta potential changes as a function of pH for P(DMAEMA-*co*-OEGMA-OH) stars. (c = 1 mg/mL, 25 °C).

Negative zeta potential values under basic conditions were also observed for stars with copolymer arms made of DMAEMA and DEGMA [11] as well as for PDMAEMA microgels [40] and PDMAEMA chains grafted onto a poly(ethylene terephthalate) (PET) substrate [41]. This behavior is probably caused by the adsorption of negative ions by the deprotonated amine groups [40]. At pH ~7.4, which is accepted as physiological conditions, the zeta potential values are between 20 and 40 mV. Therefore, the obtained P(DMAEMA-*co*-OEGMA-OH) stars can be used in future applications, e.g., as carriers of negatively charged nucleic acids.

3.5. Cytotoxicity of the Obtained Star Polymers

DMAEMA polymers of different topologies have attracted significant attention as delivery systems able to transport genetic material [42] or as antimicrobial materials [43]. The cationic charge of these polymers is responsible for their interactions with anionic species, e.g., proteins present in the membranes of the cells, which results in the apoptosis and necrosis of the cells [44,45]. In the case of gene delivery, the studies have mainly focused on the cytotoxicity of complexes of PDMAEMA (both linear and stars) with nucleic acids. On the other hand, there are fewer studies on the toxicity of PDMAEMA stars themselves. The survival of the cells in the presence of stars with DMAEMA homopolymer arms has been examined, inter alia, for Chinese hamster ovary CHO-K1 [46,47] and L929 [48] cells.

The Alamar blue reduction test was performed for HT-1080 cell line originating from a human fibrosarcoma, a malignant tumor from fibrous connective tissue. As a reference material to compare to polymers P1–P4 (Table 1), the "gold standard"—commercially available branched polyethyleneimine (PEI) was used. In all cases, it was observed that the toxicities of the stars at appropriate concentrations were much lower than that of PEI (Figure 7). The differences between the cytotoxicities of star polymers P1–P4 were mainly based on their different molar mass and chemical composition (Figure 7). For stars P1–P3 (Table 1), it may be concluded that higher concentrations of stars resulted in higher toxicities. P1 stars had the highest molar mass, as well as the highest ratio of DMAEMA:OEGMA content, and they were the most toxic (Figure 7). The introduction of the OEGMA-OH fraction resulted in a significant improvement in cell survival compared to their counterparts with PDMAEMA arms [22]. Apart from the molar mass of the star polymers, the introduction of 10% of OEGMA-OH into polymer P4 (Table 1) resulted in 100% survival of HT-1080 cells in the studied range of concentrations, and a similar effect was observed for stars with random DEGMA-*co*-OEGMA copolymer arms when the DEGMA content exceeded 40 mol% (Figure 7) [11]. The obtained results indicate that the amount of OEGMA-OH in the arms is the most important parameter for obtaining better cell viability, which should be taken into account when designing future bio-applications.

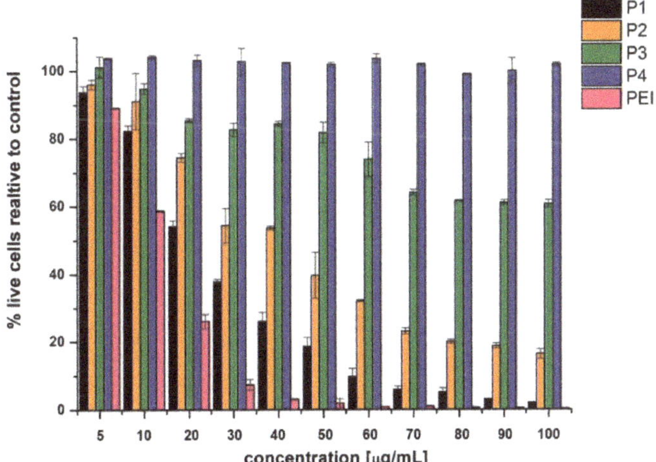

Figure 7. The cytotoxicity assay of the tested polymers. The assays were performed with HT-1080 cells. The results are presented as a % of cells surviving in the presence of the star polymers (P1–P4, Table 1) and branched polyethyleneimine (PEI) at increasing concentrations.

4. Conclusions

Star copolymers of DMAEMA and OEGMA-OH with molar masses up to 257,000 g/mol and OEGMA-OH contents up to 10 mol% were synthesized using ATRP. The introduction of the hydrophilic monomer OEGMA-OH strongly influenced the pH and thermoresponsiveness as well as the aggregation behavior of the star polymers. In organic and aqueous solutions, the stars were mainly present as isolated macromolecules, and aggregates were minor components. In general, increasing the OEGMA-OH content increased the aggregation tendency of the stars; however, as the largest hydrodynamic diameter of the stars was in the range of 100 nanometers, they still may be used in biomedical applications. The addition of 10 mol% OEGMA-OH turns off the thermoresponsive properties of the stars and limits their responsivity to changes in pH. On the other hand, the content of such a hydrophilic monomer lowers the cytotoxicity of such stars when compared to those that do not possess OEGMA-OH or have a methacrylic derivative with a methyl group instead of a hydroxyl group in their pendant ethylene glycol group. The stars obtained in this way are the least toxic at the concentrations needed for biological experiments, which is important to their applicability in biology and medicine.

Author Contributions: B.M. was responsible for synthesis of the stars, B.M. and A.K. were responsible for characterization of the obtained stars, A.F., K.K. and A.L.S. were responsible for preparation, processing and interpretation of the biological tests, M.S. delivered the hyperbranched core samples. A.D. helped with the analysis of the publication results. A.K., B.M. and A.L.S. designed the research, prepared the draft, and coordinated the work.

Funding: This work was supported by the Polish National Science Center contract no. UMO-2015/17/B/ST5/01095.

Acknowledgments: The authors thank Malwina Botor (Medical University of Silesia) for her assistance in the cytotoxicity tests.

Conflicts of Interest: The authors declare no conflict of interest.

References

1. Ren, J.M.; McKenzie, T.G.; Fu, Q.; Wong, E.H.H.; Xu, J.; An, Z.; Shanmugam, S.; Davis, T.P.; Boyer, C.; Qiao, G.G. Star polymers. *Chem. Rev.* **2016**, *116*, 6743–6836. [CrossRef] [PubMed]

2. Wu, W.; Wang, W.; Li, J. Star polymers: Advances in biomedical applications. *Prog. Polym. Sci.* **2015**, *46*, 55–85. [CrossRef]

3. Kowalczuk, A.; Trzcinska, R.; Trzebicka, B.; Müller, A.H.E.; Dworak, A.; Tsvetanov, C.B. Loading of polymer nanocarriers: Factors, mechanisms and applications. *Prog. Polym. Sci.* **2014**, *39*, 43–86. [CrossRef]

4. Markovsky, E.; Baabur-Cohen, H.; Eldar-Boock, A.; Omer, L.; Tiram, G.; Ferber, S.; Ofek, P.; Polyak, D.; Scomparin, A.; Satchi-Fainaro, R. Administration, distribution, metabolism and elimination of polymer therapeutics. *J. Controll. Release* **2012**, *161*, 446–460. [CrossRef] [PubMed]

5. Bao, Y.; De Keersmaecker, H.; Corneillie, S.; Yu, F.; Mizuno, H.; Zhang, G.; Hofkens, J.; Mendrek, B.; Kowalczuk, A.; Smet, M. Tunable ratiometric fluorescence sensing of intracellular ph by aggregation-induced emission-active hyperbranched polymer nanoparticles. *Chem. Mater.* **2015**, *27*, 3450–3455. [CrossRef]

6. Yang, K.; Liang, H.; Lu, J. Multifunctional star polymer with reactive and thermosensitive arms and fluorescently labeled core: Synthesis and its protein conjugate. *J. Mater. Chem.* **2011**, *21*, 10390–10398. [CrossRef]

7. Rezaei, S.J.T.; Nabid, M.R.; Niknejad, H.; Entezami, A.A. Folate-decorated thermoresponsive micelles based on star-shaped amphiphilic block copolymers for efficient intracellular release of anticancer drugs. *Int. J. Pharm.* **2012**, *437*, 70–79. [CrossRef] [PubMed]

8. Shi, C.; Guo, X.; Qu, Q.; Tang, Z.; Wang, Y.; Zhou, S. Actively targeted delivery of anticancer drug to tumor cells by redox-responsive star-shaped micelles. *Biomaterials* **2014**, *35*, 8711–8722. [CrossRef] [PubMed]

9. Wang, J.; Zaidi, S.S.A.; Hasnain, A.; Guo, J.; Ren, X.; Xia, S.; Zhang, W.; Feng, Y. Multitargeting peptide-functionalized star-shaped copolymers with comblike structure and a poss-core to effectively transfect endothelial cells. *ACS Biomater. Sci. Eng.* **2018**, *4*, 2155–2168. [CrossRef]

10. Cho, H.Y.; Srinivasan, A.; Hong, J.; Hsu, E.; Liu, S.; Shrivats, A.; Kwak, D.; Bohaty, A.K.; Paik, H.-J.; Hollinger, J.O.; et al. Synthesis of biocompatible peg-based star polymers with cationic and degradable core for sirna delivery. *Biomacromolecules* **2011**, *12*, 3478–3486. [CrossRef] [PubMed]

11. Mendrek, B.; Sieroń, Ł.; Żymełka-Miara, I.; Binkiewicz, P.; Libera, M.; Smet, M.; Trzebicka, B.; Sieroń, A.L.; Kowalczuk, A.; Dworak, A. Nonviral plasmid DNA carriers based on *N,N′*-dimethylaminoethyl methacrylate and di(ethylene glycol) methyl ether methacrylate star copolymers. *Biomacromolecules* **2015**, *16*, 3275–3285. [CrossRef] [PubMed]

12. Xu, F.J.; Zhang, Z.X.; Ping, Y.; Li, J.; Kang, E.T.; Neoh, K.G. Star-shaped cationic polymers by atom transfer radical polymerization from β-cyclodextrin cores for nonviral gene delivery. *Biomacromolecules* **2009**, *10*, 285–293. [CrossRef] [PubMed]

13. Chen, Q.; Han, F.; Lin, C.; Wen, X.; Zhao, P. Synthesis of bioreducible core crosslinked star polymers with *N,N′*-bis(acryloyl)cystamine crosslinker via aqueous ethanol dispersion raft polymerization. *Polymer* **2018**, *146*, 378–385. [CrossRef]

14. Wei, X.; Moad, G.; Muir, B.W.; Rizzardo, E.; Rosselgong, J.; Yang, W.; Thang, S.H. An arm-first approach to cleavable mikto-arm star polymers by raft polymerization. *Macromol. Rapid Commun.* **2014**, *35*, 840–845. [CrossRef] [PubMed]

15. Georgiou, T.K.; Vamvakaki, M.; Phylactou, L.A.; Patrickios, C.S. Synthesis, characterization, and evaluation as transfection reagents of double-hydrophilic star copolymers: Effect of star architecture. *Biomacromolecules* **2005**, *6*, 2990–2997. [CrossRef] [PubMed]

16. Li, L.; Lu, B.; Fan, Q.; Wu, J.; Wei, L.; Hou, J.; Guo, X.; Liu, Z. Synthesis and self-assembly behavior of ph-responsive star-shaped poss-(pcl-p(dmaema-co-pegma))16 inorganic/organic hybrid block copolymer for the controlled intracellular delivery of doxorubicin. *RSC Adv.* **2016**, *6*, 61630–61640. [CrossRef]

17. Lu, B.-B.; Wei, L.-L.; Meng, G.-H.; Hou, J.; Liu, Z.-Y.; Guo, X.-H. Synthesis of self-assemble ph-responsive cyclodextrin block copolymer for sustained anticancer drug delivery. *Chin. J. Polym. Sci.* **2017**, *35*, 924–938. [CrossRef]

18. Yang, Y.Q.; Zhao, B.; Li, Z.D.; Lin, W.J.; Zhang, C.Y.; Guo, X.D.; Wang, J.F.; Zhang, L.J. Ph-sensitive micelles self-assembled from multi-arm star triblock co-polymers poly(ε-caprolactone)-b-poly(2-(diethylamino)ethyl methacrylate)-b-poly(poly(ethylene glycol) methyl ether methacrylate) for controlled anticancer drug delivery. *Acta Biomater.* **2013**, *9*, 7679–7690. [CrossRef] [PubMed]

19. Forbes, D.C.; Frizzell, M.C.H.; Peppas, N.A. Polycationic nanoparticles synthesized using ARGET ATRP for drug delivery. *Eur. J. Pharm. Biopharm.* **2013**, *84*, 472–478. [CrossRef] [PubMed]

20. Kowalczuk, A.; Mendrek, B.; Żymełka-Miara, I.; Libera, M.; Marcinkowski, A.; Trzebicka, B.; Smet, M.; Dworak, A. Solution behavior of star polymers with oligo(ethylene glycol) methyl ether methacrylate arms. *Polymer* **2012**, *53*, 5619–5631. [CrossRef]

21. Mendrek, B. Zachowanie gwieździstych kopolimerów metakrylanów w roztworach. *Polimery* **2016**, *61*, 413–420. [CrossRef]

22. Mendrek, B.; Sieroń, Ł.; Libera, M.; Smet, M.; Trzebicka, B.; Sieroń, A.L.; Dworak, A.; Kowalczuk, A. Polycationic star polymers with hyperbranched cores for gene delivery. *Polymer* **2014**, *55*, 4551–4562. [CrossRef]

23. Ali, M.M.; Stöver, H.D.H. Well-defined amphiphilic thermosensitive copolymers based on poly(ethylene glycol monomethacrylate) and methyl methacrylate prepared by atom transfer radical polymerization. *Macromolecules* **2004**, *37*, 5219–5227. [CrossRef]

24. Kowalczuk, A.; Vandendriessche, A.; Trzebicka, B.; Mendrek, B.; Szeluga, U.; Cholewiński, G.; Smet, M.; Dworak, A.; Dehaen, W. Core-shell nanoparticles with hyperbranched poly(arylene-oxindole) interiors. *J. Polym. Sci. Part A Polym. Chem.* **2009**, *47*, 1120–1135. [CrossRef]

25. Hölter, D.; Burgath, A.; Frey, H. Degree of branching in hyperbranched polymers. *Acta Polym.* **1997**, *48*, 30–35. [CrossRef]

26. Huang, X.; Xiao, Y.; Lang, M. Synthesis of star-shaped pcl-based copolymers via one-pot atrp and their self-assembly behavior in aqueous solution. *Macromol. Res.* **2012**, *20*, 597–604. [CrossRef]

27. Angot, S.; Murthy, K.S.; Taton, D.; Gnanou, Y. Scope of the copper halide/bipyridyl system associated with calixarene-based multihalides for the synthesis of well-defined polystyrene and poly(meth)acrylate stars. *Macromolecules* **2000**, *33*, 7261–7274. [CrossRef]

28. Matyjaszewski, K.; Miller, P.J.; Pyun, J.; Kickelbick, G.; Diamanti, S. Synthesis and characterization of star polymers with varying arm number, length, and composition from organic and hybrid inorganic/organic multifunctional initiators. *Macromolecules* **1999**, *32*, 6526–6535. [CrossRef]

29. Newkome, G.R.; Moorefield, C.N.; Baker, G.R.; Saunders, M.J.; Grossman, S.H. Unimolekulare micellen. *Angew. Chem.* **1991**, *103*, 1207–1209. [CrossRef]

30. Paul, C.; Hiemenz, T.P.L. *Polymer Chemistry*; CRC Press Taylor and Francis Group: Boca Raton, FL, USA, 2007; ISBN 9781574447798.

31. Plamper, F.A.; Ruppel, M.; Schmalz, A.; Borisov, O.; Ballauff, M.; Müller, A.H.E. Tuning the thermoresponsive properties of weak polyelectrolytes: Aqueous solutions of star-shaped and linear poly(n,n-dimethylaminoethyl methacrylate). *Macromolecules* **2007**, *40*, 8361–8366. [CrossRef]

32. Ward, M.A.; Georgiou, T.K. Thermoresponsive polymers for biomedical applications. *Polymers* **2011**, *3*, 1215–1242. [CrossRef]

33. Dong, Z.; Wei, H.; Mao, J.; Wang, D.; Yang, M.; Bo, S.; Ji, X. Synthesis and responsive behavior of poly(*N*,*N*-dimethylaminoethyl methacrylate) brushes grafted on silica nanoparticles and their quaternized derivatives. *Polymer* **2012**, *53*, 2074–2084. [CrossRef]

34. Zhou, J.; Wang, L.; Ma, J.; Wang, J.; Yu, H.; Xiao, A. Temperature- and ph-responsive star amphiphilic block copolymer prepared by a combining strategy of ring-opening polymerization and reversible addition–fragmentation transfer polymerization. *Eur. Polym. J.* **2010**, *46*, 1288–1298. [CrossRef]

35. Cotanda, P.; Wright, D.B.; Tyler, M.; O'Reilly, R.K. A comparative study of the stimuli-responsive properties of dmaea and dmaema containing polymers. *J. Polym. Sci. Part A Polym. Chem.* **2013**, *51*, 3333–3338. [CrossRef]

36. Badi, N. Non-linear peg-based thermoresponsive polymer systems. *Prog. Polym. Sci.* **2017**, *66*, 54–79. [CrossRef]

37. Park, T.G.; Hoffman, A.S. Sodium chloride-induced phase transition in nonionic poly(n-isopropylacrylamide) gel. *Macromolecules* **1993**, *26*, 5045–5048. [CrossRef]

38. Trzebicka, B.; Szweda, D.; Rangelov, S.; Kowalczuk, A.; Mendrek, B.; Utrata-Wesołek, A.; Dworak, A. (co)polymers of oligo(ethylene glycol) methacrylates—Temperature-induced aggregation in aqueous solution. *J. Polym. Sci. Part A Polym. Chem.* **2013**, *51*, 614–623. [CrossRef]

39. Bütün, V.; Armes, S.P.; Billingham, N.C. Synthesis and aqueous solution properties of near-monodisperse tertiary amine methacrylate homopolymers and diblock copolymers. *Polymer* **2001**, *42*, 5993–6008. [CrossRef]

40. Hu, L.; Chu, L.-Y.; Yang, M.; Wang, H.-D.; Hui Niu, C. Preparation and characterization of novel cationic ph-responsive poly(*N*,*N'*-dimethylamino ethyl methacrylate) microgels. *J. Colloid Interface Sci.* **2007**, *311*, 110–117. [CrossRef] [PubMed]

41. Uchida, E.; Uyama, Y.; Ikada, Y. Zeta potential of polycation layers grafted onto a film surface. *Langmuir* **1994**, *10*, 1193–1198. [CrossRef]

42. Rinkenauer, A.C.; Schubert, S.; Traeger, A.; Schubert, U.S. The influence of polymer architecture on in vitro pdna transfection. *J. Mater. Chem. B* **2015**, *3*, 7477–7493. [CrossRef]

43. Lam, S.J.; Wong, E.H.H.; Boyer, C.; Qiao, G.G. Antimicrobial polymeric nanoparticles. *Prog. Polym. Sci.* **2018**, *76*, 40–64. [CrossRef]

44. Cai, J.; Yue, Y.; Rui, D.; Zhang, Y.; Liu, S.; Wu, C. Effect of chain length on cytotoxicity and endocytosis of cationic polymers. *Macromolecules* **2011**, *44*, 2050–2057. [CrossRef]

45. Rawlinson, L.-A.B.; O'Brien, P.J.; Brayden, D.J. High content analysis of cytotoxic effects of pdmaema on human intestinal epithelial and monocyte cultures. *J. Controll. Release* **2010**, *146*, 84–92. [CrossRef] [PubMed]

46. Schallon, A.; Jérôme, V.; Walther, A.; Synatschke, C.V.; Müller, A.H.E.; Freitag, R. Performance of three pdmaema-based polycation architectures as gene delivery agents in comparison to linear and branched pei. *React. Funct. Polym.* **2010**, *70*, 1–10. [CrossRef]

47. Synatschke, C.V.; Schallon, A.; Jérôme, V.; Freitag, R.; Müller, A.H.E. Influence of polymer architecture and molecular weight of poly(2-(dimethylamino)ethyl methacrylate) polycations on transfection efficiency and cell viability in gene delivery. *Biomacromolecules* **2011**, *12*, 4247–4255. [CrossRef] [PubMed]

48. Majewski, A.P.; Stahlschmidt, U.; Jérôme, V.; Freitag, R.; Müller, A.H.E.; Schmalz, H. Pdmaema-grafted core–shell–corona particles for nonviral gene delivery and magnetic cell separation. *Biomacromolecules* **2013**, *14*, 3081–3090. [CrossRef] [PubMed]

Article

Cell Uptake and Biocompatibility of Nanoparticles Prepared from Poly(benzyl malate) (Co)polymers Obtained through Chemical and Enzymatic Polymerization in Human HepaRG Cells and Primary Macrophages

Hubert Casajus [1,†], Saad Saba [2,†], Manuel Vlach [2], Elise Vène [2], Catherine Ribault [2], Sylvain Tranchimand [1], Caroline Nugier-Chauvin [1], Eric Dubreucq [3], Pascal Loyer [2,*], Sandrine Cammas-Marion [1,2,*] and Nicolas Lepareur [2,4,*]

[1] Ecole Nationale Supérieure de Chimie de Rennes, Univ Rennes, CNRS, ISCR, UMR 6226, F-35000 Rennes, France; casajus.hubert@gmail.com (H.C.); sylvain.tranchimand@ensc-rennes.fr (S.T.); caroline.nugier@ensc-rennes.fr (C.N.-C.)

[2] Univ Rennes, INSERM, INRA, Institut NUMECAN (Nutrition Metabolisms and Cancer) UMR_A 1341, UMR_S 1241, F-35000 Rennes, France; saad.saba@etudiant.univ-rennes1.fr (S.S.); manuel.vlach@univ-rennes1.fr (M.V.); elise.vene@univ-rennes1.fr (E.V.); catherine.ribault@univ-rennes1.fr (C.R.)

[3] Montpellier SupAgro, INRA, CIRAD, Univ Montpellier, UMR 1208 IATE, F-34060 Montpellier, France; eric.dubreucq@supagro.fr

[4] Comprehensive Cancer Center Eugène Marquis, F-35000 Rennes, France

[*] Correspondence: pascal.loyer@univ-rennes1.fr (P.L.); sandrine.marion.1@ensc-rennes.fr (S.C.-M.); n.lepareur@rennes.unicancer.fr (N.L.); Tel.: +33-223-233-873 (P.L.); +33-223-238-109 (S.C.-M.), +33-299-253-144 (N.L.)

[†] These authors contributed equally to the work.

Received: 16 October 2018; Accepted: 5 November 2018; Published: 10 November 2018

Abstract: The design of drug-loaded nanoparticles (NPs) appears to be a suitable strategy for the prolonged plasma concentration of therapeutic payloads, higher bioavailability, and the reduction of side effects compared with classical chemotherapies. In most cases, NPs are prepared from (co)polymers obtained through chemical polymerization. However, procedures have been developed to synthesize some polymers via enzymatic polymerization in the absence of chemical initiators. The aim of this work was to compare the acute in vitro cytotoxicities and cell uptake of NPs prepared from poly(benzyl malate) (PMLABe) synthesized by chemical and enzymatic polymerization. Herein, we report the synthesis and characterization of eight PMLABe-based polymers. Corresponding NPs were produced, their cytotoxicity was studied in hepatoma HepaRG cells, and their uptake by primary macrophages and HepaRG cells was measured. In vitro cell viability evidenced a mild toxicity of the NPs only at high concentrations/densities of NPs in culture media. These data did not evidence a higher biocompatibility of the NPs prepared from enzymatic polymerization, and further demonstrated that chemical polymerization and the nanoprecipitation procedure led to biocompatible PMLABe-based NPs. In contrast, NPs produced from enzymatically synthesized polymers were more efficiently internalized than NPs produced from chemically synthesized polymers. The efficient uptake, combined with low cytotoxicity, indicate that PMLABe-based NPs are suitable nanovectors for drug delivery, deserving further evaluation in vivo to target either hepatocytes or resident liver macrophages.

Keywords: enzymatic polymerization; chemical polymerization; poly(benzyl malate); biocompatible nanoparticles; cell uptake; cytotoxicity; HepaRG cells; human macrophages

---

1. Introduction

Biocompatible polymeric nanoparticles combined with the design of more specific therapeutic molecules are developed toward the goal of reducing the drug amounts administrated to patients, simplifying protocols of administration, and overcoming side effects. This research is based on the "Magic Bullet" concept first coined by Paul Ehrlich [1] at the beginning of the 20th century, which was a concept that inspired generations of scientists in the design of more efficient therapeutic molecules and vectorized therapies. Since the end of the 20th century, progress in this area has been significant, and has led to the design of nanovectors approved by the United States (US) Food and Drug Administration (FDA), or in pre-clinical or clinical phases [2–7].

The design of nanovectors for clinical applications implies very strict specifications. They must: (i) be biocompatible and/or (bio)degradable; (ii) be easy to produce in Good Manufacturing Practice (GMP) grade for clinical use; (iii) allow the encapsulation of a significant amount of active molecules such as drugs, genes, or peptides [8]; (iv) lead to the controlled release of active molecules at the targeted sites (cells, tissues, and/or organs) through the presence of targeting agents (antibodies, peptides, vitamins) [9–11]; and (v) have controlled surface characteristics to minimize their recognition by the immune system. These nanovectors can take the form of active molecule–polymers conjugates [12], nanoparticles [13], micelles [14], or polymersomes [15].

Aliphatic polyesters are produced for multiple industrial use such as the plastic bag industry instead of poly(ethylene), which is a non-biodegradable polymer derived directly from petrochemicals [16]. Their remarkable properties of biodegradability and biocompatibility make them also suitable for biomedical and pharmacological applications, including the preparation of nanoparticles (NPs) for drug delivery [11,17].

Such aliphatic polyesters are usually synthesized by the polymerization of the corresponding monomer(s) using chemically-based initiators [18]. On the other hand, some of these polyesters can be extracted from fungi, such as for instance the natural poly(malic acid) extracted from *Physarum polycephalum* [19], or may be prepared by biotechnological means using a large-scale culture of bacteria as for polyhydroxyalkanoates or (PHAs) [20]. The advantage of these biotechnological pathways relies mainly in the use of so-called "green" synthetic processes. However, it is difficult, if not impossible, to control the molar masses of the polymers, and only chiral polymers are obtained when the repeating unit of the block (co)polymers has one or more asymmetric carbon(s). Therefore, the so-called conventional "chemical" methods of synthesis cannot, at present, be totally ruled out. However, authors have pointed out the question of whether the initiators and/or catalysts that are used during these polymerization reactions might induce cytotoxicities when the corresponding polymers are in contact with cells in living organisms, even if initiators and catalysts are present only as traces in the final material [21]. Among the aliphatic polyesters' family, poly(malic acid) (PMLA) and its derivatives have raised a growing interest in the biomedical field because of their properties of biocompatibility and biodegradability [22,23] and potential applications in drug delivery. In addition to these remarkable properties, a whole family of PMLA derivatives can be obtained by modifying the synthesis of the monomers and the reactions of the homopolymerization and copolymerization of these monomers [24,25]. We and others have recently developed PMLA-based nanovectors, taking advantage of the exceptional properties of such polymers [22,24,26–28]. The first nanovector is a polymer-drug conjugate constituted of a PMLA whose lateral carboxylic acids were modified to graft different molecules of interest: doxorubicin as a drug, poly(ethylene glycol) to improve the hydrophilic properties of the conjugate, and N-acetyl galactosamine as a targeting agent [29]. Furthermore, we also prepared NPs from either hydrophobic poly(benzyl malate) (PMLABe) or amphiphilic poly(ethylene glycol)-*block*-PMLABe (PEG-*b*-PMLABe) having or not having the biotin as a targeting agent at the free end of the PEG block [22,23,30,31].

This family of polyesters is obtained by the anionic ring-opening polymerization (AROP) of alkyl malolactonates (monomers) in the presence of tetraethylammonium carboxylate salts, and more particularly tetraethylammonium benzoate [32]. It is important to note that this last initiator is a powerful neurotoxin [33]. Therefore, the use of biocatalysts, such as lipases, is an interesting alternative to avoid the presence of traces of toxic chemical initiators in polymeric materials. Lipases are enzymes belonging to the family of hydrolases, which is class 3 in the EC nomenclature (Enzyme Commission numbers). More specifically, they are serine hydrolases that have the particularity of being active in the presence of an interface, called activation, between an aqueous phase (water or buffer) and an organic phase (hydrophobic substrate and/or organic solvent). Present in plants as well as in insects, animals, and microorganisms, lipases play a key role in the degradation and synthesis of lipids [34]. The physiological reaction catalyzed by lipases is the hydrolysis of mono, di, and triglycerides. However, they are also capable of synthesizing esters, either by esterification or transesterification, when the activity of the water is low. This property is one of the main reasons for the use of this class of enzymes, along with their high selectivity during hydrolysis [34]. More than 100 three-dimensional structures are deposited on the Protein Data Bank (PDB), which has made it possible to highlight the highly conserved structural elements specific to lipases [34]. The canonical mechanism of hydrolysis of ester functions involves the three amino acids of the catalytic triad, namely, aspartate, histidine, and serine [35].

These enzymes have been studied as potential catalysts in the ring-opening polymerization of unsubstituted lactones such as ε-caprolactones [36]. The mechanism of this enzymatic catalyzed ring opening polymerization of unsubstituted lactones has been extensively studied and described in the literature: it occurs through the canonical mechanism described for the hydrolysis of esters' functions [37–39]. However, despite the interest of such enzymatic polymerization, several technological barriers still need to be overcome to completely replace the chemical initiators/catalysts. Indeed, enzymatic polymerizations are, to date, less controlled than chemical ones [40–42]. Nevertheless, we recently set up conditions to synthesize well-defined poly(benzyl malate) (PMLABe) derivatives of high molar masses through the enzymatic polymerization of benzyl malolactonate (MLABe) in the presence of pancreatic porcine lipase (PPL) [43].

Taking advantage of the availability of (co)polymers synthesized by polymerization in the presence of chemical initiators and enzymatic polymerization, the aim of this work was to compare the acute in vitro cytotoxicity of NPs prepared from PMLABe derivatives synthesized either by chemical (PMLABe chemicals) or enzymatic (PMLABe enzymatic) polymerization and their cell uptake. For that purpose, we have synthesized and characterized a series of eight PMLABe-based polymers, formulated the corresponding NPs, and studied their cytotoxicity in progenitor and differentiated hepatoma HepaRG cells, and measured their uptake by both primary macrophages and HepaRG cells.

2. Materials and Methods

2.1. Materials and Apparatus

All of the chemicals were used as received. α-methoxy,ω-carboxylic acid PEG$_{45}$ ($\overline{M}w$ = 2000 g/mol, n = 45) and α-hydroxyl,ω-methoxy-PEG$_{17}$ ($\overline{M}w$ = 736 g/mol, n = 17) were purchased from PEG Iris Biotech. Porcine pancreatic lipase (PPL) was purchased from Sigma-Aldrich Co. (Saint Louis, MO, USA).

Nuclear magnetic resonance spectroscopy: The standard temperature was adjusted to 298K. NMR spectra were recorded on a Bruker Avance III 400 spectrometer (Billerica, MA, USA) operating at 400.13 MHz for ^{1}H, equipped with a BBFO probe with a Z-gradient coil and a GREAT 1/10 gradient unit. The zg30 Bruker pulse program was used for 1D ^{1}H NMR, with a TD of 64 k, a relaxation delay d1 = 2 s, and eight scans. The spectrum width was set to 18 ppm. Fourier transform of the acquired FID was performed without any apodization in most cases.

Infrared spectroscopy: Fourier transform infrared (FT-IR) spectra were recorded on an Avatar 320FT-IR Thermo Nicolet spectrometer (ThermoFischer, Waltham, MA, USA) between 500–4000 cm^{-1} by direct measurement.

Size exclusion chromatography: Mass average molar mass ($\overline{Mw}$) and dispersity ($\text{Đ} = \overline{Mw}/\overline{Mn}$) values were measured by size exclusion chromatography (SEC) in THF at 40 °C (flow rate = 1.0 mL/min) on a GPC2502 Viscotek apparatus (Malvern Instruments, Malvern, UK) equipped with a refractive index detector Viscotek VE 3580 RI, a guard column Viscotek TGuard, Org 10 mm × 4.6 mm, a LT5000L gel column (for samples soluble in organic medium) 300 mm × 7.8 mm, and a GPC/SEC OmniSEC Software. The polymer samples were dissolved in THF (2 mg/mL). All of the elution curves were calibrated with polystyrene standards.

Polarimetry: Optical rotations were measured using a Perkin-Elmer 341 polarimeter (Waltham, MA, USA) with a sodium lamp (wavelength = 589 nm). Samples were dissolved in chloroform (concentration around 10 mg/mL) and transferred into the polarimeter microcell (length = 1 dm). Measures were performed at room temperature.

Dynamic light scattering: The size and polydispersity index (PDI) of the formulations were measured by dynamic light scattering using a Zetasizer Nano-ZR90 (Malvern Instruments, Malvern, UK) apparatus at 25 °C, an He-Ne laser at 633 nm, and a detection angle of 90°. Data were processed by the Zetasizer Software v7.11 (Malvern Instruments, Malvern, UK).

Flow cytometry: Cell monolayers were detached with trypsin and analyzed by flow cytometry (Becton Dickinson le LSRFortessa™ X-20, Franklin Lakes, NJ, USA), using the cytometry core facility of the Biology and Health Federative research structure Biosit, Rennes, France to quantify the fluorescence emitted by the DiIC18 (7) (1,1″-dioctadecyl-3,3,3″,3″-tetramethylindotricarbocyanine iodide) (DiR)-loaded NPs within cells. Cytometry data were analyzed using DIVA software (Becton Dikinson, Franklin Lakes, NJ, USA).

2.2. Synthesis of Monomers and Polymers

Synthesis of racemic and optically active MLABe: RS, R, and S-MLABe were synthesized as previously described [32]. Briefly, DL, L, or D-aspartic acid (1 eq) was reacted with sodium bromide (5.5 eq) and sodium nitrite (1.2 eq) in 2N H_2SO_4 leading, after recrystallization in acetonitrile, to the corresponding pure RS, S, or R-bromosuccinic acid. The racemic or optically active bromosuccinic acid was dissolved in anhydrous THF and reacted with trifluoroacetic anhydride (1.3 eq), leading to the corresponding racemic or optically active bromosuccinic anhydride, which was immediately reacted with benzyl alcohol (1 eq) to obtain the corresponding mixture of racemic or optically monoesters. The pH of this mixture of racemic or optically active monoesters was increased to 7.2 by addition of 2N NaOH; the resulting aqueous solution was mixed with dichloromethane, and the biphasic medium was heated at 45 °C. After purification (column chromatography and distillation under vacuum), the pure RS, R, or S-MLABe was obtained and characterized by FT-IR and ^{1}H NMR (Figure SI).

RS-MLABe (Figures SI 1.1 and SI 2.1):
FT-IR (v, cm^{-1}): 1840 (C=O, lactone) 1750 (C=O, ester)
^{1}H NMR (400 MHz, Acetone-d$_6$), δ (ppm): 7.47–7.28 (m, 5H), 5.27 (s, 2H), 5.11 (dd, J = 6.7, 4.4 Hz, 1H), 4.00–3.60 (dd, J = 16.5, 6.7 Hz, 1H), 3.72 (dd, J = 16.5, 4.4 Hz, 1H).

S-MLABe (Figures SI 1.2 and 2.2):
FT-IR (v, cm^{-1}): 1840 (C=O, lactone) 1750 (C=O, ester)
^{1}H NMR (400 MHz, Acetone-d$_6$), δ (ppm): 7.47–7.28 (m, 5H), 5.27 (s, 2H), 5.11 (dd, J = 6.7, 4.4 Hz, 1H), 4.00–3.60 (dd, J = 16.5, 6.7 Hz, 1H), 3.72 (dd, J = 16.5, 4.4 Hz, 1H).

R-MLABe (Figures SI 1.3 and 2.3):
FT-IR (v, cm^{-1}): 1840 (C=O, lactone) 1750 (C=O, ester)
^{1}H NMR (400 MHz, Acetone-d$_6$), δ (ppm): 7.47–7.28 (m, 5H), 5.27 (s, 2H), 5.11 (dd, J = 6.7, 4.4 Hz, 1H), 4.00–3.60 (dd, J = 16.5, 6.7 Hz, 1H), 3.72 (dd, J = 16.5, 4.4 Hz, 1H).

Chemical polymerization of RS and R-MLABe: The RS-PMLABe (P1) and S-PMLABe (P2) were synthesized as previously described [32] by anionic ring-opening polymerization (AROP) of the

RS-MLABe and R-MLABe, respectively, in presence of tetraethylammonium benzoate as initiator. The theoretical molar mass (30,000 g/mol) was fixed by the ratio monomer/initiator. Both polymers were purified by precipitation into ethanol. After drying under vacuum, the RS-PMLABe and S-PMLABe were analyzed by ^{1}H NMR, SEC, and polarimetry.

RS-PMLABe (P1): ^{1}H NMR (400 MHz, Acetone-d6) δ(ppm): 7.40–7.25 (m, 5nH), 5.55 (m, 1nH), 5.16 (m, 2nH), 3.24–2.67 (m, 2nH)—Figure SI 3.1. SEC (THF, 40 °C, polystyrene standards): $\overline{Mw}$ = 14,850 g/mol, Đ = 1.60. [α]$_D$ = 0;

S-PMLABe (P2): ^{1}H NMR (400 MHz, Acetone-d6) δ(ppm): 7.40–7.25 (m, 5nH), 5.55 (m, 1nH), 5.16 (m, 2nH), 3.24–2.67 (m, 2nH)—Figure SI 3.2. SEC (THF, 40 °C, polystyrene standards): impossible to measure in THF. [α]$_D$ = −10.7.

The PEG$_{45}$-*b*-PMLABe$_{73}$ (P3) was synthesized as previously described [23] through the AROP of RS-MLABe in presence of α-methoxy,ω-carboxylate PEG$_{45}$ tetraethylammonium salt as initiator, in anhydrous THF. In this case also, the PMLABe molar mass was set by the ratio monomer/initiator, and selected to be 15,000 g/mol. The resulting block copolymer was purified by precipitation in ethanol. After drying under vacuum, the pure block copolymer was analyzed by ^{1}H NMR and SEC.

PEG$_{45}$-b-PMLABe$_{73}$ (P3): ^{1}H NMR (400 MHz; Acetone-d6), δ (ppm): 7.40–7.20(m, 5nH), 5.60–5.49 (m, 1nH), 5.40 (m, 2nH), 3.60 (m, 4mH (m = 45)), 3.10–2.80 (m, 2nH)—Figure SI 3.3. M_{NMR} = 15,040 g/mol for the PMLABe block (n = 73). SEC (THF, 40 °C, polystyrene standards): $\overline{Mw}$ = 9600 g/mol, Đ = 1.30.

Enzymatic polymerization of RS, R, and S-MLABe: The RS-PMLABe P4 was synthesized as described previously [43] by ring-opening polymerization of the RS-MLABe in the presence of porcine pancreatic lipase (PPL) without toluene. Experiments were realized in a 24-multi reactor Büchi Syncore Line (Flawil, Switzerland). In a 30-mL tube, PPL (8.4 mg) was mixed with pure RS-MLABe (336 mg) and Tris/HCl buffer (125 µL, pH = 7, 105 mM). The mixture was stirred at 390 rpm, and the temperature was controlled at 60 °C. After 72 h, reaction was stopped by the addition of 3 mL of THF and 3 mL of chloroform. The mixture was then transferred into a separatory funnel. Separations were realized by adding 15 mL of distilled water and 15 mL of chloroform. Organic phases were dried over MgSO$_4$ and filtered. Finally, solvents were evaporated under reduced pressure. The RS-PMLABe P4 was dissolved in chloroform and then precipitated in a large excess of cold diethyl ether. After elimination of the supernatant, the polymer was dried under vacuum and analyzed by ^{1}H NMR, SEC, and polarimetry.

RS-PMLABe (P4): ^{1}H NMR (400 MHz, CDCl$_3$) δ(ppm): 7.25 (m, 5nH), 5.49 (m, 1nH), 5.08 (m, 2nH), 2.88 (m, 2nH)—Figure SI 3.4. SEC (THF, 40 °C, polystyrene standards): $\overline{Mw}$ = 12,250 g/mol, Đ = 1.4. [α]$_D$ = 0.

The RS-PMLABe P5, the S-PMLABe P6, and the R-PMLABe P7 were obtained as described previously [43] by ring-opening polymerization of the RS-MLABe, R-MLABe, and S-MLABe, respectively, in the presence of porcine pancreatic lipase (PPL) with toluene. Experiments were realized in a 24-multi reactor Büchi Syncore Line. In a 30-mL tube, PPL (8.4 mg) was mixed with RS-MLABe, R-MLABe, or S-MLABe (338 mg), Tris/HCl (333 µL, pH = 7, 105 mM), and toluene (667 µL). The final volume was 1 mL. The reaction mixtures were stirred at 390 rpm, and the temperature was maintained at 60 °C. After 72 h, reaction was stopped by the addition of 3 mL of THF and 3 mL of chloroform. The mixtures were then transferred into a separatory funnel. Separations were realized by adding 15 mL of distilled water and 15 mL of chloroform. Organic phases were dried over MgSO$_4$ and filtered. Finally, solvents were evaporated under reduced pressure. The RS-PMLABe P5, S-PMLABe P6, and R-PMLABe P7 were dissolved in chloroform, and then precipitated in a large excess of cold diethyl ether. After elimination of the supernatant, polymers were dried under vacuum and analyzed by ^{1}H NMR, SEC, and polarimetry.

RS-PMLABe (P5): ^{1}H NMR (400 MHz, CDCl$_3$) δ(ppm): 7.25 (m, 5nH), 5.49 (m, 1nH), 5.08 (m, 2nH), 2.88 (m, 2nH)—Figure SI 3.5. SEC (THF, 40 °C, Standards Polystyrene): $\overline{Mw}$ = 3850 g/mol, Đ = 1.5. [α]$_D$ = −3.

S-PMLABe (P6): ^{1}*H NMR (400 MHz, CDCl$_3$) δ(ppm): 7.25 (m, 5nH), 5.49 (m, 1nH), 5.08 (m, 2nH), 2.88 (m, 2nH)—Figure SI 3.6. SEC (THF, 40 °C, polystyrene standards):* $\overline{M}w$ *= 2000 g/mol, Đ = 1.50. [α]$_D$ = −11.*

R-PMLABe (P7): ^{1}*H NMR (400 MHz, CDCl$_3$) δ(ppm): 7.25 (m, 5nH), 5.49 (m, 1nH), 5.08 (m, 2nH), 2.88 (m, 2nH)—Figure SI 3.7. SEC (THF, 40 °C, polystyrene standards):* $\overline{M}w$ *= 2300 g/mol, Đ = 1.40. [α]$_D$ = +18.*

The PEG$_{17}$-b-PMLABe$_{47}$ (P8) was synthesized by enzymatic ring opening polymerization (ROP) of RS-MLABe in the presence of α-methoxy-ω-hydroxy-PEG$_{17}$ and PPL as the catalyst. Before the reaction, the PPL and the α-methoxy-ω-hydroxy-PEG$_{17}$ (M_w = 736 g/mol) were lyophilized for 12 h. In a 30-mL tube, PPL (12.5 mg) was mixed with 427 mg (2.07 mmol, 1eq.) of RS-MLABe, 77 mg (0.1 mmol, 0.05 eq) of α-methoxy-ω-hydroxy-PEG$_{17}$, and anhydrous toluene (630 μL). The reaction was performed at 60 °C at 390 rpm during 72 h. The reaction was stopped by the addition of 5 mL of chloroform. Then, the mixture was transferred in Eppendorf tubes and centrifuged for 3 min at 14,000 rpm. Supernatants were filtered, and the chloroform was evaporated under reduced pressure. Polymer P8 was dissolved in chloroform, and then precipitated in a large excess of cold diethyl ether. After elimination of the supernatant, polymers were dried under reduced pressure at room temperature. The PEG$_{17}$-b-PMLABe$_{47}$ (P8) was analyzed by ^{1}H NMR (Figure SI) and SEC.

PEG$_{17}$-b-PMLABe$_{47}$ (P8): ^{1}*H NMR (400 MHz, Acetone-d6) δ (ppm): 7.57–7.07 (m, 5nH), 5.55 (m, 1nH), 5.16 (m, 2nH), 3.60 (m, 4mH), 3.24–2.67 (m, 2nH).—Figure SI 3.8. SEC (THF, 40 °C, 1 mL/min, polystyrene standards):* $\overline{M}w$ *= 9650 g/mol Đ = 1.50.*

2.3. Formulation of PMLABe-Based Nanoparticles

Formulation of NPs for the cell uptake assays: NPs were loaded with DiR (DiIC18 (7) (1,1″-dioctadecyl-3,3,3″,3″-tetramethylindotricarbocyanine iodide)). DiR is a hydrophobic carbocyanine-type probe emitting in the near infrared (780 nm). Five mg of the selected polymer were dissolved in 100 μL of DMF, and then 50 μL of DiR (1 mg/mL in dimethylformamide (DMF), corresponding to 1 wt % with the dissolved polymer) was added. This solution was rapidly added into 1 mL of distilled water under vigorous stirring. The nanoparticles formed instantly because of the hydrophobic or amphiphilic nature of the (co)polymers. The suspension was left under stirring for 1 h. Then, DiR-loaded NPs were purified using Sephadex steric exclusion columns (PD-10 column) to eliminate free DiR. The concentration of loaded DiR was measured by UV at 670 nm in a solution prepared by mixing 80 μL of the DiR-loaded nanoparticles with 320 μL of DMF, and using a calibration curve of DiR solution in a mixture of DMF/H$_2$O (80/20). The absorbance of the solutions was measured using a Microplate Spectrophotometer powerwase XS/XS2 (Biotek, Winooski, VT, USA), equipped with Gen5 Data Analysis Software (Biotek, Winooski, VT, USA). The NPs were characterized by dynamic light scattering (DLS) in water at room temperature with a concentration in polymer of 2.5 mg/mL.

Formulation of NPs for the acute toxicity: NPs were prepared by the nanoprecipitation method as described described [44,45]. Five mg of the selected polymer were dissolved in five mL of 1,4-dioxane. This solution was rapidly added into two mL of distilled water under vigorous stirring. The suspension was left under stirring during 15 min, and the 1,4-dioxane was evaporated under reduced pressure at 40 °C. The final polymer concentration was 2.5 mg/mL. The NPs were characterized by dynamic light scattering (DLS) in water at room temperature with a concentration in polymer of 2.5 mg/mL.

2.4. In Vitro Assays

HepaRG culture: For all of the studies, proliferating progenitors HepaRG cells were seeded at a density of 3.10^4 cells/cm^2 and cultured as previously described [46], in William's E medium (Lonza, Basel, Switzerland) supplemented with 2 mM of glutamine (Gibco, ThermoFischer, Waltham, MA, USA), 5 mg/L of insulin (Sigma, Saint Louis, MO, USA), 10^{-2} mM of hydrocortisone hemisuccinate, and 10% of fetal calf serum (FCS) (Lonza, Basel, Switzerland). HepaRG cells were cultured in William's

E medium supplemented with 10% of FCS, 50 units/mL of penicillin, 50 µg/mL of streptomycin, 5 µg/mL of insulin, 2 mM of L-glutamine, and 50 µM of hydrocortisone hemisuccinate.

Primary macrophage culture: Peripheral blood mononuclear cells were purified from human buffy coat (Etablissement Français du Sang, Rennes, France) by differential centrifugation on UNI-SEP maxi U10 (Novamed, Jerusalem, Israel). The experiments were performed in compliance with the French legislation on blood donation and blood products' use and safety. Monocytes from healthy donors were enriched using a human CD14 separation kit (Microbeads; Miltenyi Biotec, Bergisch Gladbach, Germany), plated at a density of 0.5×10^6 cells per well in 24-well plates, and cultured at 37 °C with 5% humidified CO_2 in RPMI 1640 medium supplemented with 100 IU/mL of penicillin, 100 mg/mL of streptomycin, 2 mM of L-glutamine, and 10% FCS BioWhittaker®. Macrophages were obtained after differentiation from monocytes by incubation with 50 ng/mL of rhGM-CSF in RPMI 1640 medium for seven days, as previously described [47].

Cytotoxicity assays: The cytotoxicity was assessed using the human HepaRG hepatoma cell line and the Thiazolyl Blue Tetrazolium Bromide (MTT) assay. Briefly, cells were incubated with MTT (0.5 mg/mL) for 1 h at 37 °C. The formed crystals were dissolved in DMSO at room temperature for 10 min, and the absorbance was read at 490 nm with a microplate reader. The MTT values reflecting the number of viable cells were expressed in percentage relative to the absorbance determined in control cultures. The cells were exposed to the eight NPs at molar concentrations in polymers ranging from 0.5 µM to 20 µM, corresponding to different mass concentrations, since the polymers exhibit different molar masses. The polymer masses in grams per liter (L) were the following: NPs 1: 15×10^{-3} to 0.6 g/L; NPs 2: 15×10^{-3} to 0.6 g/L; NPs 3: 7.5×10^{-3} to 0.3 g/L; NPs 4: 6.125×10^{-3} to 0.245 g/L; NPs 5: 1.925×10^{-3} to 0.077 g/L; NPs 6: 10^{-3} to 0.04 g/L; NPs 7: 1.15×10^{-3} to 0.046 g/L; NPs 8: 4.8×10^{-3} to 0.193 g/L. The same range of molar concentrations in polymers was chosen, since the NPs had relatively similar diameters, indicating that the same amounts of masses of polymers would produce almost nearly identical numbers of NPs. The 5 mg of polymers that was used to prepare NPs (2 mL suspension of water) produced the following numbers of NPs: NPs 1 = 9.9×10^{11}, NPs 2 = 1.5×10^{12}, NPs 3 = 1.3×10^{13}, NPs 4 = 9.2×10^{11}, NPs 5 = 3.8×10^{12}, NPs 6 = 2.4×10^{12}, NPs 7 = 3.3×10^{12}, and NPs 8 = 1.2×10^{12}.

Cell uptake assays: For the cell uptake assay of NPs, human macrophages and HepaRG cells were plated in 24-well plates. The culture medium of macrophages and HepaRG cells was renewed, and PMLABe-based NPs loaded with DiR were added to the wells for 24 h. After incubation, culture media were discarded, and the cell monolayers were washed once with Phosphate Buffer Saline (PBS); then, they were detached with trypsin and analyzed by flow cytometry. Dot plots of forward scatter (FSC: x axis, size of events) and side scatter (SSC: y axis, structure of events) allowed to gate the viable cells prior to detecting the fluorescence emitted by DiR-loaded NPs using the BV786-1A channel. Two parameters were analyzed: the percentage of positive cells that internalized NPs and the intensity of fluorescence (mean of fluorescence) reflecting the accumulation of fluorescent NPs within the cells.

Statistical analyses: Statistical analyses were performed using a one-way Anova followed by the Kruskal–Wallis post-test or Dunn's multiple comparison tests. Statistically significant variations after treatment were compared with controls using Student's t-test with Excel software. * $p < 0.05$; ** $p < 0.01$.

3. Results and Discussion

3.1. Synthesis and Characterization of the Monomers.

The three β-substituted β-lactones involved in this study were synthesized in four steps starting from DL, L, or D-aspartic acid, as described previously (Scheme 1) [32].

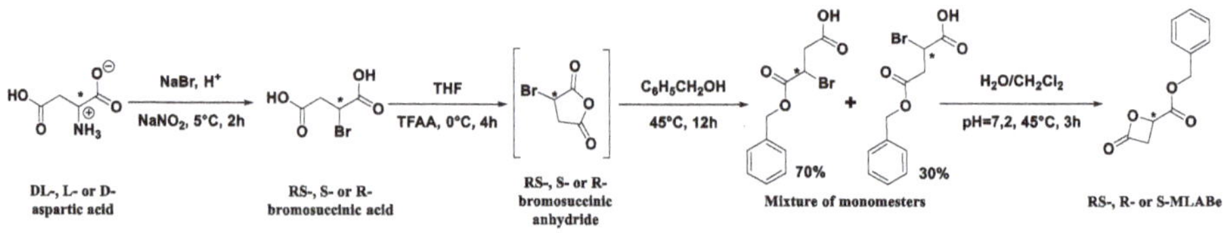

Scheme 1. Synthesis of poly(benzyl malate) (MLABe).

As shown by the results given in the Experimental part and in the Supplementary Information (Figure SI), the characteristics (yields and ^{1}H NMR and FT-IR spectra) of the synthesized RS, S, and R-MLABe were identical to those given in the literature [32]. The L-aspartic acid leads to the R-MLABe, while the D-aspartic acid leads to the S-MLABe. Indeed, the last step of the MLABe's synthesis occurs through an intramolecular nucleophilic substitution of second-order (S_N2) reaction [10]. We thus obtained pure racemic and optically active benzyl malolactonates, which can be therefore polymerized in the presence of either chemical or enzymatic initiators.

3.2. Synthesis and Characterization of the Polymers

Chemical polymerization: Two lactones, the RS and the R-MLABe, were polymerized by anionic ring-opening polymerization (AROP) in the presence of tetraethylammonium benzoate as the initiator, leading to the expected RS- and S-PMLABe, respectively (Scheme 2) [32].

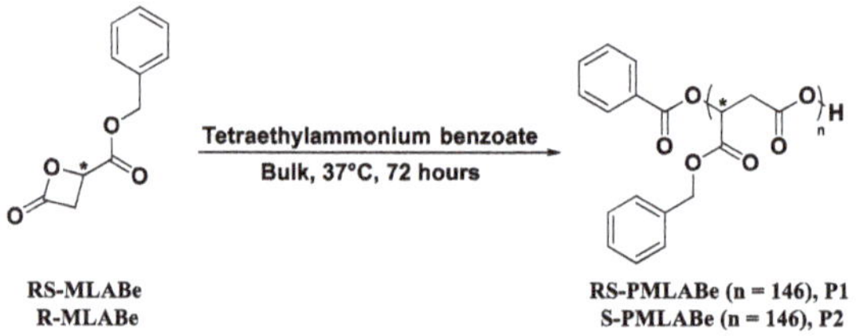

Scheme 2. Synthesis of the RS-PMLABe (P1) and S-PMLABe (P2).

The ring-opening of the monomer occurs via an O-alkyl bond cleavage, and thus with an inversion of configuration of the asymmetric carbon. Therefore, the R-MLABe leads to the S-PMLABe. Due to a fast initiation step, the molar mass of the resulting polymer is controlled by the monomer versus initiator ratio. Within this study, we chose to synthesize RS and S-PMLABe with a theoretical molar mass of 30,000 g/mol. The polymerization's reaction was followed by FT-IR and stopped when the band characteristic of the lactone ring at 1850 cm^{-1} totally disappeared from the FT-IR spectrum. After reaction, the obtained polymers were purified by precipitation in ethanol and characterized by ^{1}H NMR (structure—Figures SI 3.1 and 3.2), SEC (molar mass and dispersity) and polarimetry (Table 1). The SEC analysis on the S-PMLABe was not possible as a result of the insolubility of this polymer into THF.

Table 1. Characteristics of the RS-PMLABe (P1) and S-PMLABe (P2).

Polymers	$M_{\text{theoritical}}$ g/mol	$\overline{Mw}$ [a] g/mol	Đ [a]	$[\alpha]_D$ [b]
RS-PMLABe, P1	30,000	14,850	1.60	0
S-PMLABe, P2	30,000	n.d.	n.d.	-10.7

[a] Measured by SEC in THF, standards polystyrene, 40 °C, 1 mL/min; [b] Measured by polarimetry in chloroform. n.d.: not determined.

The synthesized RS-PMLABe (P1) and S-PMLABe (P2) have the expected characteristics and were further used for preparing the corresponding NPs.

Besides, we synthesized an amphiphilic block copolymer composed by a hydrophilic block of poly(ethylene glycol) (PEG) and a hydrophobic block of PMLABe by the AROP of the RS-MLABe in the presence of α-methoxy,ω-carboxylate-PEG tetraethylammonium salt as initiator (Scheme 3), as described previously [23]. The theoretical molar mass of the PMLABe block was fixed by the ratio monomer to the initiator and selected at 15,000 g/mol.

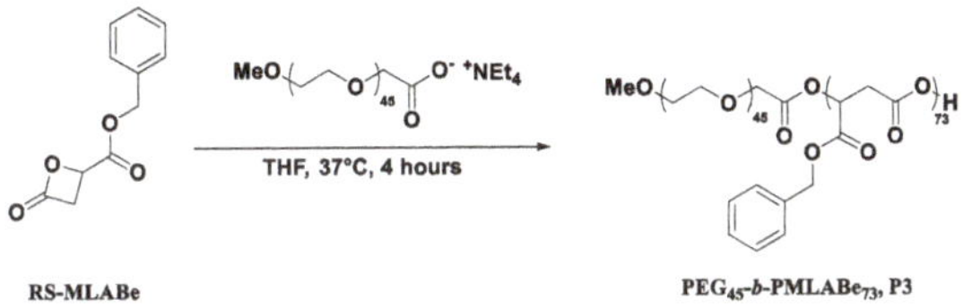

Scheme 3. Synthesis of the poly(ethylene glycol)-*block*-PMLABe (PEG$_{45}$-b-PMLABe$_{73}$) (P3).

In this case also, the polymerization's reaction was followed by FT-IR, and stopped when the band characteristic of the lactone ring at 1850 cm^{-1} totally disappeared from the FT-IR spectrum. After reaction, the obtained polymers were purified by precipitation in ethanol (elimination of unreacted PEG chains) and characterized by ^{1}H NMR (structure and PMLABe molar mass—Figure SI 3.3) and SEC (molar mass and dispersity). Due to the presence of the PEG block with a known molar mass, it is possible to determine, by ^{1}H NMR, the molar mass of the PMLABe block [23] (Table 2).

Table 2. Characteristics of the PEG$_{45}$-b-PMLABe$_{73}$ (P3).

	M_{PEG} [a] g/mol	M_{PMLABe}Theo g/mol	M_{PMLABe} NMR g/mol [b]	$\overline{M}w$ [c] g/mol	Đ [c]
PEG$_{45}$-b-PMLABe$_{73}$, P3	2000	15,000	15,040	9600	1.30

[a] Molar mass given by the supplier PEGIris Biotech. [b] Determined by ^{1}H NMR. [c] Measured by size exclusion chromatography (SEC) in THF, standards polystyrene, 40 °C, 1 mL/min.

As shown by results gathered in Table 2, the expected amphiphilic block copolymer P3 was successfully obtained.

We have included this PEG$_{45}$-b-PMLABe$_{73}$ amphiphilic block copolymer in the present study because it spontaneously forms stable core-shell NPs that are particularly suitable as a drug delivery system [22,23,30,31].

Enzymatic polymerization: The enzymatic polymerization of unsubstituted lactones (γ-butyrolactone, δ-valerolactone, ε-caprolactone, and others) has been widely studied since the 1980s–1990s [48,49]. On the other hand, the enzymatic polymerization of substituted lactones, such as MLABe, is much less described [50]. The first enzymatic polymerization of MLABe was carried out in 1996 by Matsumura et al. [51]. More recently, Panova et al. [52] have also studied the enzymatic polymerization of propyl malolactonate in the presence of Candida rugosa lipase. In this context, we have first replicated Matsumura's work, and we have next tried to improve it. Thus, the effect of various parameters (enzyme/MLABe ratio, aqueous phase/organic phase ratio, and stirring rate) was studied. However, given the large number of parameters that can influence the reaction, we have chosen to use a design of experiment approach to find the best conditions leading to well-defined high molar mass enzymatic PMLABe [43]. Using such an approach, we were able to find reaction conditions under which the expected enzymatic PMLABe were reproducibly synthesized [43].

Therefore, we used the conditions that we previously defined to synthesize racemic and optically active PMLABe as well as an amphiphilic PEG-*b*-PMLABe block copolymer by the ring-opening polymerization of racemic or optically active MLABe in the presence of PPL as a catalyst (Table 3, Scheme 4). The enzymatic polymerizations were conducted either in the presence or absence of toluene. After 72 h of reaction, all of the polymers were purified by precipitation in diethyl ether, and analyzed by ^{1}H NMR (structure, Figures SI 3.4–3.8), SEC (molar mass and dispersity) and polarimetry.

Table 3. Characteristics of RS-PMLABe (P4 and P5), S-PMLABe (P6), R-PMLABe (P7), and PEG$_{17}$-b-PMLABe$_{45}$ (P8) obtained by enzymatic polymerization in presence of porcine pancreatic lipase (PPL) as the catalyst.

Polymer	Toluene	$\overline{M}w$ [a] g/mol	Đ [a]	[a]$_D$ [c]
RS-PMLABe, P4	No	12,250	1.40	0
RS-PMLABe, P5	Yes	3850	1.50	−3
S-PMLABe, P6	Yes	2000	1.50	−11
R-PMLABe, P7	Yes	2300	1.40	+18
PEG$_{17}$-b-PMLABe$_{45}$, P8	Yes	9650	1.50	-

[a] Measured by size exclusion chromatography (SEC) in THF, standards polystyrene, 40 °C, 1 mL/min; [b] Determined by ^{1}H NMR; [c] Measured by polarimetry in chloroform.

Well-defined enzymatic PMLABe derivatives were obtained as highlighted by the relatively low dispersity values of around 1.5 (Table 3). Moreover, the molar masses of the polymers can be modulated by adding toluene or not during the enzymatic polymerization reaction. Indeed, the molar mass of the PMLABe P4 that was obtained by enzymatic polymerization of RS-MLABe in the absence of toluene is higher than the one of RS-PMLABe P5 obtained by the same reaction conducted in the presence of toluene (Table 3). The S-PMLABe P6 was obtained by enzymatic polymerization of the R-MLABe in toluene, while the R-PMLABe P7 was prepared by the ring-opening polymerization of the S-MLABe in toluene in presence of PPL as the catalyst.

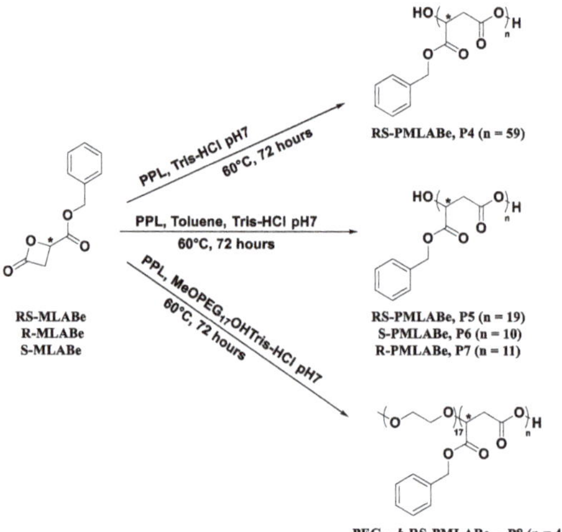

Scheme 4. Enzymatic synthesis of PMLABe P4, P5, P6, and P7, and of PEG$_{17}$-b-PMLABe$_{45}$ P8.

It is important to note that enzymatic ring-opening polymerization leads to an inversion of configuration of the asymmetric carbon. This is similar to the chemical ring-opening polymerization of the optically active MLABe [53], as demonstrated by the value of the specific rotary power for the enzymatic S-PMLABe P6 polymer, which is itself similar to the one measured by the chemical S-PMLABe P2 compound in both the value and sign. Both polymers P6 and P2 were synthesized by the ring-opening polymerization of the R-MLABe in presence of PPL and tetraethylammonium benzoate, respectively. Such an observation suggested that the MLABe ring-opening in the presence of the PPL occurs through an O-alkyl cleavage without the intervention of the serine from the catalytic triad. This result is quite surprising, especially if compared to what is usually observed for unsubstituted lactones. Indeed, the ring-opening polymerization of unsubstituted lactones in the presence of lipases has been described to occur via the canonical mechanism involving the activated serine of the catalytic triad, leading to a ring opening through an O-acyl cleavage [37–39]. Our preliminary results rather support the conclusion that the β-substituted β-lactones, as in the case of the AROP reactions, are polymerized by lipases through an O-alkyl cleavage. However, the exact mechanism needs to be further studied, and we undertake several experiments using a serine–knocked-out lipase, which is currently under synthesis, to confirm that the ring-opening polymerization of β-substituted β-lactones by lipases occurs through an O-alkyl cleavage without involvement of the catalytic serine.

Although the mechanism of polymerization is not clearly established, the enzymatic polymers obtained (P4 to P7) were used to prepare the corresponding NPs.

The last enzymatic synthesis of an amphiphilic block copolymer was realized in toluene solution in the presence of PPL as the catalyst and an α-methoxy,ω-hydroxyl PEG_{17}. The obtained polymer PEG_{17}-*b*-$PMLABe_{45}$ (P8) was characterized by ^{1}H NMR (structure, Figure SI 3.8) and SEC (molar mass and dispersity). The SEC analysis indicated the formation of a block copolymer as a result of the presence of only one peak with a quite narrow distribution (Đ = 1.50, Table 3).

3.3. Preparation and Characterization of the Nanoparticles

The methodology that was used to prepare the NPs starting from the eight polymers that were previously produced was based on the nanoprecipitation method as previously described [44,45] (see materials and methods section).

For the NPs loaded with the DiR fluorescent probe, the rate of DiR encapsulation was determined. To measure the amount of DiR encapsulated in the nanoparticles, 80 µL of the solution of nanoparticles encapsulating the DiR were added to 320 µL of DMF. The optical density of these solutions was then measured at 670 nm. The amount of encapsulated DiR was determined using a calibration curve of DiR solutions at different concentrations in DMF/water (80/20) (Figure SI 4).

Knowing the initial amount of DiR, we were able to determine the rate of DiR encapsulation that varied between 45–95%, with quantitative encapsulation rates for NPs 2, NPs 3, and NPs 8 (Table 4).

Table 4 also summarizes the hydrodynamic diameters (Dh) and dispersities (PDIs), measured by DLS, of the PMLABe-based NPs. All of the NPs have hydrodynamic diameters between 83–200 nm, which are values expected for NPs based on either hydrophobic or amphiphilic (co)polymers encapsulating a fluorescent probe such as the DiR. The samples are relatively monodisperse, since the PDIs values were all between 0.13–0.26.

Table 4. Characteristics of the prepared NPs loaded with DiR formulated for the cell uptake studies.

Entry	Polymers	Way of Synthesis	Dh nm	PDI	DiR Encapsulation Rate %
NPs 1	RS-PMLABe, P1	Chemical	195	0.19	61
NPs 2	S-PMLABe, P2	Chemical	170	0.26	Quantitative
NPs 3	PEG_{45}-*b*-$PMLABe_{73}$, P3	Chemical	83	0.22	Quantitative
NPs 4	RS-PMLABe, P4	Enzymatic	200	0.13	45
NPs 5	RS-PMLABe, P5	Enzymatic	125	0.19	83
NPs 6	S-PMLABe, P6	Enzymatic	145	0.19	95
NPs 7	R-PMLABe, P7	Enzymatic	130	0.16	94
NPs 8	PEG_{17}-*b*-$PMLABe_{45}$, P8	Enzymatic	185	0.24	Quantitative

Preparation of nanoparticles for the in vitro cytotoxicity assays. PDI: polydispersity index.

In this case, the synthesized PMLABe polymers were dissolved in 1,4-dioxane. This solution was quickly added in distilled water to trigger nanoprecipitation and auto-assembly of the polymers into NPs. The final concentration of polymers was 2.5 mg/mL. Table 5 summarizes the hydrodynamic diameters (Dh) and PDIs of the corresponding NPs used for the cytotoxicity study.

Table 5. Characteristics of the nanoparticles prepared for the in vitro cytotoxicity assays.

Entry	Polymers	Way of Synthesis	Dh nm	PDI
NPs 1	RS-PMLABe, P1	Chemical	72	0.18
NPs 2	S-PMLABe, P2	Chemical	122	0.32
NPs 3	PEG_{45}-*b*-$PMLABe_{73}$, P3	Chemical	51	0.16
NPs 4	RS-PMLABe, P4	Enzymatic	113	0.28
NPs 5	RS-PMLABe, P5	Enzymatic	96	0.17
NPs 6	S-PMLABe, P6	Enzymatic	121	0.19
NPs 7	R-PMLABe, P7	Enzymatic	107	0.18
NPs 8	PEG_{17}-*b*-$PMLABe_{45}$, P8	Enzymatic	83	0.17

The hydrodynamic diameters (Dh) of NPs were found in the range of 50 nm to 120 nm (average diameter of 102 nm $\pm$ 19 nm). The smallest Dh was measured for NPs 3 produced with the amphiphilic block copolymer PEG_{45}-*b*-$PMLABe_{73}$ P3, harboring the longest PEG chain. Such a result is in good agreement with the already observed characteristics of NPs based on PEGylated amphiphilic block copolymers [23]. All of the PDIs' values are lower than 0.3, indicating a good size distribution with the exception of the NPs 2 formulation, for which the PDI is slightly greater than 0.3. Nevertheless, we considered that all of the prepared formulations meet the conditions for the in vitro cytotoxicity assays.

3.4. In Vitro Assays

3.4.1. In Vitro Cytotoxicity Assays

In order to evaluate the biocompatibility of the PMLABe-based NPs, we incubated human hepatoma HepaRG cells with the 8 NPs in a wide range of concentrations. HepaRG cells are bipotent hepatic progenitors that can be differentiated into hepatocyte-like cells [54] expressing most of the major hepatic detoxifying enzymes [55]. These metabolically competent hepatocyte-like cells are used worldwide as an in vitro model for studying the metabolism [56] and toxicity of xenobiotics [57–59], including exposure to NPs [31,47,60].

Proliferating progenitor and quiescent differentiated HepaRG cells were incubated with NPs for 24 h and 72 h at concentrations ranging from 0.5 μM to 20 μM, and the cell viability was assayed using the MTT assay (Figure 1). The exposure of progenitor cells during 24 h triggered a moderate and partially dose-dependent decrease in MTT activity for NPs 1, 2, 3, 4, 7, and 8 compared to that measured in control cultures without nanoparticles (Figure 1A). This effect was less significant at 72 h. For these NPs, the highest concentrations in polymers (10 μM and 20 μM) reduced the cell viability by less than 30%, and the half-maximal inhibitory concentration (IC_{50}) was not reached for these amounts of NPs. The exposure of hepatocyte-like HepaRG cells that exhibit a higher ability to metabolize xenobiotics, showed a weaker decrease in MTT activities (Figure 1B).

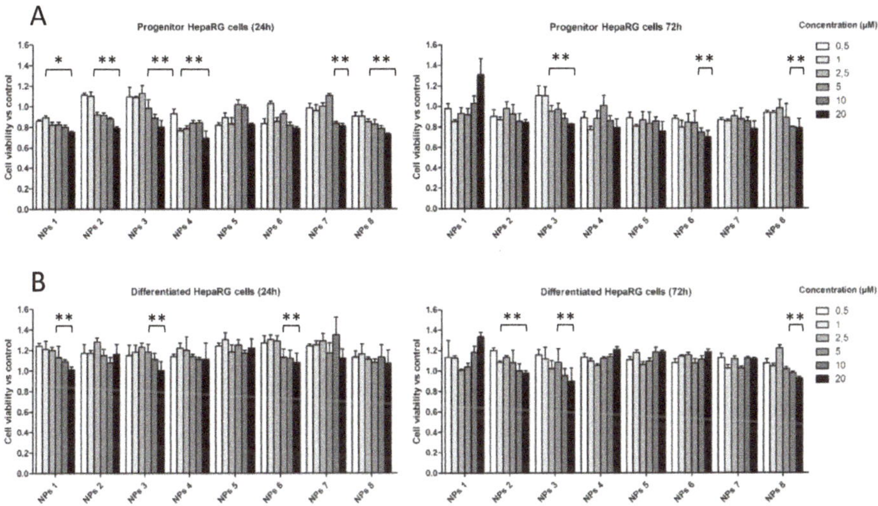

Figure 1. MTT assays (viability assay) in progenitor (**A**) and hepatocyte-like HepaRG cells (differentiated cells, **B**) incubated with PMLABe-based nanoparticles at 24 h (24 h) and 72 h (72 h). Statistical analyses: * $p < 0.05$; ** $p < 0.01$ versus non treated cells, three independent experiments.

Together, these data demonstrated the mild toxicity of these PMLABe-based NPs following incubation with HepaRG cells, which is in agreement with our previous reports describing the moderate effects on total cell numbers following the exposure of various human cell lines, including the Huh7 hepatoma cell line with NPs 1 and NPs 3 [23]. Interestingly, in this work, we compared the effects of NPs prepared from polymers synthesized by chemical (NPs 1 to 3) and enzymatic (NPs 4 to 8) polymerization. Our data indicate that mild effects on cell viability were observed with both types of NPs based on either enzymatic or chemically synthesized PMLABe. Morever, the use of NPs constituted by PMLABe derivatives prepared through enzymatic synthesis did not really improve the cell viability in comparison to what is observed with NPs based on chemical PMLABe dervatives.

3.4.2. In Vitro Cells Uptake Assays

The NPs' cell uptake by HepaRG cells (Figure 2) and human macrophages (Figure 3) of PMLABe-based nanoparticles was monitored by the detection of the cells that internalized NPs encapsulating the DiR lipophilic fluorescent dye using flow cytometry.

The flow cytometry analysis (Figures 2 and 3) was performed on single cells gated on dot plots with size of events (x axis, forward scatter, FSC-A) and structure (y axis, side scatter, SSC-A). HepaRG cells and macrophages non-incubated with NPs were used to determine the endogenous cell fluorescence and the M1 gate corresponding to negative cells. The M2 gate defined the positive cell populations that internalized fluorescent NPs.

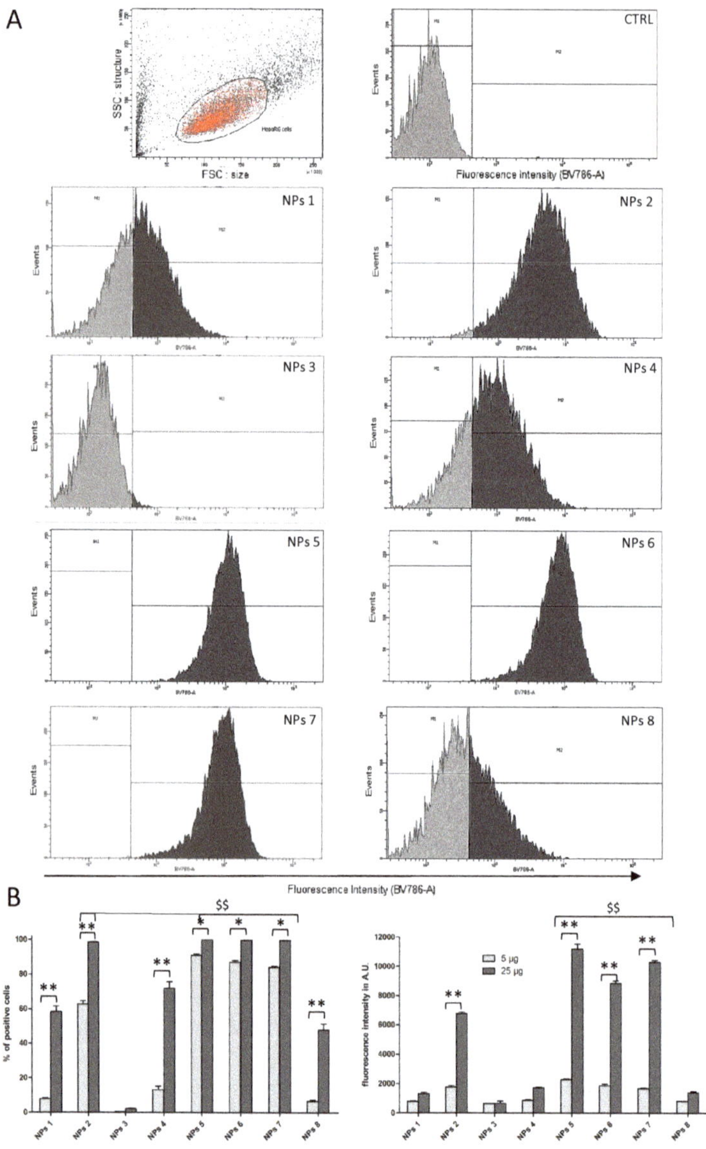

Figure 2. (**A**) Dot plot (control HepaRG cells) and histograms used to detect negative (M1 gate) and positive (M2 gate) cells and determine the mean of fluorescence in control (CTRL) and cells incubated with nanoparticles (NPs) 1 to 8; (**B**) Quantification of percentages of positive cells and mean of fluorescence in progenitor HepaRG cells incubated with PMLABe-based nanoparticles for 24 h. Statistical analyses: ** $p < 0.01$: cells incubated with 5 µg/mL of polymers versus cells incubated with 25 µg/mL; \$\$ < 0.01: highest percentages of positive cells or mean of fluorescence versus all the other conditions.

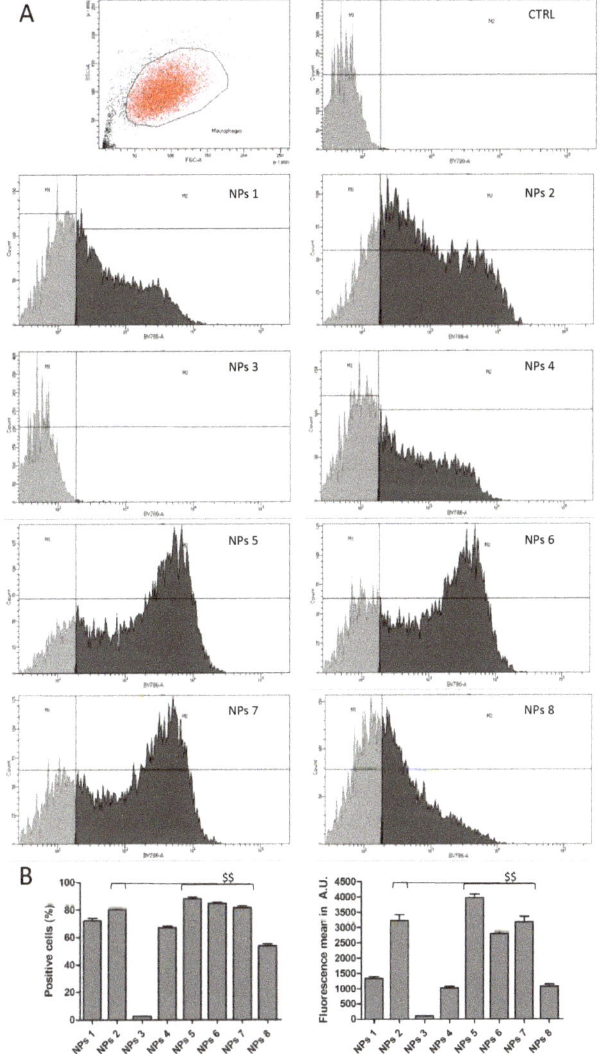

Figure 3. (**A**) Dot plot (control macrophages) and histograms used to detect negative (M1 gate) and positive (M2 gate) cells and determine the mean of fluorescence in control (CTRL) and cells incubated with NPs 1 to 8; (**B**) Quantification of percentages of positive cells and mean of fluorescence in macrophages incubated with PMLABe-based nanoparticles for 24 h. Statistical analyses: $\$\$ < 0.01$: highest percentages of positive cells or mean of fluorescence versus all the other conditions.

We first compared the HepaRG cell uptake of NPs using 5 µg/mL and 25 µg/mL of polymers (Figure 2B). At 5 µg/mL, the cell uptake was very low for NPs 1, 3, 4, and 8, with less than 10% of positive cells, while NPs 2, 5, 6, and 7 were internalized in more than 60% of the HepaRG cells. At 25 µg/mL, the percentages of positive cells and the intensity of fluorescence, reflecting the intracellular accumulation of NPs, were considerably enhanced with high levels of internalization for NPs 1, 2, 4, 5, 6, and 7, indicating that this concentration of polymers and the corresponding amounts of NPs were more appropriate to visualize the cell uptake. At this concentration, in both cultures of HepaRG cells and macrophages, cellular uptakes of NPs 2, 5, 6, and 7 prepared from PMLABe were very efficient, with at least 80% of positive cells and the highest mean of fluorescence. The efficient internalization of these NPs was not correlated to the molar masses of the corresponding homopolymers, but with small hydrodynamic diameters (Table 4). Interestingly, the cell uptake of NPs 2 was much higher than that of NPs 1, while they exhibit similar molar masses and hydrodynamic diameters. The main difference between these two NPs was that NPs 1 were prepared with RS-PMLABe, while NPs 2 were formulated with S-PMLABe, suggesting that NPs derived from this optically active polymer exhibit a physicochemical feature favoring cell uptake. Furthermore, the highest cell uptake was observed for NPs 5 to 7, which were prepared from homopolymers synthesized by enzymatic polymerization. This data led us to postulate that PMLABe homopolymers obtained by enzymatic polymerization, and with low molar masses (<4000 g/L), formed NPs that are more efficiently internalized than NPs produced from PMLABe polymers synthesized by chemical polymerization.

In contrast, the percentages of positive cells were much lower for PEGylated NPs 8 (PEG_{17}-*b*-$PMLABe_{45}$) with 45% to 55% of positive cells and almost abolished for PEGylated NPs 3 (PEG_{45}-*b*-$PMLABe_{73}$) with less than 5% of positive cells. These data are in agreement with the well-established role of the neutral PEG moiety that reduces the opsonization of PEGylated NPs by plasma proteins and their cell internalization.

4. Conclusions

In the present work, we have successfully synthesized and characterized a series of eight PMLABe-based polymers through chemical and enzymatic polymerization, formulated the corresponding NPs, studied their cytotoxicity human hepatoma HepaRG cells, and measured their uptake by both primary macrophages and HepaRG cells.

In vitro cell viability assessed by MTT on HepaRG cells evidenced a mild toxicity of the NPs formed from PMLABe polymers at high concentrations/densities of NPs in the culture media of cells. This cytotoxicity is in agreement with our previous reports, and does not evidence a higher biocompatibility of the NPs prepared from polymers synthesized by enzymatic polymerization. These data further demonstrate that the chemical polymerization and the procedure of nanoprecipitation led to the production of biocompatible PMLABe-based NPs. The efficient uptake combined with low cytotoxicity toward HepaRG cells indicates that PMLABe-based NPs are suitable nanovectors for drug delivery that will further be evaluated in vivo to target either hepatocytes or resident liver macrophages.

Supplementary Materials: The supplementary materials are available online at http://www.mdpi.com/2073-4360/10/11/1244/s1.

Author Contributions: Conceptualization, S.T., C.N.C., S.C.M., P.L. and N.L.; methodology, H.C., S.S., M.V. and E.V.; software, H.C., S.S. and M.V.; validation, S.C.M., P.L. and N.L.; formal analysis, S.C.M., P.L. and N.L.; investigation, H.C., S.S., M.V., E.V., E.D. and C.R.; resources, C.R. and E.V.; data curation, H.C., M.V., E.V., S.C.M. and P.L.; writing—original draft preparation, S.C.M., P.L. and N.L.; writing—review and editing, S.C.M., P.L. and N.L.; visualization, S.C.M., P.L. and N.L.; supervision, C.N.C., S.C.M., P.L. and N.L.; project administration, S.C.M., P.L.; funding acquisition, C.N.C., S.C.M., P.L.

Funding: This work was funded by the Institut National de la Santé et de la Recherche Médicale (Inserm), l'Ecole Nationale Supérieure de Chimie de Rennes (ENSCR, France), the Centre National de la Recherche Scientifique (CNRS) and les comités départementaux de la Ligue contre le Cancer du Grand Ouest (comités 17 et 35). Hubert Casajus and Saad Saba were recipients of fellowships from the Ministère de l'Enseignement Supérieur et de la Recherche (France) and the Islamic Center Association for Guidance and Higher Education (Lebanon), respectively.

Conflicts of Interest: The authors declare no conflict of interest.

References

1. Strebhardt, K.; Ullrich, A. Paul Ehrlich's Magic Bullet Concept: 100 Years of Progress. *Nat. Rev. Cancer* **2008**, *8*, 473–480. [CrossRef] [PubMed]

2. Torchilin, V.P. Recent Advances with Liposomes as Pharmaceutical Carriers. *Nat. Rev. Drug Discov.* **2005**, *4*, 145–160. [CrossRef] [PubMed]
3. Hoffman, A.S. The Origins and Evolution of "Controlled" Drug Delivery Systems. *J. Control. Release* **2008**, *132*, 153–163. [CrossRef] [PubMed]
4. Liu, S.; Maheshwari, R.; Kiick, K.L. Polymer-Based Therapeutics. *Macromolecules* **2009**, *42*, 3–13. [CrossRef] [PubMed]
5. Malam, Y.; Loizidou, M.; Seifalian, A.M. Liposomes and Nanoparticles: Nanosized Vehicles for Drug Delivery in Cancer. *Trends Pharmacol. Sci.* **2009**, *30*, 592–599. [CrossRef] [PubMed]
6. Misra, R.; Acharya, S.; Sahoo, S.K. Cancer Nanotechnology: Application of Nanotechnology in Cancer Therapy. *Drug Discov. Today* **2010**, *15*, 842–850. [CrossRef] [PubMed]
7. Etheridge, M.L.; Campbell, S.A.; Erdman, A.G.; Haynes, C.L.; Wolf, S.M.; McCullough, J. The Big Picture on Nanomedicine: The State of Investigational and Approved Nanomedicine Products. *Nanomed. Nanotechnol. Biol. Med.* **2013**, *9*, 1–14. [CrossRef] [PubMed]
8. Jin, S.E.; Jin, H.E.; Hong, S.S. Targeted Delivery System of Nanobiomaterials in Anticancer Therapy: From Cells to Clinics. *BioMed Res. Int.* **2014**, *2014*, 814208. [CrossRef] [PubMed]
9. Park, J.H.; Lee, S.; Kim, J.H.; Park, K.; Kim, K.; Kwon, I.C. Polymeric Nanomedicine for Cancer Therapy. *Prog. Polym. Sci.* **2008**, *33*, 113–137. [CrossRef]
10. Kumari, A.; Yadav, S.K.; Yadav, S.C. Biodegradable Polymeric Nanoparticles Based Drug Delivery Systems. *Colloids Surf. B Biointerfaces* **2010**, *75*, 1–18. [CrossRef] [PubMed]
11. Ulery, B.D.; Nair, L.S.; Laurencin, C.T. Biomedical Applications of Biodegradable Polymers. *J. Polym. Sci. Part B Polym. Phys.* **2011**, *49*, 832–864. [CrossRef] [PubMed]
12. Tucker, B.S.; Sumerlin, B.S. Poly (N-(2-Hydroxypropyl) Methacrylamide)-Based Nanotherapeutics. *Polym. Chem.* **2014**, *5*, 1566–1572. [CrossRef]
13. El-Say, K.M.; El-Sawy, H.S. Polymeric Nanoparticles: Promising Platform for Drug Delivery. *Int. J. Pharm.* **2017**, *528*, 675–691. [CrossRef] [PubMed]
14. Deshmukh, A.S.; Chauhan, P.N.; Noolvi, M.N.; Chaturvedi, K.; Ganguly, K.; Shukla, S.S.; Nadagouda, M.N.; Aminabhavi, T.M. Polymeric Micelles: Basic Research to Clinical Practice. *Int. J. Pharm.* **2017**, *532*, 249–268. [CrossRef] [PubMed]
15. Marguet, M.; Bonduelle, C.; Lecommandoux, S. Multicompartmentalized Polymeric Systems: Towards Biomimetic Cellular Structure and Function. *Chem. Soc. Rev.* **2012**, *42*, 512–529. [CrossRef] [PubMed]
16. Kale, G.; Kijchavengkul, T.; Auras, R.; Rubino, M.; Selke, S.E.; Singh, S.P. Compostability of Bioplastic Packaging Materials: An Overview. *Macromol. Biosci.* **2007**, *7*, 255–277. [CrossRef] [PubMed]
17. Seyednejad, H.; Ghassemi, A.H.; van Nostrum, C.F.; Vermonden, T.; Hennink, W.E. Functional aliphatic polyesters for biomedical and pharmaceutical applications. *J. Control. Release* **2011**, *152*, 168–176. [CrossRef] [PubMed]
18. Sauer, A.; Kapelski, A.; Fliedel, C.; Dagorne, S.; Kol, M.; Okuda, J. Structurally Well-Defined Group 4 Metal Complexes as Initiators for the Ring-Opening Polymerization of Lactide Monomers. *Dalton Trans.* **2013**, *42*, 9007–9023. [CrossRef] [PubMed]
19. Chaturvedi, D.; Mishra, S.; Tandon, P.; Portilla-Arias, J.A.; Muñoz-Guerra, S. Thermal Degradation and Theoretical Interpretation of Vibrational Spectra of Poly (β,L-Malic Acid). *Polymer* **2011**, *52*, 3118–3126. [CrossRef]
20. Philip, S.; Keshavarz, T.; Roy, I. Polyhydroxyalkanoates: Biodegradable Polymers with a Range of Applications. *J. Chem. Technol. Biotechnol.* **2007**, *82*, 233–247. [CrossRef]
21. Tanzi, M.C.; Verderio, P.; Lampugnani, M.G.; Resnati, M.; Dejana, E.; Sturani, E. Cytotoxicity of Some Catalysts Commonly Used in the Synthesis of Copolymers for Biomedical Use. *J. Mater. Sci. Mater. Med.* **1994**, *5*, 393–396. [CrossRef]
22. Loyer, P.; Cammas-Marion, S. Natural and Synthetic Poly (Malic Acid)-Based Derivates: A Family of Versatile Biopolymers for the Design of Drug Nanocarriers. *J. Drug Target.* **2014**, *22*, 556–575. [CrossRef] [PubMed]
23. Huang, Z.W.; Laurent, V.; Chetouani, G.; Ljubimova, J.Y.; Holler, E.; Benvegnu, T.; Loyer, P.; Cammas-Marion, S. New Functional Degradable and Bio-Compatible Nanoparticles Based on Poly (Malic Acid) Derivatives for Site-Specific Anti-Cancer Drug Delivery. *Int. J. Pharm.* **2012**, *423*, 84–92. [CrossRef] [PubMed]

24. Jaffredo, C.G.; Guillaume, S.M. Benzyl β-Malolactonate Polymers: A Long Story with Recent Advances. *Polym. Chem.* **2014**, *5*, 4168–4194. [CrossRef]

25. Coulembier, O.; Degée, P.; Cammas-Marion, S.; Guérin, P.; Dubois, P. New Amphiphilic Poly [(R,S)-β-Malic Acid-b-ε-Caprolactone] Diblock Copolymers by Combining Anionic and Coordination-Insertion Ring-Opening Polymerization. *Macromolecules* **2002**, *35*, 9896–9903. [CrossRef]

26. Vert, M.; Lenz, R.W. Preparation and Properties of Poly-β-Malic Acid: A Functional Polyester of Potential Biomedical Importance. *Polym. Prepr.* **1979**, *20*, 608–611.

27. Cammas, S.; Béar, M.M.; Moine, L.; Escalup, R.; Ponchel, G.; Kataoka, K.; Guérin, P. Polymers of Malic Acid and 3-Alkylmalic Acid as Synthetic PHAs in the Design of Biocompatible Hydrolyzable Devices. *Int. J. Biol. Macromol.* **1999**, *25*, 273–282. [CrossRef]

28. Cammas-Marion, S.; Guérin, P. Design of Malolactonic Acid Esters with a Large Spectrum of Specified Pendant Groups in the Engineering of Biofunctional and Hydrolyzable Polyesters. *Macromol. Symp.* **2000**, *153*, 167–186. [CrossRef]

29. Venkatraj, N.; Nanjan, M.J.; Loyer, P.; Chandrasekar, M.J.N.; Cammas-Marion, S. Poly (Malic Acid) Bearing Doxorubicin and N-Acetyl Galactosamine as a Site-Specific Prodrug for Targeting Hepatocellular Carcinoma. *J. Biomater. Sci. Polym. Ed.* **2017**, *28*, 1140–1157. [CrossRef] [PubMed]

30. Loyer, P.; Bedhouche, W.; Huang, Z.W.; Cammas-Marion, S. Degradable and Biocompatible Nanoparticles Decorated with Cyclic RGD Peptide for Efficient Drug Delivery to Hepatoma Cells in Vitro. *Int. J. Pharm.* **2013**, *454*, 727–737. [CrossRef] [PubMed]

31. Vene, E.; Jarnouen, K.; Huang, Z.W.; Bedhouche, W.; Montier, T.; Cammas-Marion, S.; Loyer, P. In Vitro Toxicity Evaluation and in Vivo Biodistribution of Polymeric Micelles Derived from Poly (Ethylene Glycol)-b-Poly(Benzyl Malate) Copolymer. *Pharm. Nanotechnol.* **2016**, *4*, 24–37. [CrossRef]

32. Cammas, S.; Renard, I.; Langlois, V.; Guerin, P. Poly (β-Malic Acid): Obtaining High Molecular Weights by Improvement of the Synthesis Route. *Polymer* **1996**, *37*, 4215–4220. [CrossRef]

33. Tetraethylammonium. Available online: https://www.drugbank.ca/drugs/DB08837 (accessed on 2 November 2018).

34. Casas-Godoy, L.; Duquesne, S.; Bordes, F.; Sandoval, G.; Marty, A. Lipases: An Overview. In *Lipases and Phospholipases*; Humana Press: New York, NY, USA, 2012; pp. 3–30.

35. Rauwerdink, A.; Kazlauskas, R.J. How the Same Core Catalytic Machinery Catalyzes 17 Different Reactions: The Serine-Histidine-Aspartate Catalytic Triad of α/β-Hydrolase Fold Enzymes. *ACS Catal.* **2015**, *5*, 6153–6176. [CrossRef] [PubMed]

36. Shoda, S.; Uyama, H.; Kadokawa, J.; Kimura, S.; Kobayashi, S. Enzymes as Green Catalysts for Precision Macromolecular Synthesis. *Chem. Rev.* **2016**, *116*, 2307–2413. [CrossRef] [PubMed]

37. MacDonald, R.T.; Pulapura, S.K.; Svirkin, Y.Y.; Gross, R.A.; Kaplan, D.L.; Akkara, J.; Swift, G.; Wolk, S. Enzyme-Catalyzed Epsilon.-Caprolactone Ring-Opening Polymerization. *Macromolecules* **1995**, *28*, 73–78. [CrossRef]

38. Henderson, L.A.; Svirkin, Y.Y.; Gross, R.A.; Kaplan, D.L.; Swift, G. Enzyme-Catalyzed Polymerizations of ε-Caprolactone: Effects of Initiator on Product Structure, Propagation Kinetics, and Mechanism. *Macromolecules* **1996**, *29*, 7759–7766. [CrossRef]

39. Johnson, P.M.; Kundu, S.; Beers, K.L. Modeling Enzymatic Kinetic Pathways for Ring-Opening Lactone Polymerization. *Biomacromolecules* **2011**, *12*, 3337–3343. [CrossRef] [PubMed]

40. Kobayashi, S. Enzymatic Polymerization: A New Method of Polymer Synthesis. *J. Polym. Sci. Part A Polym. Chem.* **1999**, *37*, 3041–3056. [CrossRef]

41. Kobayashi, S. Green Polymer Chemistry: New Methods of Polymer Synthesis Using Renewable Starting Materials. *Struct. Chem.* **2016**, *28*, 461–474. [CrossRef]

42. Albertsson, A.C.; Srivastava, R.K. Recent Developments in Enzyme-Catalyzed Ring-Opening Polymerization. *Adv. Drug Deliv. Rev.* **2008**, *60*, 1077–1093. [CrossRef] [PubMed]

43. Casajus, H.; Tranchimand, S.; Wolbert, D.; Nugier-Chauvin, C.; Cammas-Marion, S. Optimization of Lipase-Catalyzed Polymerization of Benzyl Malolactonate through a Design of Experiment Approach. *J. Appl. Polym. Sci.* **2017**, *134*, 44604. [CrossRef]

44. Thioune, O.; Fessi, H.; Devissaguet, J.P.; Puisieux, F. Preparation of Pseudolatex by Nanoprecipitation: Influence of the Solvent Nature on Intrinsic Viscosity and Interaction Constant. *Int. J. Pharm.* **1997**, *146*, 233–238. [CrossRef]

45. Martínez Rivas, C.J.; Tarhini, M.; Badri, W.; Miladi, K.; Greige-Gerges, H.; Nazari, Q.A.; Galindo Rodríguez, S.A.; Román, R.Á.; Fessi, H.; Elaissari, A. Nanoprecipitation Process: From Encapsulation to Drug Delivery. *Int. J. Pharm.* **2017**, *532*, 66–81. [CrossRef] [PubMed]
46. Laurent, V.; Glaise, D.; Nübel, T.; Gilot, D.; Corlu, A.; Loyer, P. Highly Efficient SiRNA and Gene Transfer into Hepatocyte-Like HepaRG Cells and Primary Human Hepatocytes: New Means for Drug Metabolism and Toxicity Studies. In *Cytochrome P450 Protocols*; Phillips, I.R., Shephard, E.A., Ortiz de Montellano, P.R., Eds.; Humana Press: Totowa, NJ, USA, 2013; Volume 987, pp. 295–314.
47. Vène, E.; Barouti, G.; Jarnouen, K.; Gicquel, T.; Rauch, C.; Ribault, C.; Guillaume, S.M.; Cammas-Marion, S.; Loyer, P. Opsonisation of nanoparticles prepared from poly (β-hydroxybutyrate) and poly(trimethylene carbonate)-*b*-poly(malic acid) amphiphilic diblock copolymers: Impact on the in vitro cell uptake by primary human macrophages and HepaRG hepatoma cells. *Int. J. Pharm.* **2016**, *13*, 438–452. [CrossRef] [PubMed]
48. Matsumura, S. Enzymatic Synthesis of Polyesters via Ring-Opening Polymerization. In *Enzyme-Catalyzed Synthesis of Polymers*; Kobayashi, S., Ritter, H., Kaplan, D., Eds.; Springer: Berlin/Heidelberg, Germany, 2005; Volume 194, pp. 95–132.
49. Uyama, H.; Takeya, K.; Kobayashi, S. Synthesis of Polyesters by Enzymatic Ring-Opening Copolymerization Using Lipase Catalyst. *Proc. Jpn. Acad. Ser. B* **1993**, *69*, 203–207. [CrossRef]
50. Kikuchi, H.; Uyama, H.; Kobayashi, S. Lipase-Catalyzed Ring-Opening Polymerization of Substituted Lactones. *Polym. J.* **2002**, *34*, 835–840. [CrossRef]
51. Matsumura, S.; Beppu, H.; Nakamura, K.; Osanai, S.; Toshima, K. Preparation of Poly (β-Malic Acid) by Enzymatic Ring-Opening Polymerization of Benzyl β-Malolactonate. *Chem. Lett.* **1996**, *25*, 795–796. [CrossRef]
52. Panova, A.A.; Taktak, S.; Randriamahefa, S.; Cammas-Marion, S.; Guerin, P.; Kaplan, D.L. Polymerization of Propyl Malolactonate in the Presence of *Candida Rugosa* Lipase. *Biomacromolecules* **2003**, *4*, 19–27. [CrossRef] [PubMed]
53. Guerin, P.; Francillette, J.; Braud, C.; Vert, M. Benzyl esters of optically active malic acid stereocopolymers as obtained by ring-opening polymerization of (R)-(+) and (S)-(-)-benzyl malolactonates. *Macromol. Symp.* **1986**, *6*, 305–314. [CrossRef]
54. Cerec, V.; Glaise, D.; Garnier, D.; Morosan, S.; Turlin, B.; Drenou, B.; Gripon, P.; Kremsdorf, D.; Guguen-Guillouzo, C.; Corlu, A. Transdifferentiation of hepatocyte-like cells from the human hepatoma HepaRG cell line through bipotent progenitor. *Hepatology* **2007**, *45*, 957–967. [CrossRef] [PubMed]
55. Aninat, C.; Piton, A.; Glaise, D.; Le Charpentier, T.; Langouet, S.; Morel, F.; Guguen-Guillouzo, C.; Guillouzo, A. Expression of cytochrome P450, conjugating enzymes and nuclear receptors in human hepatoma HepaRG cells. *Drug Metab. Dispos.* **2006**, *34*, 75–83. [CrossRef] [PubMed]
56. Quesnot, N.; Bucher, S.; Gade, C.; Vlach, M.; Vène, E.; Valenca, S.; Gicquel, T.; Holst, H.; Robin, M.-A.; Loyer, P. Production of chlorzoxazone glucuronides via cytochrome P4502E1 dependent and independent pathways in human hepatocytes. *Arch. Toxicol.* **2018**, *92*, 3077–3091. [CrossRef] [PubMed]
57. Jossé, R.; Aninat, C.; Glaise, D.; Dumont, J.; Fessard, V.; Morel, F.; Poul, JM.; Guguen-Guillouzo, C.; Guillouzo, A. Long-term functional stability of HepaRG hepatocytes and use for chronic toxicity and genotoxicity studies. *Drug Metab. Dispos.* **2008**, *36*, 1111–1118. [CrossRef] [PubMed]
58. Dumont, J.; Jossé, R.; Lambert, C.; Anthérieu, S.; Laurent, V.; Loyer, P.; Robin, M.-A.; Guillouzo, A. Preferential induction of the AhR gene battery in HepaRG cells after a single or repeated exposure to heterocyclic aromatic amines. *Toxicol. Appl. Pharmacol.* **2010**, *249*, 91–100. [CrossRef] [PubMed]
59. Quesnot, N.; Rondel, K.; Martinais, S.; Audebert, M.; Glaise, D.; Morel, F.; Loyer, P.; Robin, M.-A. Evaluation of genotoxicity using automated detection of gammaH2AX in metabolically competent HepaRG cells. *Mutagenesis* **2016**, *31*, 43–50. [PubMed]
60. Barouti, G.; Jarnouen, K.; Cammas-Marion, S.; Loyer, P.; Guillaume, S. Polyhydroxylakanoate-based diblock copolymers: Potential biocompatible nanovectors. *Polym. Chem.* **2015**, *6*, 5414–5429. [CrossRef]

Article

Contraction and Hydroscopic Expansion Stress of Dental Ion-Releasing Polymeric Materials

Krzysztof Sokolowski [1], Agata Szczesio-Wlodarczyk [2], Kinga Bociong [2], Michal Krasowski [2], Magdalena Fronczek-Wojciechowska [3], Monika Domarecka [4], Jerzy Sokolowski [4] and Monika Lukomska-Szymanska [4,*]

[1] Department of Restorative Dentistry, 251 Pomorska St., 92-213 Lodz, Poland; krzysztof.sokolowski@umed.lodz.pl

[2] University Laboratory of Materials Research, Medical University of Lodz, 251 Pomorska St., 92-213 Lodz, Poland; agata.szczesio@umed.lodz.pl (A.S.-W.); kinga.bociong@umed.lodz.pl (K.B.); michal.krasowski@umed.lodz.pl (M.K.)

[3] "DynamoLab" Academic Laboratory of Movement and Human Physical Performance, Medical University of Lodz, 251 Pomorska St., 92-216 Lodz, Poland; magdalena.fronczek-wojciechowska@umed.lodz.pl

[4] Department of General Dentistry, Medical University of Lodz, 251 Pomorska St., 92-213 Lodz, Poland; monika.domarecka@umed.lodz.pl (M.D.); jerzy.sokolowski@umed.lodz.pl (J.S.)

* Correspondence: monika.lukomska-szymanska@umed.lodz.pl; Tel.: +48-42-675-74-64

Received: 28 August 2018; Accepted: 29 September 2018; Published: 2 October 2018

Abstract: Ion-releasing polymeric restorative materials seem to be promising solutions, due to their possible anticaries effect. However, acid functional groups (monomers) and glass filler increase hydrophilicity and, supposedly, water sorption. The purpose of the study was to evaluate the influence of water sorption of polymeric materials on the stress state at the restoration-tooth interface. Beautifil Bulk Fill Flow, Beautifil Flow Plus F00, Beautifil Flow F02, Dyract eXtra, Compoglass Flow, Ionosit, Glasiosite, TwinkiStar, Ionolux and Fuji II LC were used for the study. The stress state was measured using photoelastic analysis after: 0.5, 24, 72, 96, 168, 240, 336, 504, 672, 1344 and 2016 h. Moreover, water sorption, solubility and absorption dynamic were assessed. The water sorption, solubility and absorption dynamic of ion-releasing restorative materials are material dependent properties. The overall results indicated that the tested restorative materials showed significant stress decrease. The total reduction in contraction stress and water expansion stress was not observed for materials with low value of water sorption (Beautifil Bulk Fill, Dyract eXtra, Glasionosit and Twinky Star). The photoelastic method turned out to be inadequate to evaluate stress changes of resin modified glass-ionomer cement (RMGI, Fuji II LC and Ionolux).

Keywords: ion-releasing materials; shrinkage stress; water sorption; hydroscopic expansion; photoelastic investigation

---

1. Introduction

Development in dentistry is driven by the desire to create a restorative material exhibiting high aesthetics, biocompatibility, durability and permanent adhesion to tooth structure [1]. Many products have been developed aiming at meeting these expectations. Dentists perceive fluoride-releasing restorative materials as being very attractive due to their advertised multifunction performance. Besides esthetic and restoring function, these materials could prevent or arrest the progression of caries lesions [2]. The glass-ionomer cements (GICs) release fluoride ions and are biocompatible, however, they exhibit inherent drawbacks such as low compressive strength, early moisture sensitivity and inadequate esthetics [3,4]. Resin modified glass-ionomer cement (RMGIC) was designed to combine chemistry of GIC with dental resins. This material shows favorable properties similar to composites while maintaining the features of the conventional GIC [5]. Acid/base reaction predominates in overall curing process of RMGIs. An additional curing process is the photo-polymerization of methacrylate monomers that are attached to molecules of the acidic liquid component [6]. Compomers provide benefits of composites (the "comp" in their name) and glass-ionomers ("omer"). These materials contain dimethacrylate monomers with acid functional group and partially salinized ion-leachable glass. These components may participate in an acid/base glass-ionomer reaction following polymerization of the resin molecule. However, this process probably does not influence the overall properties of these materials. Considering the internal structure, these materials are classified as a new type of composites [7,8]. Giomers were developed to combine the advantages of composites (good mechanical and esthetic properties) and GICs (anticariogenic activity). Being resin-based materials, they contain surface reaction-type pre-reacted glass-ionomer (S-PRG) particles. These particles are made of fluorosilicate glass that react with polyacrylic acid prior to being incorporated into the resin [9].

The polymerization shrinkage is an integral feature of all currently available resin-based/polymer materials [10,11]. Previous studies showed the reduction in the initial stress state (generated by photo-polymerization) in composite materials by hygroscopic expansion [12,13]. The value of contraction stress and the magnitude of stress reduction were proven to be material-dependent [12,14]. Moreover, acid functional groups in monomers and glass filler increased hydrophilicity and supposedly water sorption of materials that combined chemistry of GIC and resin composites. As a result, the water sorption and the change of contraction stress generated during curing might be significantly modified. The in-depth analysis of the change in stress state for compomers and giomers after ageing in water is still missing. The purpose of the study was to evaluate the influence of water sorption of selected ion-releasing polymeric restorative materials on the stress state at the restoration-tooth interface.

2. Materials and Methods

The composition of investigated materials and bonding systems is presented in Tables 1 and 2. To ensure consistent irradiance values, the light curing units (Mini L.E.D, Satelec, France) were calibrated with a radiometer system (Digital Light Meter 200 from Rolence Enterprice Inc., Taoyuan, Taiwan). Increments of 2 mm in thickness were polymerized. Materials were tested after a selected period of time (30 min, 24, 72, 96, 168, 240, 336, 504, 672, 1344 and 2016 h).

Table 1. The composition and curing time of investigated materials.

Material	Manufacturer (Country)	Type	Composition	Curing Time [s]
Beautifil Bulk Fill Flow	Shofu (Japan)	Giomer	bis-GMA, UDMA, bis-MEPEPP, TEGDMA, multi-functional glass filler and S-PRG filler based fluoro-alumino-silicate glass (73 wt %)	10
Beautifil Flow Plus F00	Shofu (Japan)	Giomer	bis-GMA, TEGDMA, multifunctional glass filler, improved S-PGR filler based on aluminofluoro-borosilicate glass, Al_2O_3 (67.3 wt %, 47.0 vol %)	10
Beautifil Flow F02	Shofu (Japan)	Giomer	bis-GMA, TEGDMA, multifunctional glass filler, improved S-PGR filler based on fluoro-boroaluminosilicate glass (54.5 wt %/34.6 vol %)	10
Dyract eXtra	Dentsply Sirona (USA)	Compomer	UDMA, carboxylic acid modified, dimethacrylate resin, TEGDMA, BHT, strontium alimino-sodium-fluoro-silicate glass (50 vol %)	10

Table 1. *Cont.*

Material	Manufacturer (Country)	Type	Composition	Curing Time [s]
Compoglass Flow	Ivoclar Vivadent (Germany)	Compomer	UDMA, PEGDMA, cycloaliphate dicarbonic acid dimethacrylate, catalysts, stabilizers and pigments, mixed oxide—silanized, ytterbiumtrifluoride, Ba-Al-fluorosilikateglass-silanized (66.8 wt %)	20
Ionosit	DMG (Germany)	Compomer	acrylic resin, glass powder, silica, aliphatic dimethacrylate, aromatic dimethacrylate, polycarboxylic polymethacrylate (72 wt % 55 vol %)	20
Glasiosite	Voco (Germany)	Compomer	bis-GMA, UDMA, TEGDMA, BHT, SiO_2, (Ba,B)AlSi, FAlSi (77.5 wt%)	20
TwinkiStar	Voco (Germany)	Compomer	bis-GMA, UDMA, carboxylic acid modified methacrylate, camphorquinone, BHT, Ba-Al-Str-fluorosilicate glass, Silicon dioxide (78 wt %)	20
Ionolux	Voco (Germany)	RMGI	polyacrylic acid, HEMA, bis-GMA, UDMA, fluoro-alumino-silicate glass	20
Fuji II LC	GC (USA)	RMGI	polyacrylic acid, HEMA, UDMA, camphorqunone, fluoro-alumino-silicate glass	20

bis-GMA—bisphenol A glycol dimethacrylate, bis-MPEPP—bisphenol A polyethoxy methacrylate, TEGDMA—triethylene glycol dimethacrylate, UDMA—urethane dimethacrylate, PEGDMA—polyethylene glycol dimethacrylate, BHT—butylated hydroxytoluene, HEMA—hydroxyethylmethacrylate.

Table 2. The composition and curing time of bonding systems.

Material	Manufacturer (Country)	Dedicated Restorative Material	Composition	Curing Time [s]
BeautiBond	Shofu (Japan)	Beautifil Flow, Beautifil Bulk Fill Flow, Beautifil Flow Plus F00	bis-GMA, TEGDMA, phosphoric acid monomer, carboxylic acid monomer	10
XP Bond	Dentsply Sirona (USA)	Dyract eXtra, Fuji II LC	TCB, PENTA, UDMA, TEGDMA, HEMA, butylated benzenediol (stabilizer), ethyl-4-dimethylaminobenzoate, camphorquinone	10
Monobond Plus	Ivoclar Vivadent (Germany)	Compoglass Flow	10-MDP, silane methacrylate, ethanol, sulfide methacrylate	10
Ecosite-Bond	DMG (Germany)	Ionosit	dental resins, ethanol, water, additives and catalysts	10
Futurabond M+	Voco (Germany)	Glasiosite, TwinkyStar, Ionolux	HEMA, bis-GMA, etanol, acidic adhesive monomer	10

2.1. Water Absorption Dynamic Study

The detailed procedure for determination of absorption dynamic has also been described in our earlier works [12,14]. In order to characterize absorbency dynamic the five cylindrical samples—15 mm in diameter and 1 mm in width—were prepared according to ISO 4049 [15]. Curing time was consistent with the manufacturers' instructions (Tables 1 and 2). Light curing units exhibited an output irradiance of 1250 mW/cm^2, as stated by the manufacturer.

Absorption is a physical or chemical phenomenon or a process in which atoms, molecules or ions enter some bulk phase—gas, liquid or solid material. This is a different process from adsorption, since molecules undergoing absorption are taken up by the volume, not by the surface (as in the case for adsorption). A more general term is sorption, which covers absorption, adsorption, and ion exchange. The main phenomenon observed in the present study is the absorption.

The samples were weighted (RADWAG AS 160/C/2, Poland) immediately after preparation and then daily for 30 days, and after 1344 h (56 days) and 2016 h (84 days).

Water absorption was calculated according to following equations:

$$A = \frac{m_i - m_0}{m_0} \cdot 100\% \tag{1}$$

where: A—water absorption, m_0—the mass of the sample in dry condition [g] and m_i—mass of the sample after storage in water for a specified (i) period of time.

2.2. Water Sorption and Solubility

Water sorption and solubility was investigated according to ISO 4049 [15]. Five samples were prepared for each dental material. The samples were prepared using the silicone mold (15 mm in diameter, 1 mm in width). Tested materials were cured in nine zones partially overlapping. Exposure time was consistent with the manufacturer instructions (Table 2). Direct contact of optical fiber with the sample surface was ensured. Specimens were placed in a vacuum desiccator (Duran®, Mainz, Germany) at a temperature of 37 ± 1 °C for 23 h, transferred to a second desiccator at the temperature of 25 ± 1 °C for 1 h, and then weighed in the balance. This cycle was repeated until a constant mass was obtained (m_1). Upon stabilization, specimens were immersed in distilled water at a temperature of 37 ± 1 °C for 7 days. Specimens were removed, gently dried with absorbent paper, and weighed again to obtain m_2. Using the same protocol as for m_1, specimens were then reconditioned in the desiccators until a constant mass was obtained (m_3).

Water sorption and solubility were calculated according to following equations:

$$W_{sp} = \frac{m_2 - m_3}{V} \tag{2}$$

$$W_{sl} = \frac{m_1 - m_3}{V} \tag{3}$$

where: W_{sp}—water sorption, W_{sl}—water solubility, m_1—initial constant mass [µg], m_2—mass after 7 days of water immersion [µg], m_3—final constant mass [µg] and V—specimen volume [mm³].

2.3. Photoelastic Study

The modified photoelastic analysis method enables analysis of the relationship between water sorption and the change of stress state (contraction or expansion) of materials with resin matrix. This test was extensively described [12,14]. Photoelastically sensitive plates of epoxy resin (Epidian 53, Organika-Sarzyna SA, Nowa Sarzyna, Poland) were used in this study. Calibrated orifices of 3 mm in diameter and of 4 mm in depth were prepared in resin plates in order to mimic an average tooth cavity. The generated strains in the plates were visualized in circular transmission polariscope FL200 (Gunt, Germany). Next, photoelastic strain calculations were based on the Timoshenko equation [16]. Three samples were prepared for each material. Curing time was consistent with the manufacturers' instructions (Tables 1 and 2).

3. Results

3.1. Absorption Dynamic Study

Figures 1–8 presented mean values and standard deviations of water absorbency and stress state. The water immersion resulted in an increase in weight of all tested materials. The water sorption (weight %) increased for Beautifil Flow F02 and Ionosit up to 3 wt % (Figures 3 and 6). Twinky Star showed the lowest value of absorption after 2016 h (84 days) (Figure 8).

3.2. Water Sorption and Solubility

Mean values of water sorption and solubility were presented in Table 3. Ionosit and Beautifil Flow F02 exhibited the highest, while Beautifil Bulk Fill Flow the lowest values of water sorption. The water sorption of the selected RMGI amounted to more than 115 μg/mm^3.

Table 3. Stress state before (0.5 h) and after 2016 h (84 days) of water immersion, contraction stress drop, absorbency and solubility of tested materials.

Material	Stress State [MPa]		Absolute Values of Stress Changes [MPa]	Contraction Stress Drop [%]	Sorption [μg/mm^3]	Solubility [μg/mm^3]
	0.5 h	2016 h				
Beautifil Flow F02	16.7 ± 0.9	-4.7 ± 0.8	21.4	128 *	45.9 ± 2.1	0.2 ± 0.1
Beautifil Bulk Fill Flow	11.1 ± 0.4	4.2 ± 0.9	6.9	62	13.6 ± 0.4	0.9 ± 0.3
Beautifil Flow Plus F00	12.8 ± 0.4	0.0 ± 0.2	12.8	100	26.4 ± 1.0	0.5 ± 0.1
Dyract eXtra	7.8 ± 0.1	1.6 ± 0.1	6.2	79	16.5 ± 0.5	2.9 ± 0.7
Compoglass Flow	10.4 ± 0.9	-0.4 ± 0.2	10.8	104 *	28.9 ± 0.8	2.3 ± 0.5
Ionosit	13.4 ± 1.1	-4.7 ± 0.2	18.1	135 *	103.8 ± 0.9	3.0 ± 0.2
Glasiosite	9.5 ± 0.2	-0.5 ± 0.2	10.0	105 *	16.4 ± 0.8	1.3 ± 0.1
TwinkiStar	7.9 ± 0.2	0.0 ± 0.2	7.9	100	17.7 ± 0.4	1.6 ± 0.8

* materials with over-compensated water expansion.

3.3. Photoelastic Study

The polymerization shrinkage stress for RMGI (Fuji II LC and Ionolux) was observed after 30 min. Based on photoelastic images, additional isochromes around restoration appeared after 24 h (Figure 9). They occurred during the transition from the contraction stress to the expansion stress (resulting from the material sorption). However, isochromes generated by contraction were still visible and the analysis of the state stress changes using this research method was impossible. It might be assumed that the material undergone remodeling process during the acid-base reaction [6] after 24 h.

In contrast, the changes in the stress state resulting from the water sorption in compomers and giomers was feasible. The results were presented in Sections 3.1–3.3.

All ion-releasing restorative materials shrank during hardening process and exhibited the associated contraction stress. The significant decrease in the stress state was observed due to hygroscopic expansion of materials (Figures 1–8). The absolute values of stress changes amounted from 6.2 up to 21.4 MPa (Table 3). Upon water aging, most of the tested materials showed total decrease in stress state (100%) and four materials showed additional stress characterized by the opposite direction of forces (expansion stress).

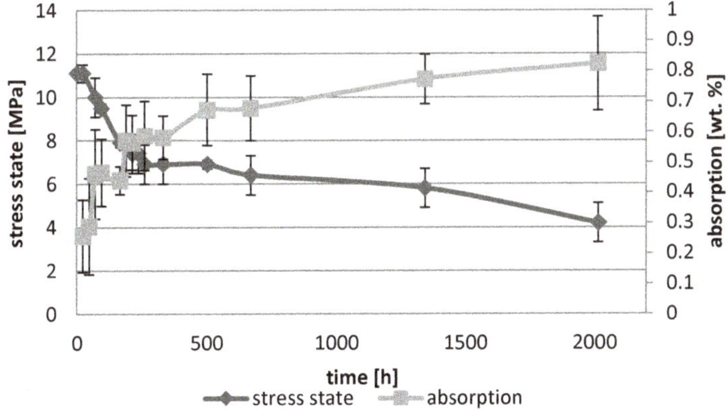

Figure 1. The relationship of absorption and stress state during water ageing (2016 h) of Beautifil Bulk Flow.

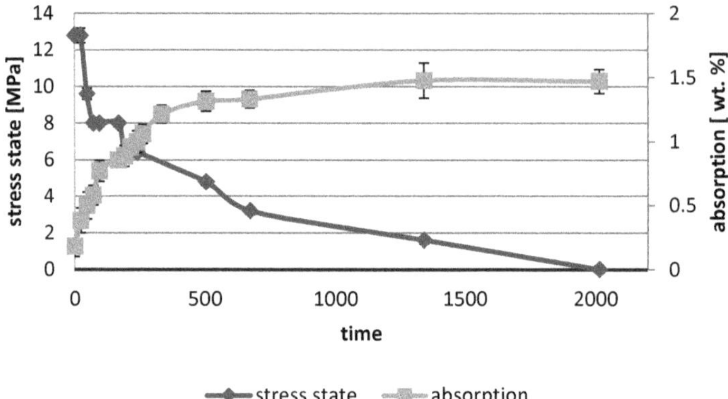

Figure 2. The relationship of absorption and stress state during water ageing (2016 h) of Beautifil Flow Plus F00.

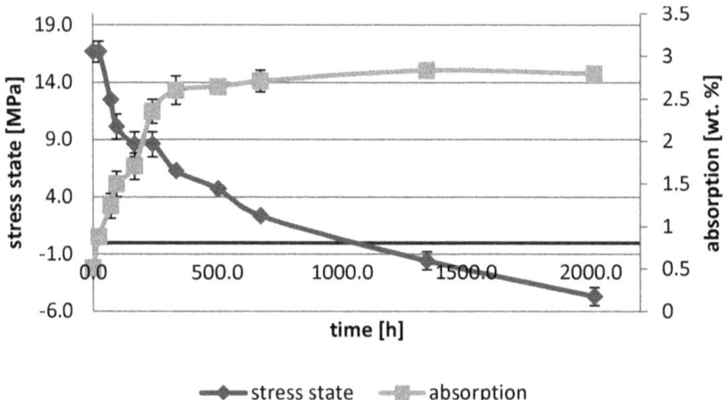

Figure 3. The relationship of absorption and stress state during water ageing (2016 h) of Beautifil Flow F02.

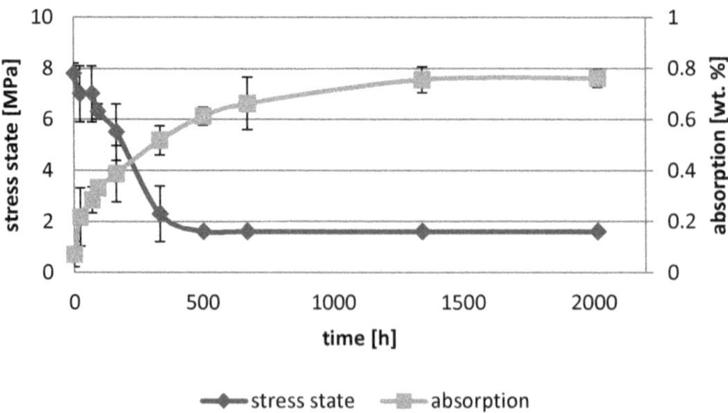

Figure 4. The relationship of absorption and stress state during water ageing (2016 h) of Dyract eXtra.

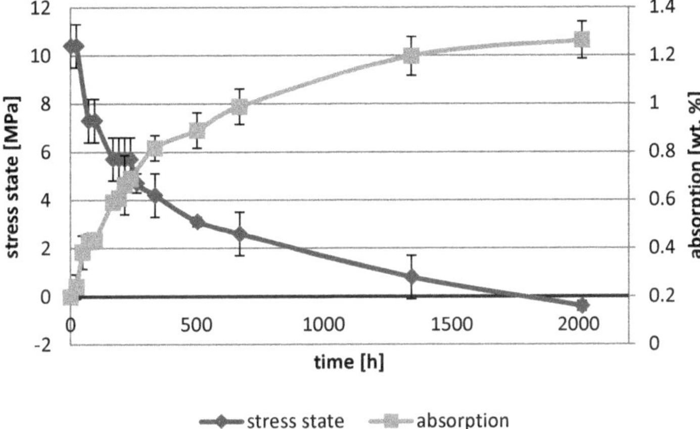

Figure 5. The relationship of absorption and stress state during water ageing (2016 h) of Compoglass Flow.

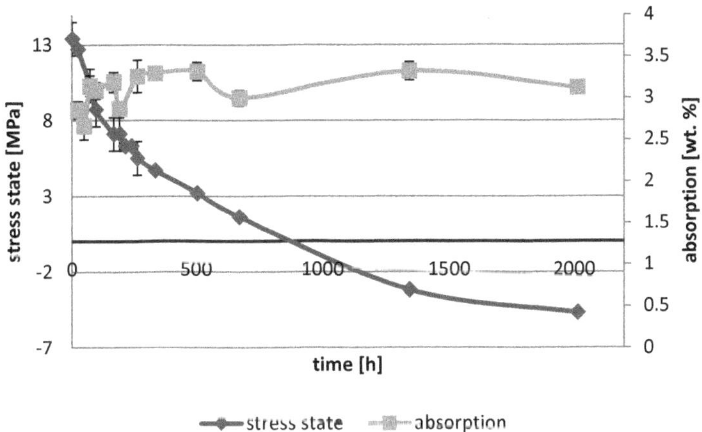

Figure 6. The relationship of absorption and stress state during water ageing (2016 h) of Ionosit.

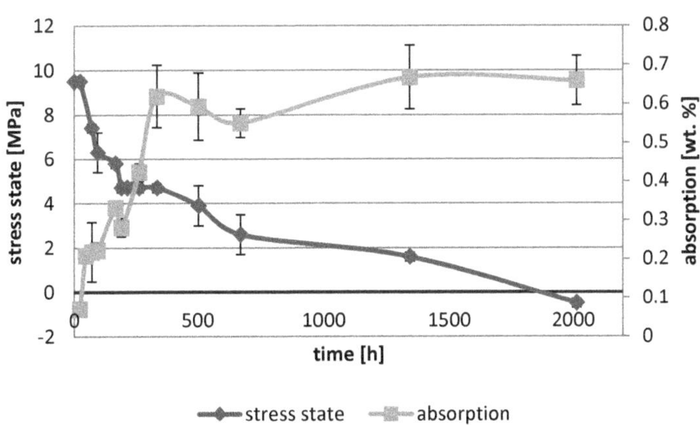

Figure 7. The relationship of absorption and stress state during water ageing (2016 h) of Glasiosite.

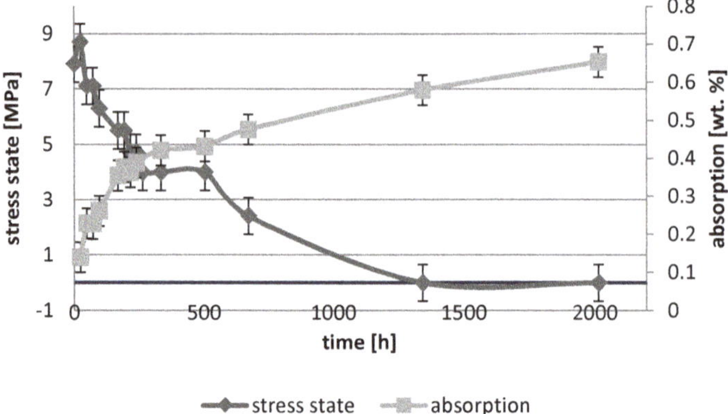

Figure 8. The relationship of absorption and stress state during water ageing (2016 h) of Twinky Star.

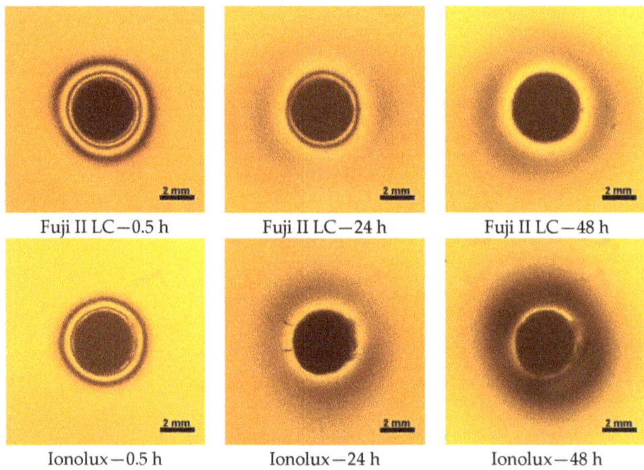

Figure 9. Isochromes in epoxy plate around Fuji II LC and Ionolux 0.5 h and after 24 and 48 h of water storage.

4. Discussion

Ion-releasing restorative materials are a substantial part of modern restorative dentistry. Released fluoride may be incorporated into the surrounding enamel or dentin with anti-cariogenic effect. These materials complement their fluoride reservoir from the oral cavity by building in fluoride delivered via rinses, toothpastes or topical application of fluoride preparations. Thus, the filling is a local fluoride storage that is used when the decrease in pH level occurs [17,18]. However, these materials exhibit significant water sorption, which may affect the formation of expansion stress endangering the integrity of tooth structure. Therefore, the profound knowledge of the rate and extent of water sorption of these materials is of paramount importance.

In the present study, the overall results indicated that the ion-releasing restorative materials showed greater stress decrease as compared with resin composites [12]. It might be associated with the process of ion release due to various types of fluoride glasses used as fillers. However, no significant difference between giomers and compomers was found. Discrepancies occurred between individual materials and depended on the material composition. Giomers consisted mainly of bis-GMA/TEGDMA resin matrix and multifunctional glass filler based on alumino-fluoro-borosilicate glass (Table 1). The differences resulted primarily from the filler amount in the materials. It is worth emphasizing, that the increase in filler content caused the decrease in sorption and solubility of materials. These phenomena might be explained by the reduction of free volume in polymer matrix [19,20]. Beautifil Bulk Fill Flow showed the lowest sorption value (13.6 $\mu g/mm^3$) due to the addition of UDMA and bis-MEPEPP resins. Both resins exhibited lower water sorption than bis-GMA and TEGDMA [21,22]. The strong relationship between the value of water sorption and the change in the stress state of giomers was observed; the higher the water sorption in the giomer, the greater the relaxation effect of stress (resulting from the plasticization effect and hydroscopic expansion). Beautifil Flow F02 showed high values of sorption and expansion stress after 1344 h of water ageing. The water sorption of Beautifil Flow Plus F00 amounted up to 26.4 $\mu g/mm^3$.

The water absorption of any material was controlled by a variety of mechanisms, but the osmotic effect led to an increased radial pressure. Since the Beautifil Flow F02 (giomer) demonstrated significant swelling along with the greatest radial pressure, it might be assumed that this material - more likely than any other studied material - was able to generate a significant osmotic effect. The main difference in microstructure between the giomers and compomers was the presence of pre-reacted glass-polyacid zones in the giomer structure (as a part of the filler) [23]. It seemed likely that these zones were responsible for generating the osmotic effect which led to swelling and pressure generation. Similar zones could be potentially formed at the surface of compomer glass particles following the delayed acid-base reaction which occurred due to water absorption. The latter reaction (in compomers) was significantly limited in extent compared with the almost total consumption of glass (in giomers). However, Ionosit (compomer) produced a significant radial pressure after 56 days of water storage. These results corresponded with the study of McCabe at al., showing compomer (F2000) exhibiting high displacement force and radial pressure due to water storage for 1 month [24].

Most compomers and giomers showed an extended process of water sorption. UDMA and a high degree of filler content were used to limit water sorption [19]. The high water sorption could be explained by the occurrence of later acid-base reaction in giomers. The setting process of compomers, in contrast to GIC, was based on photopolymerization and chemical reactions between glass particle fillers, while the reaction of monomers with acid functional group was only additional [25]. Compomers contained hydrophilic resins that were copolymerized with the more hydrophobic urethane dimethacrylate or bis-GMA [23]. The initial water uptake of these materials was lower than that of RMGICs, however, it was considerably higher than that of resin composites [12,14]. While compomers were initially anhydrous when polymerized, the dynamic synergistic interaction between the ion-leachable fillers and the acidic resin phase (resin matrix) was initiated upon water sorption [26]. The acid-base reaction in compomers invariably led to the formation of hydrogel layers around the glass fillers and an irreversible increase in the amount of 'firmly-bound' water [26]. The additional phase of resin matrix expansion to accommodate the swollen hydrogel layers was demanded. Such a phenomenon explained high water absorption of compomers and high hydroscopic expansion. Ionosit showed the highest values of water sorption among all tested materials. After polymerization the volume of material increased and this process compensated the polymerization contraction of polymer phase. This feature might be explained by the increased water sorption. Dyract eXtra differed from all tested compomers: low water absorption value and slow, consequent reduction in contraction stress only up to 1.6 MPa after 21 days of water immersion. These results could be explained by low filler content in comparison to the other compomers.

The photoelastic method turned out to be inadequate to evaluate stress changes of RMGI (Fuji II LC and Ionolux). Consequently, a different method allowing for observation of the state stress development in RMGI during water aging is highly needed. An alternative measurement technique might be a tensiometer [27,28], strain gauges [29,30] and a stress–strain analyser [31,32].

A major drawback associated with resin dental materials was polymerization shrinkage. It could cause a marginal gap formation. Water sorption might help compensate the hygroscopic expansion [33]. Moreover, it also relieved polymerization shrinkage stress generated along cavity walls during the initial setting stage. However, water absorbed into resin matrices, acts as a plasticizer adversely effecting on the materials' physical properties [34]. Furthermore, the excessive expansion could create stress that might possibly result in undesirable cuspal flexure and fracture of unsupported tooth structure [23,35].

5. Conclusions

The water sorption, solubility and absorption dynamic of ion-releasing restorative materials were material dependent properties. The overall results indicated that the tested restorative materials showed significant stress decrease. The total reduction in contraction stress or water expansion stress was not observed for materials with low value of water sorption (Beautifil Bulk Fill, Dyract eXtra, Glasionosit and Twinky Star). The photoelastic method turned out to be inadequate to evaluate stress changes of RMGI.

Author Contributions: M.L.S., K.B. and J.S. conceived and designed the experiments; A.S., M.K. performed the experiments; M.D., M.F.-W. and K.S. analyzed the data; K.S., M.L.S., A.S. and K.B. wrote the paper.

Funding: The financial support of this work by Medical University of Lodz, Poland (grant no. 502-03/2/148/04/502-24-082-18) is gratefully acknowledged.

Conflicts of Interest: The authors declare no conflict of interest. The founding sponsors had no role in the design of the study; in the collection, analyses, or interpretation of data; in the writing of the manuscript, and in the decision to publish the results.

References

1. Cramer, N.B.; Stansbury, J.W.; Bowman, C.N. Recent Advances and Developments in Composite Dental Restorative Materials. *J. Dent. Res.* **2011**, *90*, 402–416. [CrossRef] [PubMed]
2. Cury, J.A.; De Oliveira, B.H.; Dos Santos, A.P.P.; Tenuta, L.M.A. Are Fluoride Releasing Dental Materials Clinically Effective on Caries Control? *Dent. Mater.* **2016**, *32*, 323–333. [CrossRef] [PubMed]
3. Sidhu, S.; Nicholson, J. A Review of Glass-Ionomer Cements for Clinical Dentistry. *J. Funct. Biomater.* **2016**, *7*, 16. [CrossRef] [PubMed]
4. Lohbauer, U. Dental Glass Ionomer Cements as Permanent Filling Materials?—Properties, Limitations and Future Trends. *Materials* **2010**, *3*, 76–96. [CrossRef]
5. Najeeb, S.; Khurshid, Z.; Zafar, M.S.; Khan, A.S.; Zohaib, S.; Martí, J.M.N.; Sauro, S.; Matinlinna, J.P.; Rehman, I.U. Modifications in Glass Ionomer Cements: Nano-Sized Fillers and Bioactive Nanoceramics. *Int. J. Mol. Sci.* **2016**, *17*, 1134. [CrossRef] [PubMed]
6. Dinakaran, S. Sorption and Solubility Characteristics of Compomer, Conventional and Resin Modified Glass-Ionomer Immersed in Various Media. *J. Dent. Med. Sci.* **2014**, *13*, 41–45. [CrossRef]
7. Ruse, N.D. What Is a "Compomer"? *J. Can. Dent. Assoc. (Tor)* **1999**, *65*, 500–504.
8. Meyer, J.M.; Cattani-Lorente, M.A.; Dupuis, V. Compomers: Between Glass-Ionomer Cements and Composites. *Biomaterials* **1998**, *19*, 529–539. [CrossRef]
9. Abdel-karim, U.M.; El-Eraky, M.; Etman, W.M. Three-Year Clinical Evaluation of Two Nano-Hybrid Giomer Restorative Composites. *Tanta Dent. J.* **2014**, *11*, 213–222. [CrossRef]
10. Ferracane, J.L.; Hilton, T.J. Polymerization Stress—Is It Clinically Meaningful? *Dent. Mater.* **2016**, *32*, 1–10. [CrossRef] [PubMed]
11. Domarecka, M.; Sokołowski, K.; Krasowski, M.; Łukomska-Szymańska, M.; Sokołowski, J. The Shrinkage Stress of Modified Flowable Dental Composites. *Dent. Med. Probl.* **2015**, *52*, 424–433. [CrossRef]
12. Bociong, K.; Szczesio, A.; Sokolowski, K.; Domarecka, M.; Sokolowski, J.; Krasowski, M.; Lukomska-Szymanska, M. The Influence of Water Sorption of Dental Light-Cured Composites on Shrinkage Stress. *Materials* **2017**, *10*, 1142. [CrossRef] [PubMed]
13. Sokolowski, G.; Szczesio, A.; Bociong, K.; Kaluzinska, K.; Lapinska, B.; Sokolowski, J.; Domarecka, M.; Lukomska-Szymanska, M. Dental Resin Cements—The Influence of Water Sorption on Contraction Stress Changes and Hydroscopic Expansion. *Materials* **2018**, *11*, 973. [CrossRef] [PubMed]
14. Domarecka, M.; Sokołowski, K.; Krasowski, M.; Szczesio, A.; Bociong, K.; Sokołowski, J.; Łukomska-Szymańska, M. Influence of Water Sorption on the Shrinkage Stresses of Dental Composites Wpływ Sorpcji Wody Na Naprężenia Skurczowe Materiałów Kompozytowych. *J. Stomatol.* **2016**, *69*, 412–419.

15. PN-EN ISO 4049:2003. Warszawa 2003. Available online: http://sklep.pkn.pl/pn-en-iso-4049-2003p.html (accessed on 28 August 2018).
16. Müller, J.A.; Rohr, N.; Fischer, J. Evaluation of ISO 4049: Water Sorption and Water Solubility of Resin Cements. *Eur. J. Oral Sci.* **2017**, *125*, 141–150. [CrossRef] [PubMed]
17. Wiegand, A.; Buchalla, W.; Attin, T. Review on Fluoride-Releasing Restorative Materials-Fluoride Release and Uptake Characteristics, Antibacterial Activity and Influence on Caries Formation. *Dent. Mater.* **2007**, *23*, 343–362. [CrossRef] [PubMed]
18. Davis, H.B.; Gwinner, F.; Mitchell, J.C.; Ferracane, J.L. Ion Release from, and Fluoride Recharge of a Composite with a Fluoride-Containing Bioactive Glass. *Dent. Mater.* **2014**, *30*, 1187–1194. [CrossRef] [PubMed]
19. Ferracane, J.L. Hygroscopic and Hydrolytic Effects in Dental Polymer Networks. *Dent. Mater.* **2006**, *22*, 211–222. [CrossRef] [PubMed]
20. Ferracane, J.L.; Palin, W.M. 10—Effects of Particulate Filler Systems on the Properties and Performance of Dental Polymer Composites. In *Non-Metallic Biomaterials for Tooth Repair and Replacement*; Vallittu, P., Ed.; Woodhead Publishing Limited: Cambridge, UK, 2013.
21. Sideridou, I.; Tserki, V.; Papanastasiou, G. Study of Water Sorption, Solubility and Modulus of Elasticity of Light-Cured Dimethacrylate-Based Dental Resins. *Biomaterials* **2003**, *24*, 655–665. [CrossRef]
22. Kawaguchi, M.; Fukushima, T.; Horibe, T. Mechanical and Physical Properties of 2, 2'-Bis (4-Methacryloxy Polyethoxyphenyl) Propane Polymers. *Dent. Mater. J.* **1987**, *6*, 148–155, 224. [CrossRef] [PubMed]
23. Huang, C.; Kei, L.; Wei, S.H.Y.; Cheung, G.S.P.; Tay, F.R.; Pashley, D.H. The Influence of Hygroscopic Expansion of Resin-Based Restorative Materials on Artificial Gap Reduction. *J. Adhes. Dent.* **2002**, *4*, 61–71. [PubMed]
24. McCabe, J.F.; Rusby, S. Water Absorption, Dimensional Change and Radial Pressure in Resin Matrix Dental Restorative Materials. *Biomaterials* **2004**, *25*, 4001–4007. [CrossRef] [PubMed]
25. Hse, K.M.; Leung, S.K.; Wei, S.H. Resin-Ionomer Restorative Materials for Children: A Review. *Aust. Dent. J.* **1999**, *44*, 1–11. [PubMed]
26. Eliades, G.; Kakaboura, A.; Palaghias, G. Acid-Base Reaction and Fluoride Release Profiles in Visible Light-Cured Polyacid-Modified Composite Restoratives (Compomers). *Dent. Mater.* **1998**, *14*, 57–63. [CrossRef]
27. Davidson, C.L.; de Gee, A.J. Relaxation of Polymerization Contraction Stresses by Flow in Dental Composites. *J. Dent. Res.* **1984**, *63*, 146–148. [CrossRef] [PubMed]
28. Davidson, C.L.; de Gee, A.J.; Feilzer, A.; De Gee, A. The Competition between the Composite-Dentin Bond Strength and the Polymerization Contraction Stress. *J. Dent. Res.* **1984**, *63*, 1396–1399. [CrossRef] [PubMed]
29. Sakaguchi, R.L.; Peters, M.C.; Nelson, S.R.; Douglas, W.H.; Poort, H.W. Effects of Polymerization Contraction in Composite Restorations. *J. Dent.* **1992**, *20*, 178–182. [CrossRef]
30. Suiter, E.A.; Watson, L.E.; Tantbirojn, D.; Lou, J.S.B.; Versluis, A. Effective Expansion: Balance between Shrinkage and Hygroscopic Expansion. *J. Dent. Res.* **2016**, *95*, 543–549. [CrossRef] [PubMed]
31. Chen, H.Y.; Manhart, J.; Hickel, R.; Kunzelmann, K. Polymerization Contraction Stress in Light-Cured Packable Composite Resins. *Dent. Mater* **2001**, *17*, 253–259. [CrossRef]
32. Chen, H.Y.; Manhart, J.; Kunzelmann, K.H.; Hickel, R. Polymerization Contraction Stress in Light-Cured Compomer Restorative Materials. *Dent. Mater.* **2003**, *19*, 597–602. [CrossRef]
33. Versluis, A.; Tantbirojn, D.; Lee, M.S.; Tu, L.S.; Delong, R. Can Hygroscopic Expansion Compensate Polymerization Shrinkage? Part I. Deformation of Restored Teeth. *Dent. Mater.* **2011**, *27*, 126–133. [CrossRef] [PubMed]
34. Cattani-Lorente, M.A.; Dupuis, V.; Payan, J.; Moya, F.; Meyer, J.M. Effect of Water on the Physical Properties of Resin-Modified Glass Ionomer Cements. *Dent. Mater.* **1999**, *15*, 71–78. [CrossRef]
35. Momoi, Y.; McCabe, J.F. Hygroscopic Expansion of Resin Based Composites during 6 Months of Water Storage. *Br. Dent. J.* **1994**, *176*, 91–96. [CrossRef] [PubMed]

 *polymers*

Article

Design and Biophysical Characterization of Poly (L-Lactic) Acid Microcarriers with and without Modification of Chitosan and Nanohydroxyapatite

Liying Li [1], Kedong Song [1,*], Yongzhi Chen [1], Yiwei Wang [2], Fangxin Shi [3], Yi Nie [4,*] and Tianqing Liu [1,*]

[1] State Key Laboratory of Fine Chemicals, Dalian R&D Center for Stem Cell and Tissue Engineering, Dalian University of Technology, Dalian 116024, China; liyingli@mail.dlut.edu.cn (L.L.); chenyongzhi@mail.dlut.edu.cn (Y.C.)

[2] Burns Research Group, ANZAC Research Institute, Concord, University of Sydney, Sydney, NSW 2139, Australia; eweiwang@hotmail.com

[3] Zhengzhou Institute of Emerging Technology Industries, Zhengzhou 450000, China; Shifangxin1122@163.com

[4] Key Laboratory of Green Process and Engineering, Institute of Process Engineering, Chinese Academy of Sciences, Beijing 100190, China

* Correspondence: Kedongsong@dlut.edu.cn (K.S.); ynie@ipe.ac.cn (Y.N.); Liutq@dlut.edu.cn (T.L.); Tel.: +86-411-84706360 (K.S.)

Received: 25 August 2018; Accepted: 23 September 2018; Published: 25 September 2018

Abstract: Nowadays, microcarriers are widely utilized in drug delivery, defect filling, and cell culture. Also, many researchers focus on the combination of synthetic and natural polymers and bioactive ceramics to prepare composite biomaterials for tissue engineering and regeneration. In this study, three kinds of microcarriers were prepared based on physical doping and surface modification, named Poly (L-lactic) acid (PLLA), PLLA/nanohydroxyapatite (PLLA/nHA), and PLLA/nHA/Chitosan (PLLA/nHA/Ch). The physicochemical properties of the microcarriers and their functional performances in MC3T3-E1 cell culture were compared. Statistical results showed that the average diameter of PLLA microcarriers was 291.9 ± 30.7 μm, and that of PLLA/nHA and PLLA/nHA/Ch microcarriers decreased to 275.7 ± 30.6 μm and 269.4 ± 26.3 μm, respectively. The surface roughness and protein adsorption of microcarriers were enhanced with the doping of nHA and coating of chitosan. The cell-carrier cultivation stated that the PLLA/nHA microcarriers had the greatest proliferation-promoting effect, while the PLLA/nHA/Ch microcarriers performed the strongest attachment with MC3T3-E1 cells. Besides, the cells on the PLLA/nHA/Ch microcarriers exhibited optimal osteogenic expression. Generally, chitosan was found to improve microcarriers with superior characteristics in cell adhesion and differentiation, and nanohydroxyapatite was beneficial for microcarriers regarding sphericity and cell proliferation. Overall, the modified microcarriers may be considered as a promising tool for bone tissue engineering.

Keywords: microcarriers; Poly (L-lactic) acid; Chitosan; nanohydroxyapatite; osteoblasts

---

1. Introduction

Since the concept of microcarriers was first proposed by Van Wezel in 1967 [1], microcarriers with particle size in a certain micrometer range have received extensive attention and development in drug delivery, defect filling, and three-dimensional cell culture [2–8]. Compared to conventional monolayer cell culture and three-dimensional static culture with a porous scaffold, utilizing microcarriers can increase the specific surface area and achieve the uptake of cells attached on their curved surface within a short period of time. Also, microcarriers are capable of integrating with bioreactors to create a dynamic environment, and cell delivery and passage can be realized by supplementing fresh microcarriers, thereby avoiding the damage caused by enzymatic digestion [9–12]. Furthermore, some specific microcarriers can also be used as injectable materials to repair defective tissues in vivo [13–15].

The biological materials for the fabrication of these microcarriers are mainly derived from natural macromolecule materials, bioactive ceramics (hydroxyapatite, tricalcium phosphate, and glass/ceramic), and biodegradable synthetic polymers [16–18]. For example, Lai et al prepared gelatin microcarriers which were functionalized with oxidized hyaluronic acid and their study suggested that the microcarriers have good compatibility with rabbit corneal cells [19]; Song et al [20] grafted thermosensitive polymers onto the surface of glass microcarriers to improve cell adhesion and preserve osteoblasts and BM-MSCs (Rat Bone Marrow Mesenchymal Stem Cells) biological properties after consecutive passages with the thermal-liftoff method. Levato et al [21] made bone marrow mesenchymal stem cells attach to the surface of poly (L-lactic) acid microcarriers, and created an osteochondral double sphere model using gelatin acrylamide gel bio-ink with the help of 3D printing to promote the bidirectional differentiation of bone and cartilage. Among the synthetic materials used in the preparation of microcarriers, poly (L-lactic) acid has been developed rapidly in tissue engineering because of its good biodegradability and mechanical strength; usually it needs to be modified by some kind of surface treatment such as plasma and laser treatment, etching, grafting, etc. because of its poor hydrophilicity and low cytocompatibility [22–26]. As the main inorganic component of the human skeleton, HA (hydroxyapatite) can be firmly combined with bone tissue in vivo and coexist harmoniously with cells. The organic combination of poly (L-lactic) acid and nHA can lead to outstanding biological performance. On one hand, nHA can improve the osteogenic differentiation of adherent cells; on the other hand, it can partially neutralize the acidic degraded products of poly (L-lactic) acid. Studies have proved that the combination of hydroxyapatite and biopolymer materials can have complementary advantages for a specific application [27–31]. Besides, chitosan, as one of the injectable materials, also has good biocompatibility and biodegradability, while chitosan also has the ability to further strengthen the surface affinity of poly (L-lactic) acid [32–34]. For instance, Gao et al. [35] modified the surface of poly (L-lactic) acid microcarriers with chitosan, thus enhancing the adhesion and proliferation of chondrocytes.

Although utilizing microcarriers for suspension cultures can enhance cell proliferation and solve the difficulty that some anchorage-dependent cells cannot grow directly in the bioreactors, there are still some problems in the production of microcarriers, such as poor cellular affinity, high cost, and complicated preparation process. Therefore, innovative exploration and development are still needed to manufacture cost-effective and biocompatible microcarriers for practical application. In this study, PLLA, PLLA/nHA, and PLLA/nHA/Ch microcarriers were prepared based on physical doping and surface modification using gelatin as the dispersant. The differences of physicochemical properties of the three kinds of microcarriers and their functional performance in osteoblasts-like cell culture were compared. Through the improvement of PLLA microcarriers to enhance its biological function, this study may provide some reference value for the development of bone tissue engineering.

2. Materials and Methods

2.1. Materials

Murine MC3T3-E1 Subclone14 cells were purchased from the Cell bank of the Chinese Academy of Sciences. Poly (L-lactic) acid (PLLA, MW 100,000 Da) was purchased from Jinan Daigang Biomaterial Co., Ltd. (Jinan, China) Silver nitrate, gelatin (analytical agent, the purity was greater than 99.5%), chitosan (the degree of deacetylation was around 95%), and nanohydroxyapatite (nHA) were purchased from Beijing Coolaber Science & Technology Co., Ltd. (Beijing, China) 1-(3-dimethylaminopropyl)-3-ethylcarbodiimide hydrochloride (EDC), morpholine ethane sulfonic acid (MES), and N-hydroxy-succinamide (NHS) were obtained from Shanghai Macklin Biochemical Technology Co., Ltd. (Shanghai, China) Cell Counting Kit-8 (CCK-8) was purchased from Dojindo

Laboratories (Kumamoto, Japan). Calcein-AM, Hoechst 33258, and Propidium Iodide (PI) were purchased from Sigma-Aldrich Inc. (St. Louis, MO, USA). Alkaline phosphatase assay kit, BCA (Bicinchoninic Acid) protein assay kit, and Alizarin Red S reagent were purchased from Nanjing Jiancheng Bioengineering Institute (Nanjing, China). All reagents and solvents were at analytical grade. MC3T3-E1 Cells were cultured in α-Dulbecco's Modified Eagle's Medium (α-DMEM, Hyclone, Logan, UT, USA) containing 10% fetal bovine serum (FBS, Minhai, China), 100 units/mL penicillin and 100 µg/mL streptomycin (Hyclone, Logan, UT, USA) in a humidified 5% CO_2 atmosphere at 37 °C.

2.2. Preparation and Characterization of PLLA, PLLA/nHA and PLLA/nHA/Ch Microcarriers

Poly (L-lactic) acid was fully dissolved in dichloromethane at a ratio of 1:10 (W/V) prior to being added into gelatin aqueous solution which was more biocompatible than other polymer dispersants. Then, the gelatin-poly (L-lactic) acid suspension was stirred for 5 h. The liquid beads gradually solidified and precipitated at the bottom of the beaker with the volatilization of organic solvent. Finally, the wet microcarriers were collected and dried in a vacuum drying oven.

A similar protocol was used to prepare PLLA/nHA microcarriers. A certain amount of nHA was added to the polymer solution, followed by mechanical agitation for two hours with a lid on to reduce the volatilization of dichloromethane. The following steps were the same as above.

Preparation of PLLA/nHA/Ch microcarriers: The PLLA/nHA microcarriers were firstly soaked in 1 M sodium hydroxide solution, and then repeatedly rinsed with 0.1 M dilute hydrochloric acid solutions and water. The cleavage of ester bonds and exposure of carboxyl groups of PLLA can be induced by alkaline treatment. Next, the microcarriers were transferred to the crosslinking agent solution (EDC/NHS/MES) and soaked for two hours, followed by the addition of 1% chitosan acetic acid solution under 45 °C for 5 h. For this reason, the amide bonds were formed between the carboxyl groups of PLLA and the amino groups of chitosan. Then the microcarriers were soaked in EDC/NHS/MES solution again for 3 h and finally washed with 0.1 M disodium hydrogen phosphate solution and deionized water in triplicate [35].

2.3. Analysis of Particle Size Distribution

The effect of stirring speed (150 rpm, 200 rpm, 300 rpm), concentrations of PLLA (5%, 10%, 15%), gelatin (0.25%, 0.5%, 1%), and nHA (W/WPLLA: 0%, 10%, 20%) on the particle size distribution of microcarriers was analyzed. A total of 12 experiments were conducted to select suitable preparation conditions. Three different 40× images of microcarriers in each group were taken by an inverted phase contrast microscope. The diameters of 300 microspheres in each group were measured by Image-Pro Plus software (v7.0, Media Cybernetics, Inc., Bethesda, MD, USA), and the histogram of the particle size distribution of microcarriers in each group was drawn based on statistical calculation.

2.4. Morphology Observation and Component Analysis of Microcarriers

The surface profile and topography of microcarriers were observed using an inverted phase contrast microscope and scanning electron microscope (SEM, FEI Company, Hillsboro, OR, USA). Semi-quantitative analysis of specific elements of microcarriers was carried out by EDS energy spectrometry, and the content of nitrogen of microcarriers was analyzed using an elemental analyzer (VarioELIII, Elementar, Langenselbold, Germany). The specific chemical functional groups of microcarriers were detected by Fourier Transform Infrared Spectroscopy (FTIR, Nicolet 6700, Thermo Fisher Scientic Inc., Madison, WI, USA).

2.5. Protein Adsorption Determination

The BCA protein assay kit was used to determine the protein adsorption capacity of the three kinds of microcarriers. In brief, a certain volume of BSA (Bovine Serum Albumin) solution with microcarriers was set as an experimental group, while the control group was the same BSA solution without microcarriers. After incubation at 37 °C for 12 h, the supernatant was transferred to an EP (eppendorf) tube, followed by the addition of BCA working liquid. The absorbance of the assay solution at the emission wavelength of 562 nm was detected by an Enzyme-linked immune detector (Thermo scientific, Waltham, MA, USA). Then, the protein concentration was calculated according to the standard protein concentration curve. The protein adsorption capacity of microcarriers was measured by the following calculation method: the difference of protein mass between the control group and experimental group was normalized to the mass of microcarriers.

Three kinds of microcarriers of equal weight whose initial mass was taken as m_1 were immersed in phosphate buffer (PBS) of equal volume for 6 weeks. These microcarriers were collected periodically, and their dry weights were measured and regarded as m_2. The calculation formula of degradation rate was: $\frac{m_1 - m_2}{m_2} \times 100\%$.

2.6. MC3T3-E1 Cell Culture and Osteogenic Differentiation

The morphology of MC3T3-EI cells was observed by Hematoxylin-Eosin (HE) staining. One hundred microliters of cells suspension was seeded in a 24-well plate at a cell density of 2×10^4 cells/mL. After incubation for 3 h, the osteogenic medium (α-DMEM, 10%FBS, 50 µg/mL ascorbic acid, and 10 mmol/L beta sodium phosphate) was added. The osteogenic differentiation ability of MC3T3-E1 cells was determined by alkaline phosphatase (after incubation for one week), silver nitrate (Von-Kossa) (three weeks), and Alizarin Red S staining (four weeks).

2.7. Inoculation and Culture of MC3T3-E1 Cells on Microcarriers

The microcarriers were soaked in a 75% ethanol solution and simultaneously irradiated with ultraviolet light for 5 h, and finally rinsed with aseptic PBS 3 times. The sterilized microcarriers were reserved for cells culture. The MC3T3-E1 cells at passage 7 were used for the subsequent study. The three kinds of microcarriers of equal quality (50 mg/per well) were respectively placed in a low adhesion 24-well plate, followed by seeding with 50 µL of cell suspension with a density of 2×10^6 cells/mL. The medium was refreshed every 2 days. Then, the cells–microcarriers were stained with Calcein-AM/PI/Hoechst 33258 dye after cultured for 24 h and 120 h, which was observed under a fluorescence microscope. At the same time, the cells–microcarriers were immobilized after culturing for 24 h, 72 h, and 120 h with 0.25% glutaraldehyde solution and then dehydrated gradually by ethanol solution, followed by the SEM to investigate cell adhesion and growth. In addition, the process of cell adhesion and the subsequent spreading and proliferation on the surface of microcarriers was observed with an inverted phase contrast microscope after culturing for 30 min, 24 h, and 72 h. Further, trypsin digestion was used to study the process of cell desorption from the three kind microcarriers after 5 days culture. Besides, the detached cells were harvested and reseeded into 24-well plates whose viability was tested by live/dead staining after culturing for 24 h and 168 h.

2.8. Cell Adhesion, Proliferation, and Differentiation on Microcarriers

One hundred microliters of cell suspension was seeded on microcarriers in a 96-well plate at a cell density of 5×10^4 cells/well, while the cell suspension without microcarriers was set as the control group. To determine cell adhesion efficiency, in brief, supernatant and microcarriers were removed at 10 min, 20 min, and 30 min post seeding, followed by addition of 100 µL complete medium and 10 µL/well CCK-8 reagent, and incubation at 37 °C for 3 h. Then, the mixtures of culture medium and CCK-8 (100 µL) were used to measure optical density (OD) at the emission wavelengths of 562 nm and 620 nm via an Enzyme-linked immune detector. The cell adhesion rate was then calculated by comparing the difference in OD between the control group and the experimental group. Cell proliferation rate was further measured at 24 h, 72 h, and 120 h post cell seeding using the same method.

The MC3T3-E1 cells were inoculated directly onto the three kinds of microcarriers, and Alkaline Phosphatase (ALP) staining was performed after one week of osteogenic induction. Equally, MC3T3-E1 cells (1×10^5 cells/per well) were inoculated into the 24-well plate and cultured for 24 h prior to the addition of microcarriers, followed by ALP staining as above. Also, the three kinds of microcarriers with the same weights were respectively placed in a low adhesion 6-well plate, followed by the seeding of 100 μL of cell suspension with a density of 5×10^6 cells/mL. After culturing for 7 and 14 days, ALP activity of MC3T3-E1 cells was assayed using an alkaline phosphatase assay kit, and the result was normalized to the total protein content.

2.9. Statistical Analysis

All experimental results were expressed as mean ± SD (Standard Deviation). The statistical significance of differences within each group was evaluated by one-way ANOVA followed by the *t*-test analysis using OriginPro (v7.5, OriginLab Corporation, Northampton, MA, USA). It was regarded as a significant difference when the value of $p < 0.05$.

3. Results and Discussion

3.1. The Particle Size Distribution of Microcarriers

The average particle size of the microcarriers increased from 109.1 μm to 544.9 μm when the concentration of PLLA grew from 5% to 15% with the gelatin concentration at 0.25% and the stirring speed at 200 rpm (Figure $1A_1$–A_3). The reason was the oil phase became more viscous with the increasing concentration of PLLA, resulting in the formation of larger oil droplets through emulsion dispersion, thus the particle size of the microcarriers increased. This was consistent with some relevant studies [36,37]. For example, Cheng et al. [36] have examined the impact of phase viscosity on droplet size by using different octadecane-to-corn oil ratios in the oil phase and different glycerol-to-water ratios in the aqueous phase. Results showed that the mean droplet diameter plot increased with increasing viscosity ratio of the oil phase. McClements et al. [37] interpreted some potential mechanisms about the influence of oil phase viscosity on droplet size: increased droplet fragmentation due to increased disruptive shear stresses; decreased droplet re-coalescence due to decreased droplet collision frequency; increased droplet re-coalescence due to a reduction in emulsifier adsorption rate. When the PLLA concentration was fixed at 10% and the stirring speed at 200 rpm, the size drop from 316.0 μm to 129.2 μm was attributed to the increased concentration of gelatin from 0.25% to 1% (Figure $1B_1$–B_3). In general, if the solubility of PLLA remained unchanged, the viscosity and dispersion of the aqueous phase would increase with extra element of gelatin, therefore, the oil phase can be dispersed into small oil beads in a short time, and the resistance of the microcarriers to collide with each other magnified so that the probability of collision between each other was lowered. As a result, the particle size decreased with the fall of surface energy. Based on this principle, the surface energy of the oil phase drops can be further reduced by adding gelatin solution, resulting in a size drop. As shown in the diagram, increased stirring speed from 150 rpm to 300 rpm with fixed PLLA concentration at 10% and gelatin concentration at 0.25% (Figure $1C_1$–C_3), the particle size of microcarriers significantly reduced from 521 μm to 102 μm. The explanation for this finding was that the accelerated stirring speed gradually increased the shearing force of the solution, and hence the oil droplets were dispersed into a smaller size. With the addition of nHA content, at first, the diameter of microcarriers decreased slightly and later increased with the unceasing addition of nHA (Figure $1D_1$–D_3). nHA, as a hydrophilic compound, tended to cover the surface of the oil phase when added, and constantly attracted the water molecules to reduce the surface energy of the oil phase. Simultaneously, nHA can also increase the viscosity of the oil phase, resulting in the formation of larger oil droplets in the process of emulsification. Based on the above analysis, the following experimental conditions were selected for subsequent cell-microcarriers study: 10% PLLA, 0.25% gelatin, and 20% nHA with rotational speed of 200 rpm. The preparation of particles with smaller dimensions will continue to be explored in future experiments and be used for three-dimensional cell culture.

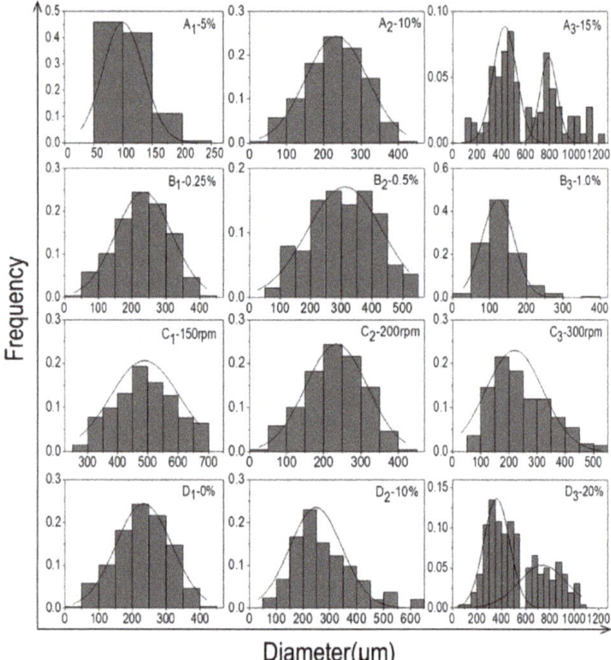

Figure 1. (A_1–A_3) Histograms of particle size distribution of PLLA microcarriers prepared with different concentrations of PLLA (A_1–A_3: 5%, 10%, 15%; the gelatin concentration was fixed at 0.25% and the stirring speed at 200 rpm); (B_1–B_3) histograms of the particle size distribution of PLLA microcarriers prepared by different concentrations of gelatin (B_1–B_3: 0.25%, 0.5%, 1%; the PLLA concentration was kept constant at 10% and stirring speed at 200 rpm); (C_1–C_3) histograms of particle size distribution of PLLA microcarriers prepared by different stirring speeds (C_1–C_3): 150 rpm, 200 rpm, 300 rpm; the concentration of PLLA and gelation was held unchanged at 10% and 0.25%, respectively); (D_1–D_3) histograms of particle size distribution of PLLA/nHA microcarriers prepared by different concentrations of nHA (D_1–D_3): 0%, 10%, 20%; the concentration of PLLA and gelation was kept unchanged at 10% and 0.25%, and stirring speed at 200 rpm).

3.2. Morphology, Component Analysis, Protein Adsorption, and Degradation of Microcarriers

The outlines of PLLA, PLLA/nHA, and PLLA/nHA/Ch microcarriers observed by inverted phase contrast microscopy are shown in Figure 2A–C. Macroscopically, the degree of sphericity of PLLA/nHA microcarriers was superior compared to the PLLA and PLLA/nHA/Ch microcarriers. The surface morphology of PLLA/nHA microcarriers with the alkaline treatment of 20 min and 10 min (Figure 2D,E) indicated that the surface erosion of microcarriers was affected by the time of alkaline treatment, also resulting in the formation of a serrated surface profile. The dissolution of microcarriers was aggravated with the extension of time. For this reason, the time of alkaline treatment was set to 10 min in the subsequent experiments. The microcarriers after crosslinking chitosan (Figure 2F) still kept the zigzag profile, which was not significantly different from that of alkaline-treated microcarriers. These microcarriers were screened in advance to remove the unformed impurities. Their average diameter was obtained by calculating the diameters of 100 particles (Table 1). It was found that the average diameter of PLLA microcarriers was 291.9 ± 30.7 µm, and the average diameter of PLLA/nHA microcarriers was 275.7 ± 30.6 µm when doped with a certain amount of nHA, while the average diameter of PLLA/nHA/Ch microcarriers turned to be 269.4 ± 26.3 µm. These phenomena explained that the addition of hydroxyapatite improved the sphericity and reduced the average diameter of

PLLA microcarriers to a certain extent, and the average diameter of PLLA/nHA/Ch microcarriers was further decreased and their surface roughness was increased by alkaline treatment.

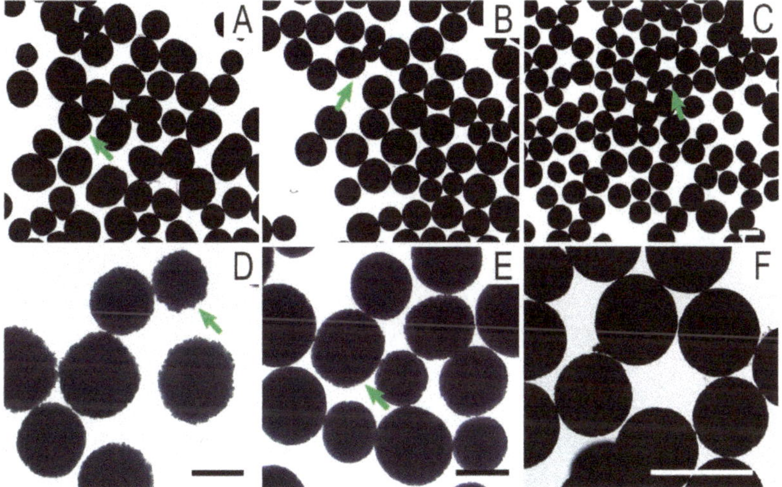

Figure 2. Outline drawings of microcarriers (**A**): PLLA microcarriers; (**B**): PLLA/nHA microcarriers; (**C,F**): PLLA/nHA/Ch microcarriers; (**D**): the PLLA/nHA microcarriers treated with sodium hydroxide for 20 min; (**E**): the PLLA/nHA microcarriers treated with sodium hydroxide for 10 min. (**A–C, D,E, F**: 40×, 100×, 200×; Scale: 250 μm).

Table 1. Analysis of mean diameter and sphericity of microcarriers.

Microcarriers	PLLA	PLLA/nHA	PLLA/nHA/Ch
Mean diameter/μm	291.9 ± 30.7	275.7 ± 30.6	269.4 ± 26.3
Sphericity	++	+++	+

Note: the sphericity of microcarriers was artificially judged by three viewers who were not involved in this experiment. The three plus signs indicated that the sphericity of the microcarriers was the best.

SEM observation of PLLA and PLLA/nHA microcarriers (Figure 3A,B) indicated that the presence of nHA changed the surface morphology of PLLA microcarriers resulting in a porous surface. The mechanism of emulsion solvent evaporation has been investigated by Rosca et al. [38]. It was concluded in their research that emulsion solvent evaporation (ESE) was mainly a two-step process: the emulsification of a polymer solution containing the encapsulated substance, followed by particle hardening through solvent evaporation and polymer precipitation according to the optical microscopic observations corroborated with laser diffractometry analysis. The second step, with the solvent transporting out from the emulsion droplets, determined the particle morphology. In addition, Hong et al. [39] thought the possible reason for the morphology of the PLLA microcarriers in this manuscript was that the shear stress might create defects during the solvent evaporation process. Also, small pores appeared on the surface of the microcarriers when they mixed the n-hexane with PLLA/dichloromethane solution. They believed that it was caused by the removal of the non-solvent. Therefore, in this study, the morphological differences between PLLA and PLLA/nHA microcarriers should be attributed to the hydrophilicity of nHA and its effect on the volatilization of dichloromethane, thereby leading to the formation of many uniformly distributed small and closed pores. The internal structure of PLLA/nHA microcarriers was more compact with prominent white luminous spots of agglomerated nHA (Figure 3B$_4$). The topography of alkaline-treated microcarriers (10 min) and chitosan cross-linking in Figure 3C,D showed that the surface of PLLA/nHA/Ch microcarriers became uneven, and the closed pores formed by solvent evaporation were interconnected, which increased the specific area of the microcarriers and enhanced the surface roughness.

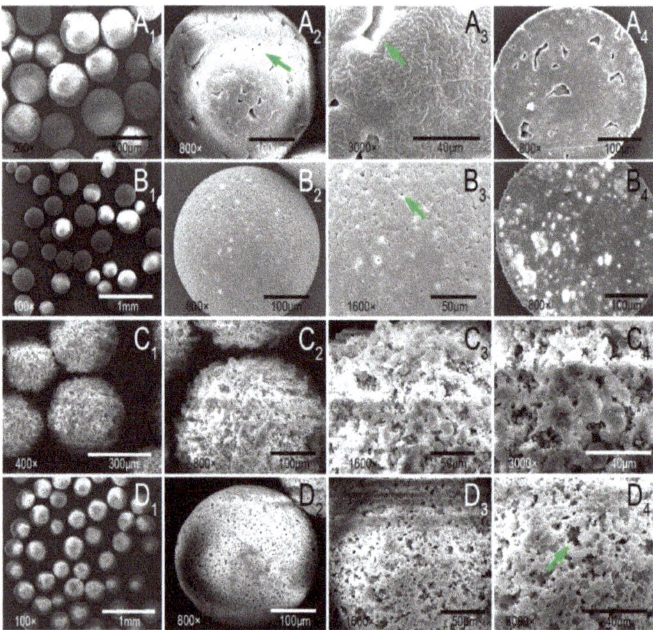

Figure 3. (A_1–A_3): The surface morphology of PLLA microcarriers; (B_1–B_3): The surface morphology of PLLA/nHA microcarriers; (A_4,B_4): The cross-sectional morphology of PLLA and PLLA/nHA microcarriers; (C_1–C_4): The surface morphology of PLLA/nHA microcarriers after alkaline treatment of 10 min; (D_1–D_4): The surface morphology of PLLA/nHA/Ch microcarriers coated with chitosan.

The EDS spectrometer provided a semi-quantitative elemental analysis of the specific area of a sample whose accuracy was from 1% to 5% and depth from 1 μm to 5 μm. It cannot guarantee that the same square box for PLLA and PLLA/nHA/Ch samples was drawn. Hence, each sample was determined in triplicate, and only one set of the three tests was displayed in Figure 4A. Meanwhile, as shown in the table of Figure 4A, the mean ± SD values of the specific elements of the intercepted region of microcarriers, which was calculated based on the triplicate tests of each sample, illustrated that the addition of nHA and crosslinking of chitosan had been successfully realized because calcium, phosphorus, and nitrogen were detected in the PLLA/nHA/Ch microcarriers, but not in the PLLA microcarriers. A quantitative analysis of nitrogen is listed in Table 2, which further stated that the presence of chitosan as nitrogen was not detected in either PLLA or PLLA/nHA microcarriers. The Fourier transform infrared spectrum of the characteristic chemical groups of microcarriers is shown in Figure 4B. At 568 cm^{-1} and 608 cm^{-1}, the characteristic absorption peaks of phosphate groups appeared, indicating that doping with nHA was achieved. Meanwhile, the enhanced C=O peak (acrylamide band I) and the absorption peak of acrylamide band II and III at 1636 cm^{-1} and 1092 cm^{-1} also demonstrated that chitosan was successfully coated on the surface of microcarriers. Protein adsorption of materials was influenced by factors such as surface charge, roughness, or the hydrophilicity of the material, which affected cell interaction with materials. In Figure 4C, the amount of protein adsorption of the microcarriers increased when nHA was added, owing to the calcium ions in nHA which can bind to negatively charged protein. Although the surface roughness of PLLA/nHA/Ch microcarriers was the largest, the protein adsorption of the PLLA/nHA/Ch microcarriers had no obvious significance with that of PLLA/nHA microcarriers. This was mainly due to the partial loss of nHA caused by the dissolution of microcarriers in the alkaline solvent. Some studies have investigated the correlation between nHA and protein adsorption. For example, Kandori et al. [40,41] stated that calcium atoms would be exposed on the Hap surface by dissolution of OH ions at the particle surface to produce calcium ions or positively charged sites to bind to acidic groups of proteins after dispersing nHA particles in aqueous media. In Figure 4D, it was found that the degradation of PLLA microcarriers increased with time, and it was slowed down by the addition of hydroxyapatite. The decreased degradation ratio of PLLA/nHA in the 2-week period was believed to be caused by the mineralization of nHA in the PBS buffer solution. Kanjwal et al. [42] studied the mineralization of nanohydroxyapatite/PCL in SBF, a simulated body fluid. Its composition was somewhat similar to PBS, including NaCl, KCl, K_2HPO_4. The principle of mineralization was that the incubation of the introduced PCL/HAp in the SBF led to the formation of excessive apatite due to stimulating the crystallization of the biological apatite. However, with the extension of time, the degradation of PLLA eventually dominated, resulting in the decline of the weight of microcarriers. But for PLLA/nHA/Ch microcarriers this condition did not happen. This was mainly because of the alkaline treatment of the PLLA/nHA/Ch microcarriers, which resulted in the loss of part of nHA. Meanwhile, although chitosan also had the advantage of retarding the degradation of PLLA/nHA/Ch microcarriers, the surface erosion caused by alkaline treatment promoted the degradation to a certain extent [43]. As Lyu et al. [44] pointed out, the degradation mechanism of PLLA can be either surface or bulk erosion depending on the competition between the diffusion of water molecules into PLLA and the rate of the hydrolysis reactions. It was also confirmed by elemental analysis that chitosan was successfully crosslinked with microcarriers (Table 2).

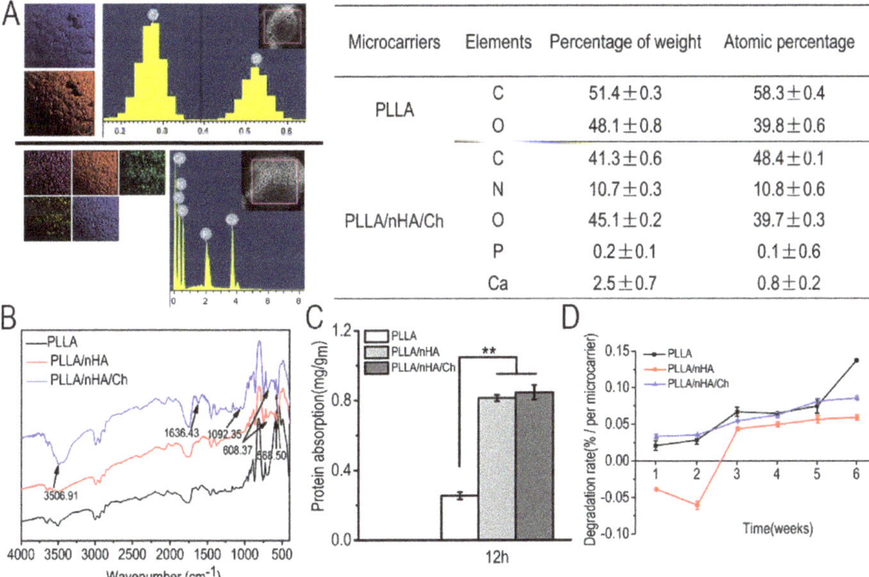

Microcarriers	Elements	Percentage of weight	Atomic percentage
PLLA	C	51.4±0.3	58.3±0.4
	O	48.1±0.8	39.8±0.6
PLLA/nHA/Ch	C	41.3±0.6	48.4±0.1
	N	10.7±0.3	10.8±0.6
	O	45.1±0.2	39.7±0.3
	P	0.2±0.1	0.1±0.6
	Ca	2.5±0.7	0.8±0.2

Figure 4. (**A**): Semi-quantitative analysis of surface elements of PLLA microcarriers (including carbon and oxygen) and PLLA/nHA/Ch microcarriers (including carbon, nitrogen, calcium, phosphorus, and oxygen); (**B**): Analyses of characteristic chemical functional groups of the three kinds of microcarriers by Fourier transform infrared spectroscopy; (**C**): Amount of protein adsorbed on the surfaces of microcarriers after incubation for 12 h at 37 °C. Data were expressed as mean ± SD ($n = 6$, ** $p < 0.01$); (**D**): The degradation of the three kinds of microcarriers in PBS solution.

Table 2. Nitrogen content of microcarriers measured by an elemental analyzer.

Microcarriers	Elemental Fraction (wt %)		
	C	N	H
PLLA	41.93 ± 0.11	0	4.80 ± 0.03
PLLA/nHA/Ch	46.96 ± 0.13	0.03 ± 0.01	5.22 ± 0.01

3.3. The Morphology and Osteogenic Differentiation of MC3T3-E1 Cells

Figure 5A–C shows the adhesion and spreading pattern of MC3T3-E1 cells by light microscopy, Calcein-AM staining, and HE staining. After one week of osteogenic induction, alkaline phosphatase staining revealed that the cells had high expression of alkaline phosphatase, characteristic of the first stage of differentiation (Figure 5D). Also, red calcium nodules were noted by Alizarin Red S staining (Figure 5E) and a brown-black substance was observed with Von-Kossa staining (Figure 5F). It was suggested from the above tests that MC3T3-E1 cells with good proliferation and high osteogenic differentiation ability in vitro could be used for subsequent cell–microcarrier complex cultivation.

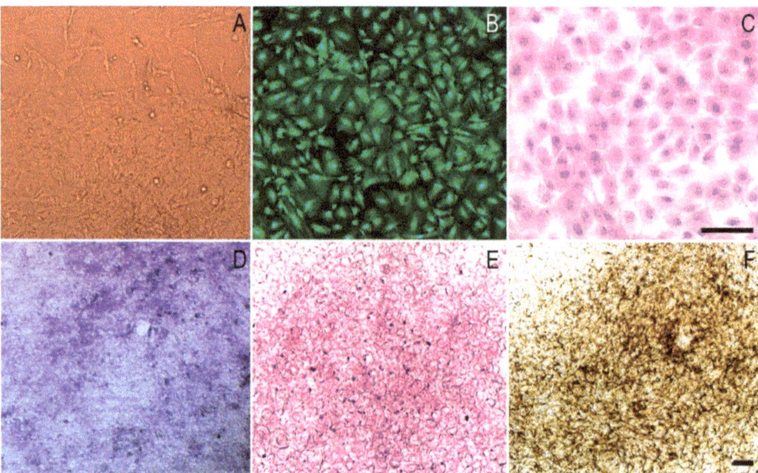

Figure 5. Phase contrast microscopy images of the morphology and differentiation of MC3T3-E1 cells. (**A**): cell morphology of MC3T3-E1cells; (**B**): Calcein-AM staining of the seventh generation of cells; (**C**): Hematoxylin-eosin (HE) staining of the seventh generation of cells; (**D**): Alkaline phosphatase staining after one week of osteogenic induction; (**E**): Alizarin red S staining after three weeks of osteogenic induction; (**F**): Von-Kossa staining after four weeks of osteogenic induction. Scale: 250 μm.

3.4. Live/Dead Staining of Cells on Microcarriers

The live/dead staining of cells cultured on the surface of microcarriers conducted at 24 h (Figure 6) showed that MC3T3-E1 cells had adhered and spread on the surface of microcarriers and maintained high cell viability, especially on the PLLA/nHA microcarriers. The observation of cells on PLLA/nHA/Ch microcarriers was difficult because the microcarriers were also stained due to the surface-modified chitosan. After 120 h (Figure 7), it was found that originally dispersed PLLA and PLLA/nHA microcarriers were packaged by a large number of cells. However, this did not appear on the PLLA/nHA/Ch microcarriers, which was speculated to be due to the fact that the rough surface of the PLLA/nHA/Ch microcarriers did not significantly promote cell proliferation at the same culturing time compared with the PLLA/nHA microcarriers.

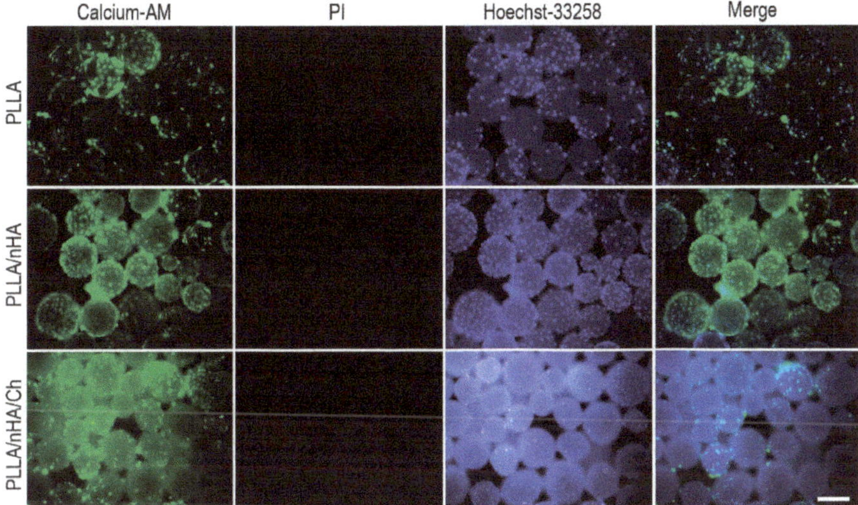

Figure 6. Live/dead staining of MC3T3-E1 cells of P7 passage cultured for 24 h on the surface of the three microcarriers, which were respectively the Calcein-AM, PI and Hoechst33258 staining, and their overlay composite graph. Scale: 250 μm.

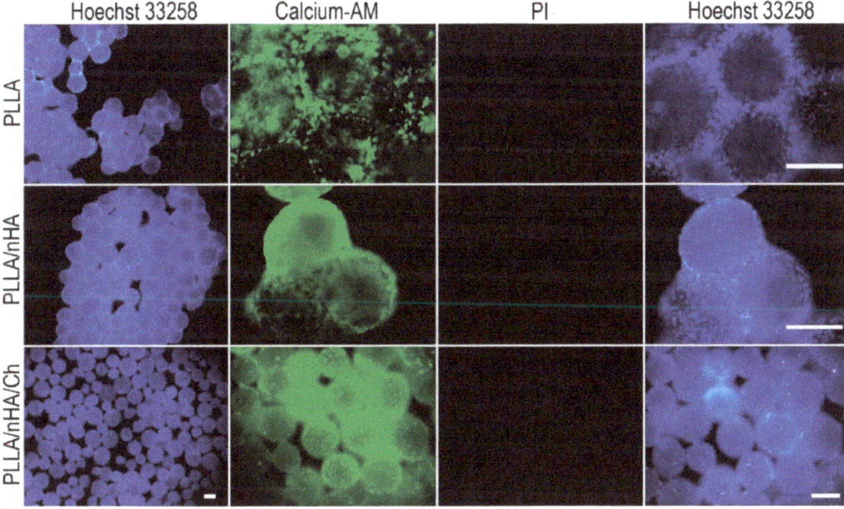

Figure 7. Live/dead staining of MC3T3-E1 cells of P7 passage cultured for 120 h on the surface of the three microcarriers. Scale: 250 μm.

3.5. Cell Adhesion, Detachment, and Differentiation on Microcarriers

The process of cell adhesion and spreading on the microcarriers was exhibited in Figure 8A. In the first 30 min, the cells initially came into contact with the surface of the three kinds of microcarriers in the shape of sphere, and then gradually became ellipsoid, flattened, and stretched out to adhere to the surface of microcarriers within 24 h. After 120 h, cells had proliferated in large numbers and encapsulated the entire surface of the microparticles. However, this was not found on the PLLA/nHA/Ch microcarriers, implying that they did not produce an active effect on the expansion of

MC3T3-E1 cells. After cultured for 7 days, the desorption process of MC3T3-E1 cells in three minutes under the action of trypsin (Figure 8B) showed easy cell detachment from PLLA microcarriers, but not from PLLA/nHA/Ch microcarriers in the same time interval, indicating that the adhesion between MC3T3-E1 cells and PLLA/nHA/Ch microcarriers was the strongest, and that of PLLA microcarriers was the weakest. The cells which were reseeded into the 24-well plate were found to have a good proliferation state after a week of culture, and show an obvious difference with the morphology of cells in 2D (two-dimensional) culture (Figure 5A,B), illustrating that the cells detaching from the three kinds of microcarriers retained good adhesion and division capability.

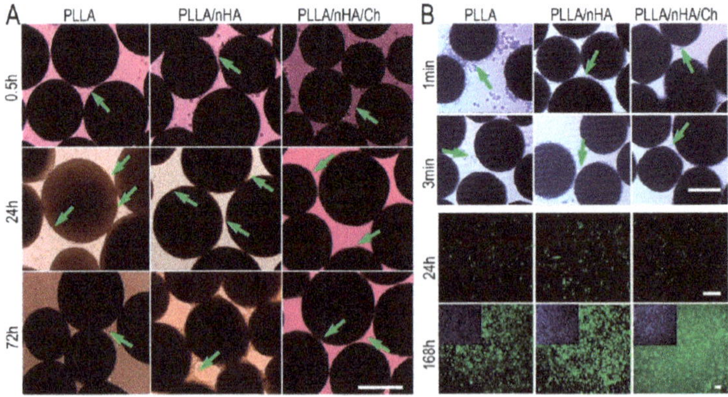

Figure 8. (**A**) The time-course of adhesion and spread of MC3T3-E1 cells of P7 passage on the surface of the three microcarriers with the inoculation density of 1×10^5 cells/mL containing the stages of contact, adhesion, spreading, and proliferation; (**B**) The detachment process of MC3T3-E1 from microcarriers and live/dead staining of the detached cells which were reseeded into 24-well plates and cultured for one week. Scale: 250 μm.

The spread morphology of cells on the surface of microcarriers observed by SEM (Figure 9A) showed that the proliferation of cells on the first two kinds of microcarriers was significantly better than on PLLA/nHA/Ch microcarriers after culturing for 72 h. The cells were seeded on the three kind microcarriers by two methods, as mentioned above. In Figure 9B, the green arrows indicate the cell–microcarrier complexes in the 24-well plate stained by alkaline phosphatase after cultivation for seven days in osteogenic medium. It can be seen all of them performed an obvious expression with a dark blue-violet color which was similar to Figure 5D. It was also found that the cells growing in the 24-well plate would migrate to the microcarriers, because many cells had appeared on the top surface of the microcarriers, suggesting that these three kinds of microcarriers prepared using gelatin as dispersant had a certain cellular affinity. As shown in Figure 9C, the cell adhesion rate was found to be significantly different on these three kinds of microcarriers in the first 10 min post cell seeding, followed by a slight increase until they reached the stable condition. The highest cell adhesion efficiency was found in the group of PLLA/nHA/Ch. This could be explained by the increased surface roughness, which was conducive to the retention of cells. The cell proliferation rate in Figure 9D demonstrated that the PLLA/nHA microcarriers had the optimum proliferation-promoting characteristics, whereas the cross-linked modified microcarriers were not favorable. As it was found by the front observation of SEM, although the modification of the microcarriers increased the surface roughness and created interconnected microchannels, these microchannels were too small to help the growth of MC3T3-E1 cells, with evidence that all cells spread out flat on the surface. The quantitative detection of ALP (Figure 9E) showed that the cells on the PLLA/nHA/Ch microcarriers had the highest ALP activity. Although the cells grew slowly on the surface of PLLA/nHA/Ch microcarriers, the cells could still fully coat the microcarriers after a certain period of culturing. Some studies have found that surface roughness plays an extremely subtle role in cell proliferation and differentiation, and only in a suitable roughness range can the best effect be achieved. For example, Borsari et al. [45] concluded in their study that values of roughness that overcome specific limits were not correctly seen by cells and HA coating had a synergic positive effect on cells only when the proper roughness is present. Zan et al. [46] made β-TCP/chitosan composite microspheres with four kinds of surface roughness and it was found that the optimum roughness of the surface was 2.0 μm. Hu et al. [47] have found that a combination of low surface roughness and high stiffness of the substrate appeared to be the most favorable for proliferation and myogenic-differentiation of C2C12 cells, while in contrast, hMSCs demonstrated a preference for higher surface roughness and stronger micro/nano-scale surface patterns. Zhang et al. [48] demonstrated that a rough stainless steel surface improved cell adhesion and morphology, while HA coating which had a higher roughness contributed to superior cell adhesion, but inhibited cell proliferation. In a word, a rough surface has both its pros and cons on cell adhesion and proliferation.

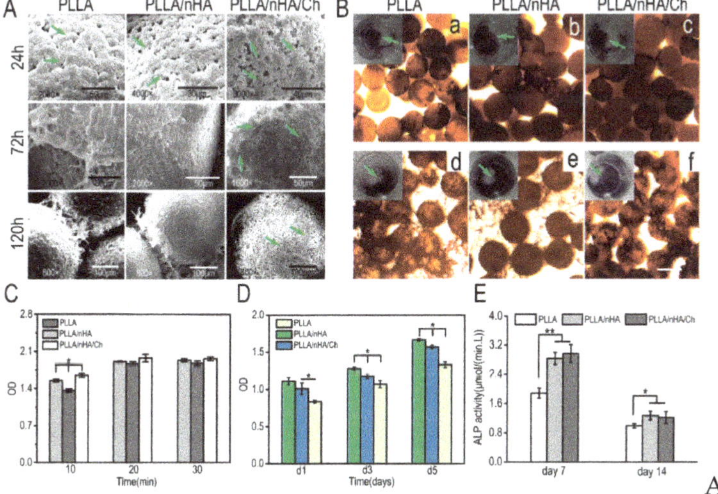

Figure 9. (**A**): Adhesion, spread, and proliferation of MC3T3-E1 cells cultured on the surface of microcarriers for 24 h, 72 h, and 120 h observed by scanning electron microscopy; (**B**): a–c: Alkaline phosphatase staining of MC3T3-E1 cells inoculated onto the surface of microcarriers after one-week osteogenic induction (the green arrows indicated the cells-microcarriers complex in the 24-well plate stained by alkaline phosphatase). d–f: Alkaline phosphatase staining of MC3T3-E1 cells firstly inoculated into the 24-well plate and cultured for 3 days, then adhered to the microcarriers. Scale: 250 μm; (**C**): The adhesion determination of MC3T3-E1 cells on the surface of microcarriers; (**D**): The proliferation detection of MC3T3-E1 cells on the surface of microcarriers ($n = 3$, * $P < 0.05$); (**E**): ALP activity assay of MC3T3-E1 cells on the three kinds of microcarriers ($n = 3$, * $p < 0.05$, ** $p < 0.01$).

Currently, many researchers focus on combinations of synthetic polymer materials, natural polymer materials, and bioactive ceramics to prepare composite materials for tissue engineering and regeneration [49–51], not only making up for hydrophobicity and the lack of adhesive ligands of synthetic polymers, but also resolving the readily degradable and structurally unstable disadvantages of natural polymers, and most importantly, enhancing the proliferation and differentiation of cells. In this study, gelatin with high biocompatibility was utilized as the dispersant, different from other dispersants such as polyvinyl alcohol, toluene, and Tween, which made the preparation process safer and more concise. Even if the dispersant remained on the surface of the microcarriers, it would not cause toxicity to the cells. Considering the hydrophobicity of PLLA, chitosan was further coated on the surface of the microcarriers after finishing the construction of PLLA/nHA composite microcarriers. It was found that the PLLA/nHA microcarriers had the greatest proliferation-promoting effect, while the PLLA/nHA/Ch microcarriers had the strongest attachment with MC3T3-E1 cells. The cells on the PLLA/nHA/Ch microcarriers exhibited optimal osteogenic expression. Generally, the function of PLLA microcarriers on osteoblast-like cell cultivation has been greatly enhanced by the doping of nanohydroxyapatite and further modification with chitosan.

4. Conclusions

In this study, three kinds of microcarriers were prepared based on physical doping and surface modification, namely Poly (L-lactic) acid (PLLA), PLLA/nanohydroxyapatite (PLLA/nHA), and PLLA/nHA/Chitosan (PLLA/nHA/Ch) using gelatin as the dispersant. The physicochemical properties of the three kinds of microcarriers and their functional performance in MC3T3-E1 cell culture were compared. The surface roughness and protein adsorption of the microcarriers were enhanced with the doping of nHA and coating of chitosan. Murine MC3T3-E1 cells with demonstrated osteogenic differentiation activity were seeded onto the surface of microcarriers to investigate the cytocompatibility, and the results showed that MC3T3-E1 cells could adhere and proliferate on the three kinds of microcarriers, and the PLLA/nHA microcarriers had the greatest proliferation-promoting effect, while the PLLA/nHA/Ch microcarriers exhibited the strongest attachment with MC3T3-E1 cells. Besides, the cells on the PLLA/nHA/Ch microcarriers exhibited optimal osteogenic expression. Generally, chitosan improved the microcarrier characteristics of cell adhesion and differentiation, and nanohydroxyapatite improved their sphericity and promotion of cell proliferation. In conclusion, the modified microcarriers provide a promising strategy for 3D cultivation of osteoblast-like cells, and might be good candidates for bone tissue engineering. Future work will study the impact of relevant experimental parameters, such as alkaline treatment time and concentration of chitosan, on the biocompatibility and morphology of PLLA/nHA/Ch microcarriers. Three-dimensional dynamic experiments utilizing the modified PLLA microcarriers will also be carried out to seek the appropriate ex vivo environment for cell survival and propagation.

Author Contributions: Data Curation, L.L. and Y.C.; Formal Analysis, K.S.; Methodology, L.L.; Software, L.L. and Y.C.; Validation, Y.N.; Writing-Original Draft Preparation, L.L.; Writing-Review & Editing, K.S., Y.W., F.S. and T.L.; Project Administration, T.L.

Funding: This research was funded by [National Natural Science Foundation of China] grant number [31670978/31370991/21676041], [Fok Ying Tung Education Foundation] grant number [132027], [State Key Laboratory of Fine Chemicals] grant number [KF1111], [Natural Science Foundation of Liaoning] grant number [20180510028], [Fundamental Research Funds for the Central Universities] grant number [2016ZD210] and [SRF for ROCS, SEM] grant number [42].

Acknowledgments: This work was supported by the National Natural Science Foundation of China (31670978/31370991/21676041), the Fok Ying Tung Education Foundation (132027), the State Key Laboratory of Fine Chemicals (KF1111), the Natural Science Foundation of Liaoning (20180510028) and the Fundamental Research Funds for the Central Universities (2016ZD210) and SRF for ROCS, SEM (42).

Conflicts of Interest: The authors have no conflicts of interest.

References

1. Wezel, A.L.V. Chapter 2–Microcarrier cultures of animal cells. *Tissue Cult.* **1973**, 372–377. [CrossRef]
2. Freiberg, S.; Zhu, X.X. Polymer microspheres for controlled drug release. *Int. J. Pharm.* **2004**, *282*, 1–18. [CrossRef] [PubMed]
3. Zhang, Y.; Sun, T.; Jiang, C. Biomacromolecules as carriers in drug delivery and tissue engineering. *Acta Pharm. Sin. B* **2018**, *8*, 34–50. [CrossRef] [PubMed]
4. Ventini, D.C.; Damiani, R.; Sousa, A.P.; De Oliveira, J.E.; Peroni, C.N.; Ribela, M.T.; Pereira, C.A. Improved bioprocess with CHO-hTSH cells on higher microcarriers concentration provides higher overall biomass and productivity for rhTSH. *Appl. Biochem. Biotechnol.* **2011**, *164*, 401–409. [CrossRef] [PubMed]

5. Chung, T.W.; Huang, Y.Y.; Liu, Y.Z. Effects of the rate of solvent evaporation on the characteristics of drug loaded PLLA and PDLLA microspheres. *Int. J. Pharm.* **2001**, *212*, 161–169. [CrossRef]

6. Liggins, R.T.; Burt, H.M. Paclitaxel loaded poly (L-lactic acid) (PLLA) microspheres: II. The effect of processing parameters on microsphere morphology and drug release kinetics. *Int. J. Pharm.* **2004**, *281*, 103–106. [CrossRef] [PubMed]

7. Aubert-Pouëssel, A.; Bibby, D.C.; Venier-Julienne, M.C.; Hindré, F.; Benoît, J.P. A novel in vitro delivery system for assessing the biological integrity of protein upon release from PLGA microspheres. *Pharm. Res.-Dordr.* **2002**, *19*, 1046–1051. [CrossRef]

8. Sart, S.; Agathos, S.N.; Li, Y. Engineering stem cell fate with biochemical and biomechanical properties of microcarriers. *Biotechnol. Progr.* **2013**, *29*, 1354–1366. [CrossRef]

9. Olmos, E.; Martin, C.; Loubiere, K.; Delaplace, G.; Marc, A. Critical agitation for microcarriers suspension in orbital shaken bioreactors: Experimental study and dimensional analysis. *Chem. Eng. Sci.* **2015**, *122*, 545–554. [CrossRef]

10. Perez, R.A.; Riccardi, K.; Altankov, G.; Ginebra, M.P. Dynamic cell culture on calcium phosphate microcarriers for bone tissue engineering applications. *J. Tissue Eng.* **2014**, *5*. [CrossRef]

11. Ferrari, C.; Olmos, E.; Balandras, F.; Tran, N.; Chevalot, I.; Guedon, E.; Marc, A. Investigation of growth conditions for the expansion of porcine mesenchymal stem cells on microcarriers in stirred cultures. *Appl. Biochem. Biotechnol.* **2014**, *172*, 1004–1017. [CrossRef]

12. Rafiq, Q.A.; Brosnan, K.M.; Coopman, K.; Nienow, A.W.; Hewitt, C.J. Culture of human mesenchymal stem cells on microcarriers in a 5 l stirred-tank bioreactor. *Biotechnol. Lett.* **2013**, *35*, 1233–1245. [CrossRef] [PubMed]

13. Chun, K.W.; Yoo, H.S.; Yoon, J.J.; Park, T.G. Biodegradable PLGA microcarriers for injectable delivery of chondrocytes: Effect of surface modification on cell attachment and function. *Biotechnol. Prog.* **2010**, *20*, 1797–1801. [CrossRef] [PubMed]

14. Sivandzade, F.; Mashayekhan, S. Design and fabrication of injectable microcarriers composed of acellular cartilage matrix and chitosan. *J. Biomater. Sci.-Polym. E* **2018**, *29*, 1–28. [CrossRef] [PubMed]

15. Arora, A. Synthesis and Characterization of Novel Fluorescent Injectable Micro-Carriers for Tissue Regeneration (*Doctoral Dissertation*). 2014. Available online: http://oaktrust.library.tamu.edu/handle/1969.1/152694 (accessed on 28 April 2014).

16. Anderson, J.M.; Shive, M.S. Biodegradation and biocompatibility of PLA and PLGA microspheres. *Adv. Drug Deliv. Rev.* **2012**, *64*, 72–82. [CrossRef]

17. Park, J.H.; Lee, E.J.; Knowles, J.C.; Kim, H.W. Preparation of in situ hardening composite microcarriers: Calcium phosphate cement combined with alginate for bone regeneration. *J. Biomater. Appl.* **2014**, *28*, 1079. [CrossRef] [PubMed]

18. Li, W.; Han, Y.; Yang, H.; Wang, G.; Lan, R.; Wang, J.Y. Preparation of microcarriers based on zein and their application in cell culture. *Mater. Sci. Eng. C-Mater.* **2016**, *58*, 863–869. [CrossRef] [PubMed]

19. Lai, J.Y.; Ma, D.H. Ocular biocompatibility of gelatin microcarriers functionalized with oxidized hyaluronic acid. *Mater. Sci. Eng. C-Mater.* **2017**, *72*, 150–159. [CrossRef]

20. Song, K.; Yang, Y.; Wu, S.; Zhang, Y.; Feng, S.; Wang, H.; Wang, Y.; Wang, L.; Liu, T. In vitro culture and harvest of BMMSCs on the surface of a novel thermo-sensitive glass microcarriers. *Mater. Sci. Eng. C* **2016**, *58*, 324–330. [CrossRef]

21. Levato, R.; Visser, J.; Planell, J.A.; Engel, E.; Malda, J.; Mateostimoneda, M.A. Biofabrication of tissue constructs by 3d bioprinting of cell-laden microcarriers. *Biofabrication* **2014**, *6*, 035020. [CrossRef]

22. Singh, B.; Singh, P.; Sutherland, A.J.; Pal, K. Control of shape and size of poly (lactic acid) microspheres based on surfactant and polymer concentration. *Mater. Lett.* **2017**, *195*, 48–51. [CrossRef]

23. Chen, J.P.; Su, C.H. Surface modification of electrospun PLLA nanofibers by plasma treatment and cationized gelatin immobilization for cartilage tissue engineering. *Acta Biomater.* **2011**, *7*, 234. [CrossRef] [PubMed]

24. Szustakiewicz, K.; Stępak, B.; Antończak, A.J.; Maj, M.; Gazińska, M.; Kryszak, B.; Pigłowski, J. Femtosecond laser-induced modification of PLLA/hydroxypatite composite. *Polym. Degrad. Stab.* **2018**, *149*, 152–161. [CrossRef]

25. Neděla, O.; Slepička, P.; Švorčík, V. Surface modification of polymer substrates for biomedical applications. *Materials* **2017**, *10*, 1115. [CrossRef] [PubMed]

26. Rimpelová, S.; Peterková, L.; Kasálková, N.S.; Slepička, P.; Švorčík, V.; Ruml, T. Surface modification of biodegradable poly (l-lactic acid) by argon plasma: Fibroblasts and keratinocytes in the spotlight. *Plasma Process. Polym.* **2015**, *11*, 1057–1067. [CrossRef]

27. Turon, P.; Valle, L.J.D.; Alemán, C.; Puiggalí, J. Chapter 2—Grafting of hydroxyapatite for biomedical applications. *Biopolym. Graft.* **2018**, 45–80. [CrossRef]

28. Pramanik, N.; Mishra, D.; Banerjee, I.; Maiti, T.K.; Bhargava, P.; Pramanik, P. Chemical synthesis, characterization, and biocompatibility study of hydroxyapatite/chitosan phosphate nanocomposite for bone tissue engineering applications. *Int. J. Polym Mater.* **2014**. [CrossRef]

29. Ravichandran, R.; Venugopal, J.R.; Sundarrajan, S.; Mukherjee, S.; Ramakrishna, S. Precipitation of nanohydroxyapatite on PLLA/PBLG/collagen nanofibrous structures for the differentiation of adipose-derived stem cells to osteogenic lineage. *Biomaterials* **2012**, *33*, 846–855. [CrossRef]

30. Wang, X.; Song, G.; Lou, T. Fabrication and characterization of nano-composite scaffold of PLLA/silane modified hydroxyapatite. *Med. Eng. Phys.* **2010**, *32*, 391–397. [CrossRef]

31. Wang, J.L.; Chen, Q.; Du, B.B.; Cao, L.; Lin, H.; Fan, Z.Y.; Dong, J. Enhanced bone regeneration composite scaffolds of PLLA/β-TCP matrix grafted with gelatin and hap. *Mat. Sci. Eng. C* **2018**, *87*, 60–69. [CrossRef]

32. Jiao, Y.; Liu, Z.C. Fabrication and characterization of PLLA-chitosan hybrid scaffolds with improved cell compatibility. *J. Biomed. Mater. Res. A* **2007**, *80A*, 820–825. [CrossRef] [PubMed]

33. Kean, T.; Thanou, M. Biodegradation, biodistribution and toxicity of chitosan. *Adv. Drug Deliv. Rev.* **2010**, *62*, 3–11. [CrossRef] [PubMed]

34. Guo, Z.; Bo, D.; He, Y.; Luo, X.; Li, H. Degradation properties of chitosan microspheres/poly(L-lactic acid) composite in vitro and in vivo. *Carbohydr. Polym.* **2018**, *193*, 1–8. [CrossRef] [PubMed]

35. Lao, L.; Tan, H.; Wang, Y.; Gao, C. Chitosan modified poly (l-lactide) microspheres as cell microcarriers for cartilage tissue engineering. *Colloid Surf. B* **2008**, *66*, 218–225. [CrossRef] [PubMed]

36. Qian, C.; McClements, D.J. Formation of nanoemulsions stabilized by model food-grade emulsifiers using high-pressure homogenization: Factors affecting particle size. *Food Hydrocoll.* **2011**, *25*, 1000–1008. [CrossRef]

37. McClements, D.J. Food emulsions: Principles, practice, and techniques. In *CRC Series in Contemporary Food Science Boca Raton*, 2nd ed.; CRC Press: Boca Raton, FL, USA, 2005.

38. Rosca, I.D.; Watari, F.; Uo, M. Microparticle formation and its mechanism in single and double emulsion solvent evaporation. *J. Control. Release* **2004**, *99*, 271–280. [CrossRef] [PubMed]

39. Hong, Y.; Gao, C.; Shi, Y.; Shen, J. Preparation of porous polylactide microspheres by emulsion-solvent evaporation based on solution induced phase separation. *Polym. Adv. Technol.* **2005**, *16*, 622–627. [CrossRef]

40. Kandori, K.; Oda, S.; Fukusumi, M.; Morisada, Y. Synthesis of positively charged calcium hydroxyapatite nanocrystals and their adsorption behavior of proteins. *Colloid Surf. B* **2009**, *73*, 140–145. [CrossRef]

41. Kandori, K.; Kuroda, T.; Togashi, S.; Katayama, E. Preparation of calcium hydroxyapatite nanoparticles using microreactor and their characteristics of protein adsorption. *J. Phys. Chem. B* **2011**, *115*, 653–659. [CrossRef] [PubMed]

42. Kanjwal, M.A.; Sheikh, F.A.; Nirmala, R.; Macossay, J.; Kim, H.Y. Fabrication of poly (caprolactone) nanofibers containing hydroxyapatite nanoparticles and their mineralization in a simulated body fluid. *Fiber Polym.* **2011**, *12*, 50–56. [CrossRef]

43. He, Y.; Qian, Z.; Zhang, H.; Liu, X. Alkaline degradation behavior of polyesteramide fibers: Surface erosion. *Colloid Polym. Sci.* **2004**, *282*, 972–978. [CrossRef]

44. Lyu, S.P.; Schley, J.; Loy, B.; Lind, D.; Hobot, C.; Sparer, R.; Untereker, D. Kinetics and time−temperature equivalence of polymer degradation. *Biomacromolecules* **2007**, *8*, 2301–2310. [CrossRef] [PubMed]

45. Borsari, V.; Fini, G.W. Physical characterization of different-roughness titanium surfaces, with and without hydroxyapatite coating, and their effect on human osteoblast-like cells. *J. Biomed. Mater. Res. B* **2005**, *75B*, 359–368. [CrossRef] [PubMed]

46. Zan, Q.; Wang, C.; Dong, L.; Cheng, P.; Tian, J. Effect of surface roughness of chitosan-based microspheres on cell adhesion. *Appl. Surf. Sci.* **2008**, *255*, 401–403. [CrossRef]

47. Hu, X.; Park, S.H.; Gil, E.S.; Xia, X.X.; Weiss, A.S.; Kaplan, D.L. The influence of elasticity and surface roughness on myogenic and osteogenic-differentiation of cells on silk-elastin biomaterials. *Biomaterials* **2011**, *32*, 8979–8989. [CrossRef] [PubMed]

48. Zhang, H.; Han, J.; Sun, Y.; Huang, Y.; Zhou, M. MC3T3-E1 cell response to stainless steel 316l with different surface treatments. *Mat. Sci. Eng. C* **2015**, *56*, 22–29. [CrossRef] [PubMed]

49. Sobczak-Kupiec, A.; Pluta, K.; Drabczyk, A.; Włoś, M.; Tyliszczak, B. Synthesis and characterization of ceramic-polymer composites containing bioactive synthetic hydroxyapatite for biomedical applications. *Ceram. Int.* **2018**, *44*, 13630–13638. [CrossRef]
50. Swetha, M.; Sahithi, K.; Moorthi, A.; Srinivasan, N.; Ramasamy, K.; Selvamurugan, N. Biocomposites containing natural polymers and hydroxyapatite for bone tissue engineering. *Int. J. Biol. Macromol.* **2010**, *47*, 1–4. [CrossRef] [PubMed]
51. Shahin-Shamsabadi, A.; Hashemi, A.; Tahriri, M.; Bastami, F.; Salehi, M.; Abbas, F.M. Mechanical, material, and biological study of a PCL/bioactive glass bone scaffold: Importance of viscoelasticity. *Mater. Sci. Eng. C* **2018**, *90*, 280–288. [CrossRef]

 *polymers*

Article

Synthesis and Properties of Silk Fibroin/Konjac Glucomannan Blend Beads

Carla Giometti França [1], Vicente Franco Nascimento [1], Jacobo Hernandez-Montelongo [2,3,*], Daisy Machado [4], Marcelo Lancellotti [4] and Marisa Masumi Beppu [1]

[1] Faculdade de Engenharia Química, Universidade Estadual de Campinas, Campinas 13083-852, São Paulo, Brazil; carlagfranca@gmail.com (C.G.F.); vicente.qmc@gmail.com (V.F.N.); beppu@feq.unicamp.br (M.M.B.)

[2] Departamento de Ciencias Matemáticas y Físicas, Facultad de Ingeniería, Universidad Católica de Temuco, Temuco 4813302, Chile

[3] Núcleo de Investigación en Bioproductos y Materiales Avanzados (BioMa), Facultad de Ingeniería, Universidad Católica de Temuco, Temuco 4781312, Chile

[4] Laboratório de Biotecnologia, Instituto de Biologia, Universidade Estadual de Campinas, Campinas 13083-862, São Paulo, Brazil; daisy.machado@gmail.com (D.M.); mlancell@unicamp.br (M.L.)

* Correspondence: jacobo.hernandez@uct.cl

Received: 4 June 2018; Accepted: 7 August 2018; Published: 18 August 2018

Abstract: Silk fibroin (SF) and konjac glucomannan (KGM) are promising materials in the biomedical field due to their low toxicity, biocompatibility, biodegradability and low immune response. Beads of these natural polymers are interesting scaffolds for biomedical applications, but their fabrication is a challenge due to their low stability and the necessary adaptation of their chemical and mechanical properties to be successfully applied. In that sense, this study aimed to synthesize a blend of silk fibroin and konjac glucomannan (SF/KGM) in the form of porous beads obtained through dripping into liquid nitrogen, with a post-treatment using ethanol. Intermolecular hydrogen bonds promoted the integration of SF and KGM. Treated beads showed higher porous size, crystallinity, and stability than untreated beads. Characterization analyses by Fourier-transform infrared spectroscopy (FTIR), thermogravimetric (TGA), and X-ray diffraction (XDR) evidenced that ethanol treatment allows a conformational transition from silk I to silk II in SF and an increase in the KGM deacetylation. Those chemical changes significantly enhanced the mechanical resistance of SF/KGM beads in comparison to pure SF and KGM beads. Moreover, samples showed cytocompatibility with HaCaT and BALB/c 3T3 cells

Keywords: biopolymers; silk fibroin; konjac glucomannan; porous beads; scaffolds; tissue engineering

---

1. Introduction

Global challenges in scaffolds technology offer opportunities for further research that involves new types of polymers in different forms such as films, membranes, and beads. In the last decades, natural polymers have been investigated to be used as cell scaffolds [1]. However, to be successfully used these materials should be modified to improve different properties, such as in-vivo stability, biocompatibility, biodegradability, and potential interaction with cells. For that, the scaffolds should also have a cell-friendly surface, and interconnected pores for the transport of cells, nutrients, and metabolites [2].

One of the most promising materials to be used as cell scaffolds, among other applications in biomedicine is silk fibroin (SF), which is a biodegradable polymer, and biocompatible with human tissues. Its chemical composition consists of simple amino acids, such as glycine, alanine, serine, and tyrosine [3,4]. SF is a molecule with high tensile strength, elasticity, and resilience. It has been used in diverse applications, such as hydrogels synthesis, supports for enzymes immobilization, drug delivery, and etc. [3,5,6]. Konjac glucomannan (KGM) is also a biodegradable and biocompatible material [1]. Its chemical structure is composed of D-mannose and D-glucose monomers linked by $\beta(1\rightarrow4)$ bonds. KGM is a hydrophilic polymer with high viscoelasticity and mechanical resistance. Over the last years, it was mainly used as a bioactive polymer. KGM, combined with other polymers, can be used as a controlled delivery system for drugs, and antimicrobial materials [7,8].

In spite of the mentioned properties of SF and KGM, their mechanical properties could be improved in order to be used as a scaffold. Blending two or more polymers is a usual technique to overcome challenges in polymer properties and resulting in the new materials. Several studies of SF-based blends have been reported in the literature. Park et al. (1999) verified that the mechanical tensile strength of SF and chitosan (CHI) membranes depended on CHI concentration. Moreover, the resistance of membranes presented higher value than pure polymers, due to the increase in β-sheet conformational transition [9]. Kweon et al. (2001) studied films of SF and CHI to understand the effect of fibroin and chitosan ratio on the physical characteristics of blends. They verified that the mechanical properties the films were improved when blends were formed with 10–40% chitosan [10]. Lee et al. (2004) studied the physical and structural properties of SF and sodium alginate blends and concluded that the structural characteristics of SF in blends were not affected by the incorporation of sodium alginate. Also, the mechanical properties of these blends were improved because sodium alginate is a hard polysaccharide [11]. She et al. (2008) reported that the mechanical properties of SF/CHI scaffolds could be controlled by adjusting their composition [12].

In that sense, this study was focused on the synthesizes and characterization of porous blend beads obtained from two natural polymers, SF and KGM, with suitable characteristics to be applied to tissue engineering or drug release field [13]. A surface morphological analysis was performed by scanning electron microscope (SEM), physicochemical characterizations were obtained by Fourier-transform infrared spectroscopy (FTIR), thermogravimetric (TGA), and X-ray diffraction (XDR), and a validation of mechanical resistance was performed using compression tests. Finally, cytocompatibility assays were carried out using two kinds of cell lines: human keratinocytes (HaCaT) and murine fibroblast (BALB/c 3T3).

2. Materials and Methods

2.1. Beads Synthesis

2.1.1. Solutions Preparation

Bombyx mori silkworm cocoons were supplied by Fiação Bratac (Bastos-SP, Brazil). Cocoons were degummed three times by soaking in Na_2CO_3 1 g/L solution at 85 °C for 30 min. Afterward, the obtained SF fibers were rinsed with distilled water to remove Na_2CO_3 residues and dried at room temperature (25 °C) for 24 h. A ternary solvent of $CaCl_2$:ethanol:H_2O in a molar ratio of 1:2:8 was used to dissolve the SF fibers to 10% (w/v) at 85 °C for 1.5 h. The SF salt solution was dialyzed by a cellulose membrane (MWCO 3500, Viscofan®, Spain) against distilled water for 72 h at 4 °C, with changes every 24 h [5,14]. The obtained concentration of SF solution was 5% (w/v), which was determined by casting the solution in a petri dish and weighing the dried mass after solvent evaporation. Finally, the dialyzed solution was diluted in deionized water to a concentration of 2.5% (w/v).

KGM (Konjac Glucomannan Powder®, Sunnyvale, CA, USA) was dissolved in deionized water to a concentration of 1 g/L. After homogenization, $Ca(OH)_2$ (Synth®, Diadema, Brazil) was added at a concentration of 1.5 mmol/L to pre-stabilize KGM [15–17]. Calcium was used to crosslink the KGM, to make it insoluble in water.

The blends were prepared by adding the solution of SF 2.5% (w/v) into KGM 1% (w/v) at a weight ratio of 50% (w/w) under continuous stirring, and this maintained for 5 min for complete homogenization.

2.1.2. Beads Synthesis

SF, KGM, and SF/KGM beads were prepared by using a peristaltic pump (7518-00, Masterflex®, Gelsenkirchen, Germany) engaged in a silicone hose (diameter: 0.3 mm, Masterflex®, Gelsenkirchen, Germany) and a needle (diameter: 0.70 mm × 30 mm, BD SoloMed, Curitiba, Brazil). Each solution was pumped to the needle and dripping into liquid nitrogen (fast freezing) for the formation of beads. Figure 1 shows the used experimental setup. Finally, the obtained beads were freeze-dried (L101, Liobras, São Carlos, Brazil) for 24 h.

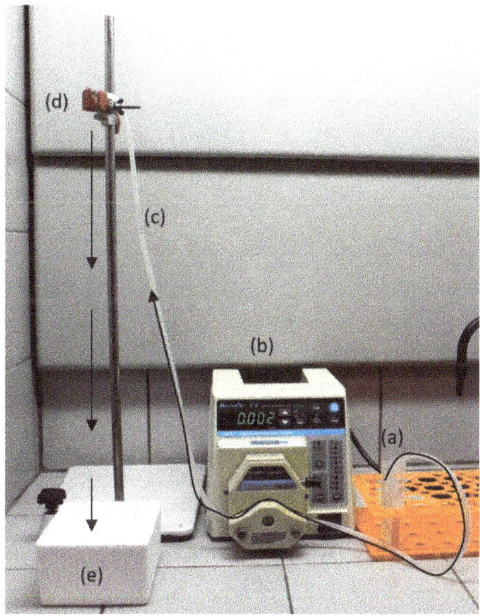

Figure 1. Experimental setup for beads syntheses: (**a**) natural polymer solution; (**b**) peristaltic pump; (**c**) silicone hos; (**d**) needle and (**e**) liquid nitrogen.

To study the effect of a post-treatment with ethanol, SF, KGM, and SF/KGM beads were treated with ethanol 95% (*v*/*v*) at room temperature for 24 h. Ethanol was used to induce SF structural transformation and KGM deacetylation [15–17]. After treatment, beads were rinsed with deionized water, frozen at 4 °C and lyophilized for 24 h.

2.2. Physicochemical Characterizations

2.2.1. Surface Tension and Size Beads

The surface tension of solutions was determined by a tensiometer (K12, KRUSS, Hamburg, Germany) using the ring method [18]. The size of beads was obtained analyzing fifteen images of each kind of sample using Image J® software (National Institutes of Health, Bethesda, USA).

2.2.2. Morphological Characterization

The morphology of the surface beads was examined using a scanning electron microscope (FEG 250, Thermo Scientific Quanta™, Waltham, MA, USA) with a current and voltage of 50 pA and 10 kV, respectively. The average pore size of the beads was determined from SEM images by ImageJ® software.

The samples were fixed on aluminum supports using double-sided carbon tape, and in sequence, the samples were metallized using a 200 Å thick gold layer. Metallization was performed in a vacuum chamber (K-450, Sputter Coater EMITECH, Quorum technologies, Lewes, UK) under argon flow for 15 min. After metallization was completed, the supports containing the metalized samples were placed under a microscope, and the samples were photographed for recording purposes.

2.2.3. Chemical Characterization

Attenuated total reflection Fourier-transform infrared spectroscopy (ATR-FTIR) was used for the chemical characterization of the samples. Analyses were carried out with an ATR-FTIR-6100 (JASCO, Easton, MD, USA) spectrometer. Each measurement was acquired in transmittance mode by an accumulation of 128 scans at a resolution of 4 cm^{-1} in the range of 650–4000 cm^{-1}. Deconvolution of FTIR spectra was obtained by a Gaussian function using the Fityk 0.9.8 software (Marcin Wojdyr, Warsaw, Poland). The percentage of β-sheet conformation was evaluated by integrating the curve area of the whole studied region.

2.2.4. Thermogravimetric Characterization

Thermogravimetric characterization was performed by TGA equipment (TGA/DSC1, Mettler Toledo, Barueri, Brazil) in a temperature range of 25–600 °C with a ramp rate of 10 °C/min, and an N_2 flow of 50 mL/min. The glass transition temperature (T_g) was obtained by a differential scanning calorimeter (DSC1, Mettler-Toledo, Barueri, Brazil). All measurements were carried out from 25 to 300 °C under a nitrogen atmosphere using a heating rate of 10 °C/min.

2.2.5. Crystallinity

The crystallinity of beads was studied by X-ray diffraction (X'Pert-MPD, Philips panalytical X-ray, Malvern Panalytical, Almelo, The Netherlands) with Cu-Kα radiation (λ = 1.54056 Å). The X-ray source was operated at 40 kV and 40 mA. The measurements were performed in a range of 2θ = 5–80° using a scanning speed of 0.033°/s with a step size of 0.02°.

2.3. Mechanical Assays

The mechanical properties of freeze-dried beads were measured by confined compression test using a TA.XT2 texturometer (Stable Micro Systems, Godalming, UK) with a 50 N load cell at room temperature. The speed test was 0.2 mm/s. The beads were subjected to a compressive force of up to 50%.

2.4. Cytotoxicity Assays

2.4.1. Cell Culture

Normally established cell lines HaCaT (human keratinocytes) and BALB/c 3T3 (murine fibroblast) were used. The cells were grown in plastic flasks (25 cm^2) with Roswell Park Memorial Institute (RPMI) 1640 medium, supplemented with 10% inactivated fetal bovine serum (FBS) and 1% antibiotic solution (penicillin and streptomycin). The cultures were incubated at 37 °C in an atmosphere containing 5% CO_2. The medium was changed every 48 h, and when the culture reached confluence, the subculture was treated with trypsin-EDTA, until complete release of the cells. The released cells were transferred to a new plastic flask or 24-well plates.

2.4.2. MTT Reduction Assay

HaCaT and BALB/c 3T3 cells were distributed in 24-well plates using a density of the 4×10^4 cell/mL and 6×10^4 cell/mL, respectively, and were incubated at 37 °C with 5% of CO_2 for 24 h. Later, the cells were treated with different concentrations of SF, KGM and SF/KGM for 24 h. After incubation, the medium was removed from the cells and 1 mL of

3-(4,5-dimethylthiazol-2-yl)-2,5-diphenyltetrazolium bromide (MTT) solution (0.5 mg/mL in FBS free culture medium) was added to each well. After incubation for 2 h at 37 °C, the MTT solution was removed and the formazan crystal was solubilized in 1 mL of ethanol. The plate was shaken for 5 min and the absorbance of each well was read in the spectrophotometer (Infinity M200Pro, Tecan, Männedorf, Switzerland). The measured absorbance at $\lambda = 570$ nm was normalized to % of control [19]. This value was calculated by multiplying the absorbance of a treated well by 100 and dividing it by the average absorbance of control well, which was considered 100%.

2.5. Statistical Analysis

Each experiment was carried out in triplicate unless otherwise specified. All results are presented as the mean ± standard deviation (SD). The experimental data from all the studies were analyzed using Analysis of Variance (ANOVA). Statistical significance was set to p-value ≤ 0.05.

3. Results and Discussions

3.1. Size Beads and Surface Tension

Figure 2 shows images of the beads treated without and with ethanol, and Table 1 presents the surface tension of solutions and diameters of beads. Beads under ethanol treatment showed smaller diameters than beads without treatment: SF, KGM and SF_{50}/KGM_{50} presented a decrease of 8.9%, 7.6% and 23.4%, respectively. In the case of SF_{50}/KGM_{50} beads, they showed an intermediate size between both pure natural polymers beads, due to the intermolecular interactions between both substances. According to the literature, higher surface tensions should produce higher diameters of drops, because the drop increases its size adding mass until the surface tension cannot keep it together [20]. However, in our case, the opposite was, solutions with higher surface tensions produced smaller diameter of beads (See Table 1 and Figure 2). These results suggest that other factors, in addition to the cohesive force between molecules, affect the diameter of the beads. An example is the molar mass of the natural polymer: higher molar mass would produce more compacted beads. In native form, the molar mass of SF is around 25 kDa for light chains and 325 kDa for heavy chain [21], while KGM is so much higher, around 1.50×10^6 Da [22].

Table 1. The surface tension of SF, KGM, and SF_{50}/KGM_{50} solutions and diameters of SF, KGM, and SF_{50}/KGM_{50} beads.

Samples	Surface Tension of Solutions (mPa) *	The Diameter of Beads without Ethanol Treatment (mm) *	The Diameter of Beads with Ethanol Treatment (mm) *
SF	45.2 ± 0.2 [a]	3.4 ± 0.1 [d]	3.1 ± 0.1 [h]
KGM	63.7 ± 0.8 [b]	2.0 ± 0.2 [f]	1.8 ± 0.1 [e]
SF_{50}/KGM_{50}	50.7 ± 0.7 [c]	3.2 ± 0.15 [d]	2.5 ± 0.1 [g]

* Mean ± SD ($n = 3$), mean with the same letter indicate that there is no significant difference ($p < 0.05$) by the Tukey test.

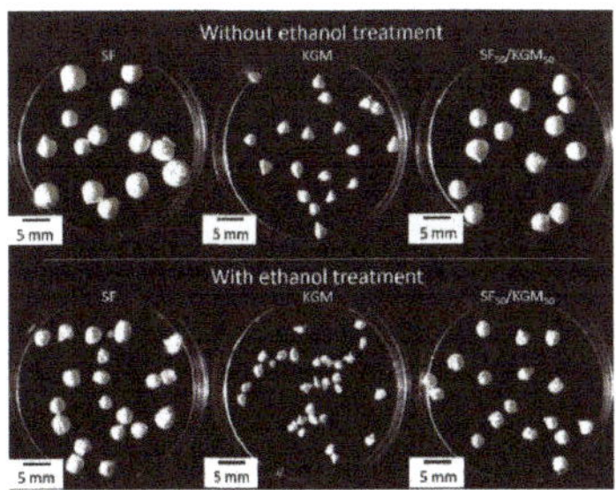

Figure 2. Images of the synthesized beads.

3.2. Morphological Characterization

Figure 3 shows the SEM images of beads surface. All cases presented pores and rough surfaces. The fast freezing with liquid nitrogen and the freeze-drying in vacuum-process used during the syntheses, generated these particular characteristics. In the case of SF beads, it is possible to observe that both samples, without and with ethanol treatment, presented homogenous surfaces. However, the untreated samples exhibited pores of 0.97 ± 0.2 µm, and the treated samples presented pores of 76.0 ± 26.5 µm. The KGM beads were the most affected by the ethanol treatment. Samples without treatment showed a spongy, homogeneous surface and many pores with a size of 1.3 ± 0.3 µm. The treated beads presented a heterogeneous, texturized surface and higher pores with a size of 32.5 ± 7.0 µm. The untreated samples of SF_{50}/KGM_{50} exhibited a rough and homogeneous surface with pores of 0.5 ± 0.2 µm. On the contrary, the treated beads presented large pores of 64.2 ± 13.4 µm. According to studies of Teimouri et al. (2015), the range of the pore size should be from 50 to 150 µm for successfully grown fibroblast cells and keratinocytes, since this size favors the transport of nutrients and oxygen [23]. In the case of the studies of Sepulveda et al. (2000), authors indicated that human osteoblasts could penetrate pores with a size between 100 and 200 µm, and a porosity above 30% is needed to facilitate oxygenation and nutrient transport [24]. Then, in the case of the synthesized beads, SF and SF_{50}/KGM_{50} treated with ethanol should provide the requirements for the cellular growth of fibroblasts and keratinocytes.

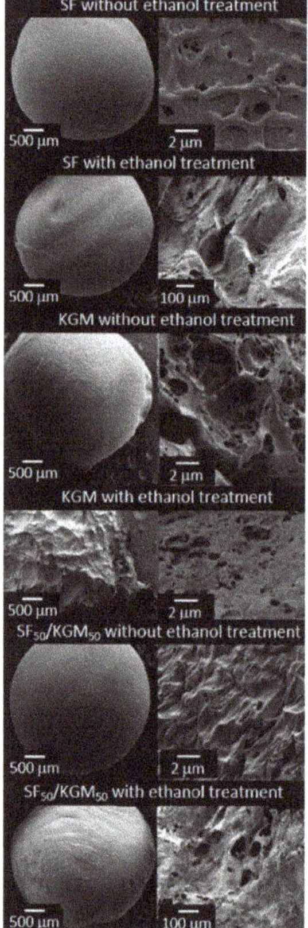

Figure 3. SEM images of the beads surface.

3.3. Chemical Characterization

To chemically characterize the synthesized beads, FTIR-ATR analyses were performed (Figure 4). Figure 4a presents the spectra of the untreated and treated beads with ethanol. In the case of the untreated SF beads, amide bands corresponding to the silk I conformation (α-helix or random coil) were identified [25]: amide I (1638–1655 cm^{-1}), amide II (1535–1540 cm^{-1}) and amide III (1230–1240 cm^{-1}). The amide bands I, II and III are attributed to C=O stretching, N–H deformation and O–C–N folding, respectively [25]. After the ethanol treatment the amide bands of the silk I were shifted to the amide bands corresponding to silk II (β-sheet): amide I (1616–1637 cm^{-1}), amide II (1515–1530 cm^{-1}) and amide III (1250–1260 cm^{-1}), which is a more stable conformation of the SF [26]. Both KGM spectra, untreated and treated beads, presented their typical bands [27] in the range of 1000–1300 cm^{-1}: at 1024 (C–O and C–OH), 1066 (C–H), 1153 (C–O), and 1247 cm^{-1} (CH$_2$OH and C(=O)O). The main difference in the treated bead appears in the 1643 cm^{-1} band, which is attributed to the deformation in the plane of the water molecule due to crystallization. This is an intermolecular hydrogen bonding interaction originated by treatment in an ethanol and alkali agent [27]. Finally, a slight band at 1730 cm^{-1} (C=O) was faded, which is related to an increase of deacetylation, causing an increase of hydrogen bonds [28]. As expected for the SF$_{50}$/KGM$_{50}$ beads, previously discussed bands corresponding to SF and KFM were identified. In fact, as the case of SF, SF$_{50}$/KGM$_{50}$ spectra exhibited a shift of higher energies: from amides of silk I to amides of silk II. The ethanol treatment made a more stable conformation of beads. Yang et al. (2000) reported the same tendency for SF/cellulose films [29]. Also, a shift in the amide bands was observed when comparing the spectra of SF with SF$_{50}$/KGM$_{50}$, which is usually accompanied by the formation of hydrogen bonds between polymers [30].

To achieve a better understanding of the chemical conformation of the blend SF/KGM, FTIR-ATR analysis was also performed on the treated beads using different proportions of SF/KGM (v/v) (Figure 4b). As KGM is added to the blend, the conformational transition of amides from α-helix conformation and random coil to β-sheet is increased. Bands shift from 1650, 1540 and 1250 cm^{-1} to 1630, 1522 and 1230 cm^{-1}. These results may indicate that a part of the random coil formation was converted to β-sheet with the addition of KGM, by intermolecular hydrogen bonds formed between the natural polymers [31]. According to the literature, when SF was mixed with other natural polymers containing hydroxyl groups (cellulose, alginate, heparin, and others), cross-linking between the polymers may benefit the conformational transition of SF [30]. Chen et al. (1997) performed structural analyzes on SF and chitosan films [32]. Authors found that SF exhibited β-sheet structure, indicating that hydrogen bonding was formed between SF and chitosan in the blend films. Also, they asserted that SF could use the rigid molecules of chitosan to extend SF molecules, resulting in the induction of the conformational β-sheet formation caused by hydrogen bonding. This study is consistent with the obtained results. Deconvolution corresponding to amide bands in FTIR-ATR spectra of SF/KGM was performed to analyze the evolution of β-sheet when KGM was added to the blend (Figure S1, Supplementary Materials). Table S1 (Supplementary Materials) shows the percent of β-sheet structure in each case of the amide. Results indicated that, as the proportion of KGM in the blend is increased, the percentage of β-sheet increase: 29.07%, 1.94% and 13.32% for amides I, II and III, respectively. These results show that part of the random coil was converted to β-sheet with the addition of KGM. The groups of KGM that can interact with SF are CH$_2$OH, COOH and OH. If the CH$_2$OH and COOH do not participate in the hydrogen bonding, the domain of the random coil conformation should be increased. However, as results show through the increasing of β-sheet, such groups may participate in the intermolecular bonds with the NH group of SF [33–35].

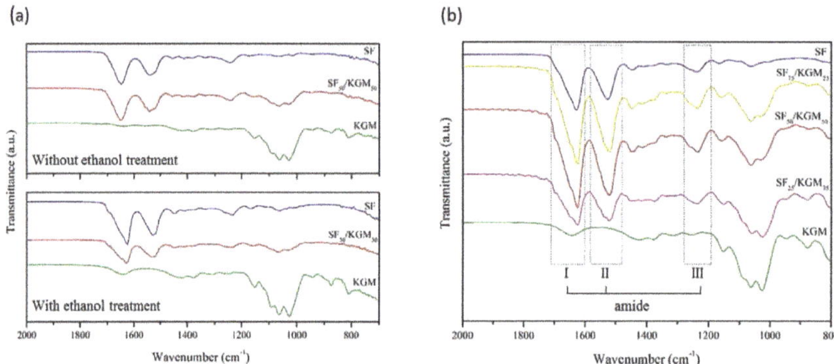

Figure 4. (**a**) FTIR-ATR spectra of beads, and (**b**) FTIR-ATR spectra of SF/KGM beads treated with ethanol obtained at different ratios of SF and KGM (*v/v*).

3.4. Thermal Characterization

Thermal stability of the beads was studied by thermogravimetric analysis (Figure 5A). The experiments involved three steps: (1) vaporization and removal of moisture in beads (to 100 °C), (2) material decomposition (at a different range for each case), and (3) thermal degradation and carbonization of the compounds (375 °C for all cases). In the case of SF samples, they presented a loss mass related to the breakdown of peptide bonds and side-chains of the residual amino acids [14,36]. Untreated and treated beads presented similar loss of mass in the second step: 36% and 33%, respectively. However, the decomposition temperature range of the beads treated with ethanol (235–297 °C) was higher than the untreated beads (216–283 °C). This suggests that ethanol post-treatment provides thermal stability, possibly due to the structural transition from silk I to silk II, which is more stable. The loss of mass of KGM is related to the decomposition of saccharide rings and the disintegration of the molecule chains [37]. In the second step, untreated beads showed 84% of loss mass, and treated beads just reached 55% of losing mass. The decomposition temperature range of untreated and treated beads were 223–322 °C and 260–296 °C, respectively. Although the maximum decomposition temperature was lower in treated samples than untreated samples, the initial temperature of decomposition was increased due to the higher deacetylation [38]. As ethanol decreases the acetyl groups of KGM, the intermolecular interactions are strengthened, and this consequently improves the thermal stability of the material [39]. In the case of SF$_{50}$/KGM$_{50}$ beads, the loss of mass was very similar between untreated and treated samples: 43% and 46%, respectively. Regarding the decomposition temperatures ranges, both cases were similar but with a slight increase in the maximum decomposition temperature: 233–299 °C for untreated beads, and 233–303 °C for treated beads.

The interaction between the polymers can also be analyzed by the glass transition temperature (T_g) of the beads, which were obtained from the DSC (Figure 5B). The calculated T_g was 116, 101 and 104 °C for the untreated samples of SF, KGM and SF$_{50}$/KGM$_{50}$, respectively. These results confirm that SF$_{50}$/KGM$_{50}$ was a miscible mixture because its T_g is an intermediate value between both T_g of pure polymers [40]. The same tendency was observed in the treated samples: 119, 107, and 108 °C for SF, KGM, and SF$_{50}$/KGM$_{50}$ beads, respectively. One point to highlight is that T_g of treated samples was higher than untreated ones, confirming that the ethanol treatment enhanced the stability of beads.

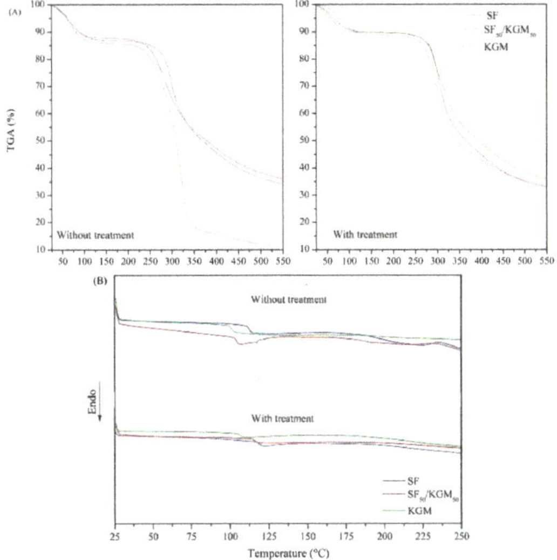

Figure 5. (A) TGA and (B) DSC analysis of beads.

3.5. Crystallinity

Figure 6 shows the X-ray diffractograms of samples. In the case of SF, values of 2θ at 23° and 45° are attributed to silk I (α-helix) while values at 21° are attributed to silk II (β-sheet) conformation. After ethanol treatment, the matrix becomes more crystalline in the silk II (β-sheet) structure, which shows that the protein chains of SF are more organized and energetically stable when they are treated with solvents or temperature [14,25,41]. According to the previous study of KGM cristallinaty [28], KGM beads presented their typical peak around 2θ = 20°. However, the increment of the peak around 12°, after the ethanol treatment, exhibited a higher deacetylation degree produced by the solvent [28]. As SF_{50}/KGM_{50} is a blend, their diffractograms presented lower crystallinity than diffractograms of pure SF and KGM [42]. However, in this case, significant changes in crystallinity were not caused by the ethanol treatment, as the cases of pure polymers.

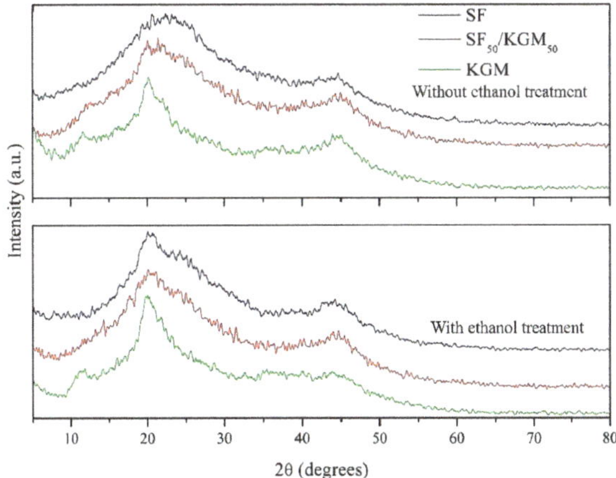

Figure 6. XRD diagrams of beads.

3.6. Mechanical Assays

Figure 7 presents the mechanical tests applied to samples obtained at different ratios of SF and KGM (v/v). For all cases (pure polymer beads and different blends) the ethanol treatment significantly improved the strength of samples. Regarding the SF component, the organic solvent dehydrated its chains favoring hydrogen bonds and Van der Waals forces. It, thus, changes to a more stable structural conformation [3]. According to the literature [16,43], the increasing of the mechanical resistance is caused by higher amounts of β-sheet structure, which is more stable. In the case of KGM, ethanol assisted in the deacetylation of the molecule [16], facilitating the formation of hydrogen bonds and hydrophobic interactions between chains composing a gel network structure [7,42]. This result was consistent with those obtained by ATR-FTIR, TGA and XDR analyses. When KGM was incorporated into SF, the mechanical strength of beads was increased. This is strongly related to intermolecular interactions of hydrogen bonds formed between both polymers [30,31,44], providing a more stable structure than pure polymer beads.

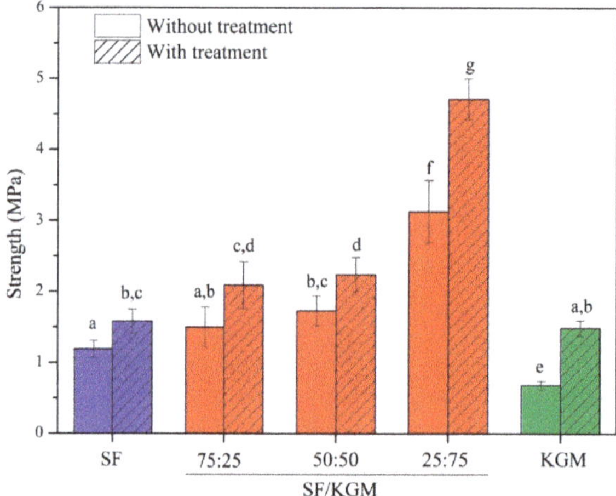

Figure 7. Mechanical properties of beads obtained at different ratios of SF and KGM (v/v). Mean ± SD (n = 15), means with the same letter indicate that there is no significant difference ($p < 0.05$) by the Tukey test.

Park et al. (1999) found that the mechanical tensile strength of SF/CHI membranes depended on CHI content. In fact, blends presented higher resistance than pure polymers, possibly due to the increase in the conformational transition β-sheet [9]. Other works in the literature have reported the increase of strength when one of the components is slightly added. For example, Yin et al. (2012) reported that scaffolds of collagen:chitosan:P(LLACL) were stronger when the ratio was 20:5:75 in comparison to pure P(LLACL) and collagen-chitosan blended scaffold (80:20:0) [45].

3.7. Cytotoxicity Assays

The cytotoxicity of solutions (Figure 8) and beads (Figure 9) was tested with two kinds of cell lines: human keratinocytes (HaCaT) and murine fibroblast (BALB/c 3T3).

Figure 8 shows that the natural polymers did not present cytotoxicity up to of 1 mg/mL. The IC50 value (reduction of 50% of the cell population) was not found in any of lineages. Also, it is noted the presence of SF in the SF/KGM solution favored the cytotoxicity of the SF since the profiles were very similar.

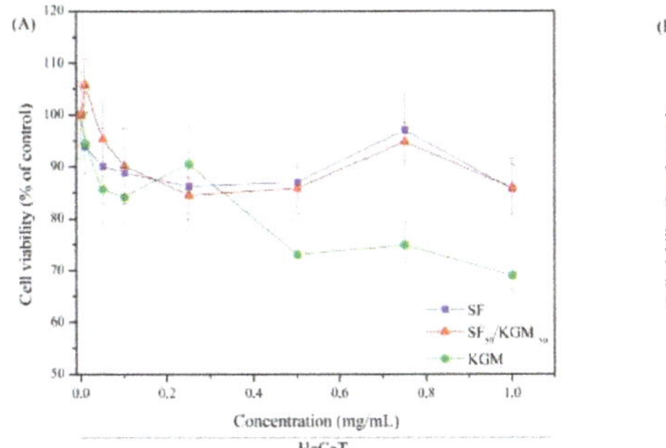
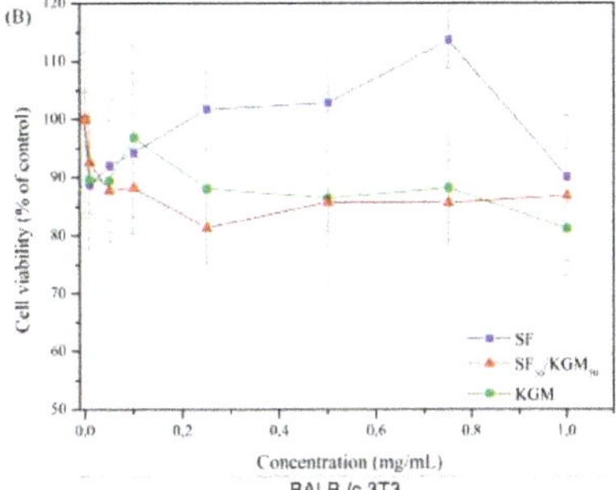

Figure 8. Cytotoxicity of solutions to (**A**) HaCaT and (**B**) BALB/c 3T3 cells. Mean ± SD (n = 3).

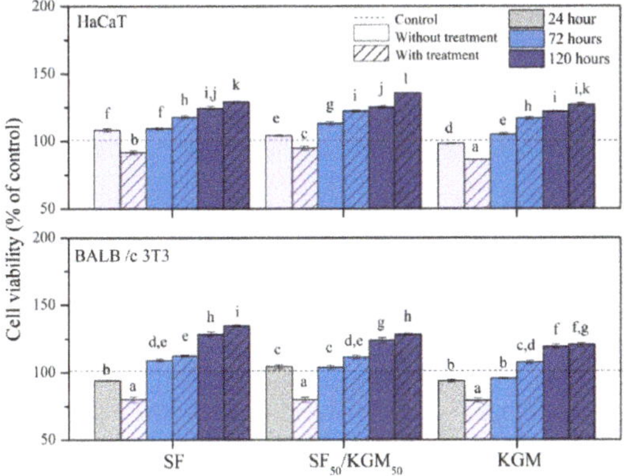

Figure 9. Cytotoxicity of beads to HaCaT and BALB/c 3 T3 cells. Mean ± SD (n = 3), means with the same letter indicate that there is no significant difference ($p < 0.05$) by the Tukey test.

KGM showed a tendency towards low cytotoxic profile for HaCaT in the concentration of 1 mg/mL since it is within the limit of the recommended in ISO 10993-5 (2009) [46]. Based on the MTT tests, it was not possible to estimate how the polymer acted in the cells, so it is necessary to perform signaling pathways, apoptosis (programmed cell death) and inflammatory process to verify the mechanism of decay of the cytotoxic profile.

In general, the decrease in cytotoxicity at the highest concentration tested was approximately 30% for HaCaT and 20% for BALB/c 3T3. When the cells were exposed to the combination of both natural polymers, it did not show cytotoxicity to the studied strains when compared to MTT reduction. Thus, the obtained beads could be applied in the field of tissue engineering.

Regarding the beads (Figure 9), in general, untreated and treated beads did not present cytotoxicity for both kinds of cell lines, since the cell viability for all cases and times was higher than 50%. Materials are considered cytotoxic only if it presents <50% of viable cells at any time [46,47]. However, during first 24 h of cell culture, treated beads showed a slight reduction of cell viability. This could be due to some residual traces of ethanol. Concentrations of this solvent above 0.1% may already be considered toxic to cells: dividing cells, trypsinization, and apoptosis. At 72 and 120 h of culture, all samples showed higher cell proliferation than control, probably due to natural selection occurring during the first 24 h. Nevertheless, the treated beads exhibited better cell viability than untreated samples. After 72 h, this increase can lead it to be assumed that treated beads acted as a dose-response system. Although in some cases, cell viability was similar to pure polymers, treated samples of SF₅₀/KGM₅₀ reached the highest values of cell viability: 136% $\pm$ 0.1% for HaCaT, and 135% $\pm$ 1% for BALBc/3T3, both at 120 h.

4. Conclusions

SF, KGM, and SF/KGM beads were obtained by a viable protocol that produced porous beads by dripping into liquid nitrogen, with a post-treatment using ethanol. Ethanol treatment reduced the diameter of beads but significantly increased the size of surface pores; making them viable for cell adhesion and growth. Physicochemical analysis of samples showed that SF/KGM beads were integrated due to the formation of hydrogen bonds between both polymers. Also, beads treated with ethanol exhibited more crystallinity and thermal stability, which were generated by a conformational transition from silk I (α-helix or random coil) to silk II (β-sheet) in SF, and an increase in the deacetylation of KGM. Due to these chemical changes, treated SF/KGM beads showed higher mechanical resistance than pure SF and KGM beads under compression tests. Finally, all beads did no present cytotoxicity effect for HaCaT and BALBc/3T3 cells up to 120 h and treated SF/KGM beads reached the highest values of cell viability. In general, the obtained porous blend beads presented improved characteristics compared to pure polymer beads, which could be used in the field of biomedicine.

Supplementary Materials: The following are available online at http://www.mdpi.com/2073-4360/10/8/923/s1, Figure S1: Deconvolution of FTIR-ATR spectra of SF/KGM beads treated with ethanol obtained at different ratios of SF and KGM (v/v), Table S1: β-sheet percent obtained from the deconvolution of FTIR-ATR spectra of SF/KGM beads treated with ethanol obtained at different ratios of SF and KGM (v/v) for the amide I, II and III. In red the percentage referring to the conformation β-sheet.

Author Contributions: C.G.F. and M.M.B. conceived and designed the research presented in this manuscript. M.L. and M.M.B. facilitated their labs for the experiments. C.G.F. and V.F.N. performed the synthesis of samples. C.G.F. and V.F.N. carried out the physicochemical characterization. D.M. and M.L. performed the biological characterization. C.F.G. and J.H.M. analyzed the data. C.F.G. and J.H.M. wrote the manuscript and made figures. All authors revised and approved the final version of the manuscript.

Funding: This study was financially supported by the Brazilian National Council for Scientific and Technological Development (CNPq).

Acknowledgments: The authors acknowledge Maria Helena Santana and Gilson Maia Júnior for granting access to analysis of surface tension. Bertran and lab team for providing the use of scanning microscopy equipment. Eduardo Gemis for granting access to infrared spectroscopy technique. The authors also kindly acknowledge Marcelle Spera and Denise Villalva for their valuable review and suggestions made to this work.

Conflicts of Interest: The authors declare no conflicts of interest.

References

1. Chen, X.; Wang, S.; Lu, M.; Chen, Y.; Zhao, L.; Li, W.; Yuan, Q.; Norde, W.; Li, Y. Formation and characterization of light-responsive TEMPO-oxidized konjac glucomannan microspheres. *Biomacromolecules* **2014**, *15*, 2166–2171. [CrossRef] [PubMed]

2. Wang, X.; Yucel, T.; Lu, Q.; Hu, X.; Kaplan, D.L. Silk nanospheres and microspheres from silk/pva blend films for drug delivery. *Biomaterials* **2010**, *31*, 1025–1035. [CrossRef] [PubMed]

3. Mori, H.; Tsukada, M. New silk protein: Modification of silk protein by gene engineering for production of biomaterials. *Rev. Mol. Biotecnol.* **2000**, *74*, 95–103. [CrossRef]

4. Hakimi, O.; Knight, D.P.; Vollrath, F.; Vadgama, P. Spider and mulberry silkworm silks as compatible biomaterials. *Compos. Part. B Eng.* **2007**, *38*, 324–337. [CrossRef]

5. Li, M.Z.; Lu, S.Z.; Wu, Z.Y.; Tan, K.; Minoura, N.; Kuga, S. Structure and properties of silk fibroin-poly(vinyl alcohol) gel. *Int. J. Biol. Macromol.* **2002**, *30*, 89–94. [CrossRef]

6. Altman, G.H.; Diaz, F.; Jakuba, C.; Calabro, T.; Horan, R.L.; Chen, J.S.; Lu, H.; Richmond, J.; Kaplan, D.L. Silk-based biomaterials. *Biomaterials* **2003**, *24*, 401–416. [CrossRef]

7. Alonso-Sande, M.; Teijeiro-Osorio, D.; Remunan-Lopez, C.; Alonso, M.J. Glucomannan, a promising polysaccharide for biopharmaceutical purposes. *Eur. J. Pharm. Biopharm.* **2009**, *72*, 453–462. [CrossRef] [PubMed]

8. Alvarez-Mancenido, F.; Landin, M.; Lacik, I.; Martinez-Pacheco, R. Konjac glucomannan and konjac glucomannan/xanthan gum mixtures as excipients for controlled drug delivery systems. Diffusion of small drugs. *Int. J. Pharm.* **2008**, *349*, 11–18. [CrossRef] [PubMed]

9. Park, S.J.; Lee, K.Y.; Ha, W.S.; Park, S.Y. Structural changes and their effect on mechanical properties of silk fibroin/chitosan blends. *J. Appl. Polym. Sci.* **1999**, *74*, 2571–2575. [CrossRef]

10. Kweon, H.; Ha, H.C.; Um, I.C.; Park, Y.H. Physical properties of silk fibroin/chitosan blend films. *J. Appl. Polym. Sci.* **2001**, *80*, 928–934. [CrossRef]

11. Lee, K.G.; Kweon, H.Y.; Yeo, J.H.; Woo, S.O.; Lee, J.H.; Park, Y.H. Structural and physical properties of silk fibroin/alginate blend sponges. *J. Appl. Polym. Sci.* **2004**, *93*, 2174–2179. [CrossRef]

12. She, Z.D.; Zhang, B.F.; Jin, C.R.; Feng, Q.L.; Xu, Y.X. Preparation and in vitro degradation of porous three-dimensional silk fibroin/chitosan scaffold. *Polym. Degrad. Stab.* **2008**, *93*, 1316–1322. [CrossRef]

13. De Alteriis, R.; Vecchione, R.; Attanasio, C.; De Gregorio, M.; Porzio, M.; Battista, E.; Netti, P.A. A method to tune the shape of protein-encapsulated polymeric microspheres. *Sci. Rep.* **2015**, *5*. [CrossRef] [PubMed]

14. Um, I.C.; Kweon, H.Y.; Park, Y.H.; Hudson, S. Structural characteristics and properties of the regenerated silk fibroin prepared from formic acid. *Int. J. Biol. Macromol.* **2001**, *29*, 91–97. [CrossRef]

15. Qu, J.Q.; Wang, L.W.; Hu, Y.; Wang, L.; You, R.Y.; Li, M.L. Preparation of silk fibroin microspheres and its cytocompatibility. *Biomat. Nanobiotechnol.* **2013**, *4*, 84–90. [CrossRef]

16. Li, J.; Ye, T.; Wu, X.; Chen, J.; Wang, S.; Lin, L.; Li, B. Preparation and characterization of heterogeneous deacetylated konjac glucomannan. *Food Hydrocoll.* **2014**, *40*, 9–15. [CrossRef]

17. Sun, L.; Xiong, Z.; Zhou, W.; Liu, R.; Yan, X.; Li, J.; An, W.; Yuan, G.; Ma, G.; Su, Z. Novel konjac glucomannan microcarriers for anchorage-dependent animal cell culture. *Biochem. Eng. J.* **2015**, *96*, 46–54. [CrossRef]

18. Wang, L.; Xie, H.G.; Qiao, X.Y.; Goffin, A.; Hodgkinson, T.; Yuan, X.F.; Sun, K.; Fuller, G.G. Interfacial rheology of natural silk fibroin at air/water and oil/water interfaces. *Langmuir* **2012**, *28*, 459–467. [CrossRef] [PubMed]

19. Mosmann, T. Rapid colorimetric assay for cellular growth and s urvival: Application to proliferation and c ytotoxicity assays. *J. Immunol. Methods* **1983**, *65*, 55–63. [CrossRef]

20. Lautrup, B. *Physics of Continuous Matter: Exotic and Everyday Phenomena in the Macroscopic World*; CRC Press: Boca Raton, FL, USA, 2011; p. 696.

21. Gong, Z.; Huang, L.; Yang, Y.; Chen, X.; Shao, Z. Two distinct beta-sheet fibrils from silk protein. *Chem. Commun.* **2009**, 7506–7508. [CrossRef] [PubMed]

22. Chua, M.; Chan, K.; Hocking, T.J.; Williams, P.A.; Perry, C.J.; Baldwin, T.C. Methodologies for the extraction and analysis of konjac glucomannan from corms of Amorphophallus konjac K. Koch. *Carbohydr. Polym.* **2012**, *87*, 2202–2210. [CrossRef]

23. Teimouri, A.; Ebrahimi, R.; Chermahini, A.N.; Emadi, R. Fabrication and characterization of silk fibroin/chitosan/nano gamma-alumina composite scaffolds for tissue engineering applications. *RSC Adv.* **2015**, *5*, 27558–27570. [CrossRef]

24. Sepulveda, P.; Binner, J.G.P.; Rogero, S.O.; Higa, O.Z.; Bressiani, J.C. Production of porous hydroxyapatite by the gel-casting of foams and cytotoxic evaluation. *J. Biomed. Mater. Res.* **2000**, *50*, 27–34. [CrossRef]

25. Rusa, C.C.; Bridges, C.; Ha, S.W.; Tonelli, A.E. Conformational changes induced in Bombyx mori silk fibroin by cyclodextrin inclusion complexation. *Macromolecules* **2005**, *38*, 5640–5646. [CrossRef]

26. Lv, Q.; Cao, C.B.; Zhang, Y.; Ma, X.L.; Zhu, H.S. Preparation of insoluble fibroin films without methanol treatment. *J. Appl. Polym. Sci.* **2005**, *96*, 2168–2173. [CrossRef]

27. Zhang, H.; Yoshimura, M.; Nishinari, K.; Williams, M.A.K.; Foster, T.J.; Norton, I.T. Gelation behaviour of konjac glucomannan with different molecular weights. *Biopolymers* **2001**, *59*, 38–50. [CrossRef]

28. Jin, W.; Song, R.; Xu, W.; Wang, Y.; Li, J.; Shah, B.R.; Li, Y.; Li, B. Analysis of deacetylated konjac glucomannan and xanthan gum phase separation by film forming. *Food Hydrocoll.* **2015**, *48*, 320–326. [CrossRef]

29. Yang, G.; Zhang, L.N.; Liu, Y.G. Structure and microporous formation of cellulose/silk fibroin blend membranes I. Effect of coagulants. *J. Membr. Sci.* **2000**, *177*, 153–161. [CrossRef]

30. Shang, S.; Zhu, L.; Fan, J. Intermolecular interactions between natural polysaccharides and silk fibroin protein. *Carbohydr. Polym.* **2013**, *93*, 561–573. [CrossRef] [PubMed]

31. Liang, C.X.; Hirabeyashi, K. Improvemets of the physical properties of fibroin membranes with sodium alginate. *J. Appl. Polym. Sci.* **1992**, *45*, 1937–1943. [CrossRef]

32. Chen, X.; Li, W.J.; Yu, T.Y. Conformation transition of silk fibroin induced by blending chitosan. *J. Polym. Sci. Part B Polym. Phys.* **1997**, *35*, 2293–2296. [CrossRef]

33. Hu, X.; Kaplan, D.; Cebe, P. Determining beta-sheet crystallinity in fibrous proteins by thermal analysis and infrared spectroscopy. *Macromolecules* **2006**, *39*, 6161–6170. [CrossRef]

34. Matsumoto, A.; Chen, J.; Collette, A.L.; Kim, U.-J.; Altman, G.H.; Cebe, P.; Kaplan, D.L. Mechanisms of silk fibroin sol-gel transitions. *J. Phys. Chem. B* **2006**, *110*, 21630–21638. [CrossRef] [PubMed]

35. Ling, S.; Qi, Z.; Knight, D.P.; Shao, Z.; Chen, X. Synchrotron FTIR microspectroscopy of single natural silk fibers. *Biomacromolecules* **2011**, *12*, 3344–3349. [CrossRef] [PubMed]

36. Markovic, M.; Fowler, B.O.; Tung, M.S. Preparation and comprehensive characterization of a calcium hydroxyapatite reference material. *J. Res. Natl. Inst. Stand. Technol.* **2004**, *109*, 553–568. [CrossRef] [PubMed]

37. Yu, H.; Huang, Y.; Ying, H.; Xiao, C. Preparation and characterization of a quaternary ammonium derivative of konjac glucomannan. *Carbohydr. Polym.* **2007**, *69*, 29–40. [CrossRef]

38. Enomoto-Rogers, Y.; Ohmomo, Y.; Iwata, T. Syntheses and characterization of konjac glucomannan acetate and their thermal and mechanical properties. *Carbohydr. Polym.* **2013**, *92*, 1827–1834. [CrossRef] [PubMed]

39. Jin, W.; Mei, T.; Wang, Y.; Xu, W.; Li, J.; Zhou, B.; Li, B. Synergistic degradation of konjac glucomannan by alkaline and thermal method. *Carbohydr. Polym.* **2014**, *99*, 270–277. [CrossRef] [PubMed]

40. Canevarolo, S.V., Jr. *Técnicas de Caracterização de Polímeros*; Artliber: São Paulo, Brazil, 2003; 448p.

41. Teimouri, A.; Ebrahimi, R.; Emadi, R.; Beni, B.H.; Chermahini, A.N. Nano-composite of silk fibroin-chitosan/nano ZrO$_2$ for tissue engineering applications: Fabrication and morphology. *Int. J. Biol. Macromol.* **2015**, *76*, 292–302. [CrossRef] [PubMed]

42. Xu, X.; Li, B.; Kennedy, J.F.; Xie, B.J.; Huang, M. Characterization of konjac glucomannan-gellan gum blend films and their suitability for release of nisin incorporated therein. *Carbohydr. Polym.* **2007**, *7*, 192–197. [CrossRef]

43. Lawrence, B.D.; Omenetto, F.; Chui, K.; Kaplan, D.L. Processing methods to control silk fibroin film biomaterial features. *J. Mater. Sci.* **2008**, *43*, 6967–6985. [CrossRef]

44. Guang, S.; An, Y.; Ke, F.; Zhao, D.; Shen, Y.; Xu, H. Chitosan/silk fibroin composite scaffolds for wound dressing. *J. Appl. Polym. Sci.* **2015**, *132*. [CrossRef]

45. Yin, A.L.; Zhang, K.H.; McClure, M.J.; Huang, C.; Wu, J.L.; Fang, J.; Mo, X.M.; Bowlin, G.L.; Al-Deyab, S.S.; El-Newehy, M. Electrospinning collagen/chitosan/poly(L-lactic acid-co-epsilon-caprolactone) to form a vascular graft: Mechanical and biological characterization. *J. Biomed. Mater. Res. Part A* **2013**, *101*, 1292–1301. [CrossRef] [PubMed]

46. International Organization for Standardization. *Biological Evaluation of Medical Devices—Part 5 Tests for In Vitro Cytotoxicity*; ISO 10993-5; ISO: Geneva, Switzerland, 2009.

47. Hernandez-Montelongo, J.; Lucchesi, E.G.; Nascimento, V.F.; Franca, C.G.; Gonzalez, I.; Macedo, W.A.A.; Machado, D.; Lancellotti, M.; Moraes, A.M.; et al. Antibacterial and non-cytotoxic ultra-thin polyethylenimine film. *Mater. Sci. Eng. C Mater. Biol. Appl.* **2017**, *71*, 718–724. [CrossRef] [PubMed]

Article

Sustained Delivery of Analgesic and Antimicrobial Agents to Knee Joint by Direct Injections of Electrosprayed Multipharmaceutical-Loaded Nano/Microparticles

Yung-Heng Hsu [1,2], **Dave Wei-Chih Chen** [3], **Min-Jhan Li** [2], **Yi-Hsun Yu** [1,2], **Ying-Chao Chou** [1,2] and **Shih-Jung Liu** [1,2,*]

[1] Department of Orthopedic Surgery, Chang Gung Memorial Hospital-Linkou, Tao-Yuan 33305, Taiwan; laurencehsu.hsu@gmail.com (Y.-H.H.); alanyu1007@gmail.com (Y.-H.Y.); enjoycu@ms22.hinet.net (Y.-C.C.)

[2] Department of Mechanical Engineering, Chang Gung University, Tao-Yuan 33302, Taiwan; liminjan127001@gmail.com

[3] Department of Orthopedic Surgery, Chang Gung Memorial Hospital-Keelung, Keelung 20401, Taiwan; mr5181@cgmh.org.tw

* Correspondence: shihjung@mail.cgu.edu.tw

Received: 30 May 2018; Accepted: 7 August 2018; Published: 9 August 2018

Abstract: In this study, we developed biodegradable lidocaine–/vancomycin–/ceftazidime–eluting poly(D,L–lactide–*co*–glycolide) (PLGA) nano/microparticulate carriers using an electrospraying process, and we evaluated the release behaviors of the carriers in knee joints. To prepare the particles, predetermined weight percentages of PLGA, vancomycin, ceftazidime, and lidocaine were dissolved in solvents. The PLGA/antibiotic/lidocaine solutions were then fed into a syringe for electrospraying. After electrospraying, the morphology of the sprayed nano/microparticles was elucidated by scanning electron microscopy (SEM). The in vitro antibiotic/analgesic release characteristics of the nano/microparticles were studied using high-performance liquid chromatography (HPLC). In addition, drug release to the synovial tissues and fluids was studied in vivo by injecting drug-loaded nano/microparticles into the knee joints of rabbits. The biodegradable electrosprayed nano/microparticles released high concentrations of vancomycin/ceftazidime (well above the minimum inhibition concentration) and lidocaine into the knee joints for more than 2 weeks and for over 3 days, respectively. Such results suggest that electrosprayed biodegradable nano/microcarriers could be used for the long-term local delivery of various pharmaceuticals.

Keywords: electrospraying; biodegradable nano/microparticles; drug delivery; septic arthritis; release characteristics

---

1. Introduction

Painful swollen joints remain a challenge for physicians and surgeons. The most severe cases are caused by septic arthritis, which occurs in approximately 2–10 out of 100,000 cases per year [1–4]. Delayed diagnosis without adequate treatment usually leads to permanent and even life-threatening joint damage. Indeed, the mortality rate of monoarticular sepsis is approximately 11% [5].

The most commonly found organisms in septic arthritis include Gram-positive staphylococci and streptococci in various populations, and Gram-negative organisms in older populations [2,6]. Both surgical intervention with arthroscopic irrigation or open arthrotomy to remove purulent material and serial irrigation have been employed as effective treatments for native septic knee arthritis [7]. In 2009, Ravindran et al. demonstrated that medical treatment (closed serial needle aspiration and intravenous antibiotics treatment) has satisfactory therapeutic outcomes for most cases of uncomplicated native septic arthritis [8].

Although debates continue over whether medical treatment or surgical intervention should be the standard of care for native septic arthritis, patients with septic arthritis generally must be admitted to a hospital for advanced assessment and treatment with intravenous antibiotics and/or surgical intervention (i.e., radical joint debridement). Broad-spectrum antibiotics are usually administered after bacterial culture and identification. The consensus for mainstay treatment includes the removal of purulent material combined with a 6-week course of adequate antibiotics, beginning with intravenous drugs followed by oral antibiotics [5]. However, possible alternatives are under study.

Knee injection of pharmaceuticals is commonly employed for the relief of knee discomfort. For that reason, we believe that knee injections of antimicrobial-and analgesic-loaded bioabsorbable nano/microcarriers could be a viable option for the treatment of septic arthritis. The procedure potentially provides high localized concentrations of broad-spectrum pharmaceuticals to knee joints for the treatment of infection and discomfort. Thus, we considered a method for creating such nano/microcarriers.

Electrospraying [9,10] is a simple and versatile electro-hydrodynamic technique for the production of nano/microparticulates [11–13] that has been widely employed to produce nano/microcarriers for drug delivery [14–17]. The process involves the application of a high electric field. When the electrostatic force is sufficiently high, solution at the end of a metallic needle stretches and forms a conical jet that eventually breaks into droplets. As the droplets travel toward the collector, the solvent evaporates and solid particles are collected. Core-shell type nano/microparticles for the encapsulation of susceptible protein-based therapeutic agents such as growth factor can be prepared using coaxial [18–21] or triaxial [22] electrospraying techniques.

Various studies have been performed on electrospraying and the production of drug-eluting nano/microparticles. Enayati et al. [12] used electrospraying to prepare particles, capsules, and bubbles for biomedical engineering applications. Hao et al. [15] employed electrospraying technology to formulate porous poly(lactic–*co*–glycolic acid) nano/microparticles. They found that the electrospraying method has a considerably shorter production time than the multi-emulsion method. Prabhakaran et al. [14] used electrospraying to fabricate metronidazole-loaded PLGA particles that provide a minimum of 41 days of sustained drug release in phosphate-buffered saline (PBS). Songsurang et al. [16] adopted electrospraying for the preparation of doxorubicin-loaded chitosan nanoparticles. They proposed that high encapsulation efficiency for doxorubicin-loaded nanoparticles was attainable. Malik et al. [17] reported a formulation recipe for generating spherical poly(lactic–*co*–glycolic acid) microspheres. Their technique employed electrospraying coupled with a novel thermally-induced phase separation process combined with a tailored liquid nitrogen collection scheme for the production of controlled-release medication.

Among the various polymers available for the formulation of local release systems, poly(D,L–lactide–*co*–glycolide) (PLGA) is one of the most thoroughly researched biodegradable materials due to its favorable properties including its highly controlled and consistent degradation characteristics and outstanding reproducible mechanical and physical properties [23,24].

The present study used electrospraying to develop biodegradable lidocaine–/vancomycin–/ceftazidime–eluting PLGA nano/microcarriers, which were subsequently evaluated for in vitro and in vivo release behaviors. Vancomycin was chosen because it is widely used for the treatment of infections complicated by Gram-positive methicillin-resistant *Staphylococcus aureus*. Meanwhile, ceftazidime exhibits strong activity against susceptible Gram-negative bacteria. Lidocaine is a cardiac depressant used as an antiarrhythmic agent [25], but it is also used as a local anesthetic.

2. Materials and Methods

To prepare the biodegradable drug-eluting particles, predetermined weight percentages of PLGA, vancomycin, ceftazidime, and lidocaine were dissolved in solvents. The PLGA/antibiotic/lidocaine solutions were then fed into a syringe for electrospraying. After electrospraying, the morphology of the sprayed microparticles was examined by scanning electron microscopy (SEM). In addition, the in vitro microparticle antibiotic/analgesic release characteristics were elucidated via high-performance liquid chromatography (HPLC). To characterize in vivo medication release to synovial tissues and fluids, drug-loaded nano/microparticles were injected into the knee joints of New Zealand white rabbits.

2.1. Fabrication of Biodegradable Drug-Eluting Microparticles

All materials utilized in this study, including PLGAs (50:50 with a molecular weight of 33,000 Da, and 75:25 with a molecular weight of 70,000 Da), vancomycin hydrochloride, ceftazidime hydrate, and lidocaine hydrochloride, were purchased from Sigma-Aldrich (St. Louis, MO, USA). Three types of solvents, namely dichloromethane (DCM), chloroform (CHF), and 1,1,1,3,3,3-hexafluoro-2-propanol (HFIP; Sigma-Aldrich), were used in the experiments.

Antibiotic–/analgesic–eluting nano/microparticles were prepared using the electrospraying device shown in Figure 1, which consisted of a syringe and needle (internal diameter of 0.60 mm), ground electrode, collector, and high voltage supply. PLGA and pharmaceuticals (360 mg PLGA, 60 mg vancomycin, 60 mg ceftazidime, and 60 mg lidocaine) [25] were first dissolved in 6 mL of solvent and mixed with a magnetic stirrer for 12 h. Three different types of solvents were employed to determine their feasibility for electrospraying. During electrospraying, the solution was transported utilizing a syringe pump that possessed a flow rate of 0.6 mL per hour. The journey distance between the needle and the collection was 20 cm.

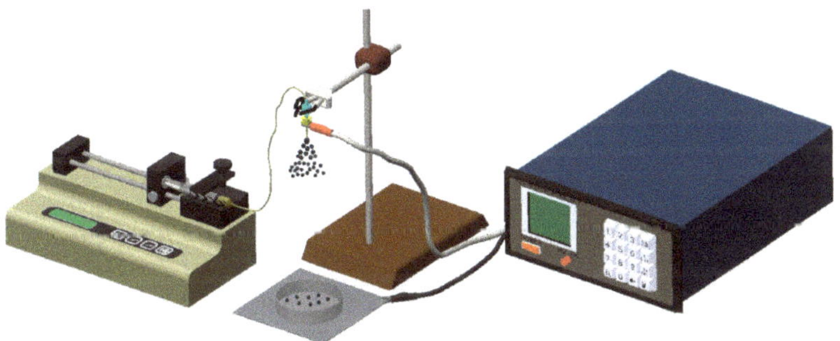

Figure 1. Schematic of the experimental setup to electrospray the drug-eluting nano/microcarriers.

2.2. Scanning Electron Microscopy

The morphology of electrosprayed nano/microparticles was observed on a field emission scanning electron microscope (FESEM; JEOL Model JSM-7500F, Tokyo, Japan) after coating with gold.

2.3. Infrared Spectrometry

Fourier transform infrared (FTIR) spectrometry was employed to examine the spectra of pure PLGA nano/microparticles and drug-loaded PLGA nano/microparticles. The FTIR analysis was conducted on a Nicolet iS5 spectrometer (Thermo Fisher Scientific, Waltham, MA, USA) at a resolution of 4 cm^{-1} (32 scans). Electrosprayed particles were pressed as KBr discs, and spectra were recorded over the range of 400–4000 cm^{-1}.

2.4. In Vitro Liberation of Antibiotics and Analgesic

The in vitro release of antibiotics and analgesics from the nano/microparticles was evaluated by an elution method. Particles of predetermined weights were added to the test tubes. Each tube contained 1 mL PBS and was incubated at 37 °C for 24 h. The eluent was then collected and replaced by 1 mL new PBS. This elution process was duplicated for 30 days. The concentrations of lidocaine and antibiotics in the collected solutions were quantified by HPLC, which was conducted on a Hitachi L-2200 Multisolvent Delivery System.

A Symmetry C_8 column (3.9 cm $\times$ 150 mm) was used to characterize vancomycin and ceftazidime (Waters, Milford, MA, USA), while an Atlantis dC18, 4.6 cm $\times$ 150 mm column (Waters Corp., Milford, MA, USA) was employed to evaluate lidocaine. The mobile phase used to separate vancomycin contained 0.01 mol heptanesulfonic acid (Fisher Scientific, Loughborough, UK) and acetonitrile (85/15, *v/v*; Mallinckrodt, St. Louis, MO, USA). The absorbance wavelength and flow rate were 280 nm and 1.4 mL/min, respectively. The mobile phase utilized for ceftazidime was composed of methanol and PBS (5 mM, pH 7.5) with a volumetric ratio of 10:90. The absorbance was monitored at 254 nm with a flow rate of 0.6 mL/min. The mobile phase used to quantify lidocaine consisted of ammonium formate and methanol (20/80, *v/v*; Sigma-Aldrich). The absorbance wavelength was set at 210 nm, while the flow rate was maintained at 1.0 mL/min. All experiments were performed in triplicate ($N = 3$).

2.5. In Vivo Study

Six New Zealand white rabbits weighing 2.8–3.2 kg were enrolled in the experiment. All animal-related procedures were approved by an institutional animal care and use committee (IACUC Approval No. CGU106-058). All animals received general anesthesia by inhaling isoflurane through a vaporizer (Matrx, Pompano Beach, FL, USA) in a plastic box (40 cm $\times$ 20 cm $\times$ 28 cm). Anesthesia was maintained during the entire injected procedure via mask inhalation of isoflurane. After the rabbits were sedated, the injected sites were depilated, washed with soft soap, and disinfected with 70% ethanol directly before the injection procedure. The posterior part of the animals was covered with a sterile blanket. The test solution (1 mL PBS + 200 mg drug-loaded nano/microparticles) was then directly injected into their bilateral knee joints (Figure 2).

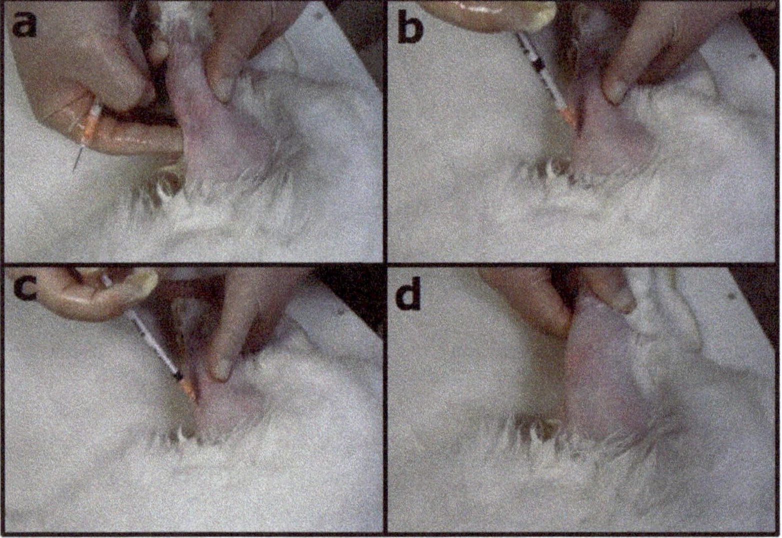

Figure 2. Knee injection with biodegradable microparticles to a New Zealand white rabbit. (**a**) The surgical site was shaved and the aseptic procedure was performed at the injection site; (**b**) the supra-patella pouch of the knee joint of rabbit was identified and penetrated with a needle; (**c**) 1 mL phosphate-buffered solution (PBS) + 200 mg drug-loaded nano/microparticles was injected; (**d**) free range of motion after injection was confirmed.

The vivo drug concentrations were determined by sampling specimens (synovial tissue and synovial fluid) on Days 1, 3, 7, and 14 after the injected procedure. Surgery of open biopsy was completed to obtain the specimens (Figure 3). The vancomycin, ceftazidime, and lidocaine concentrations of the specimens were determined by the same HPLC assay described in Section 2.4.

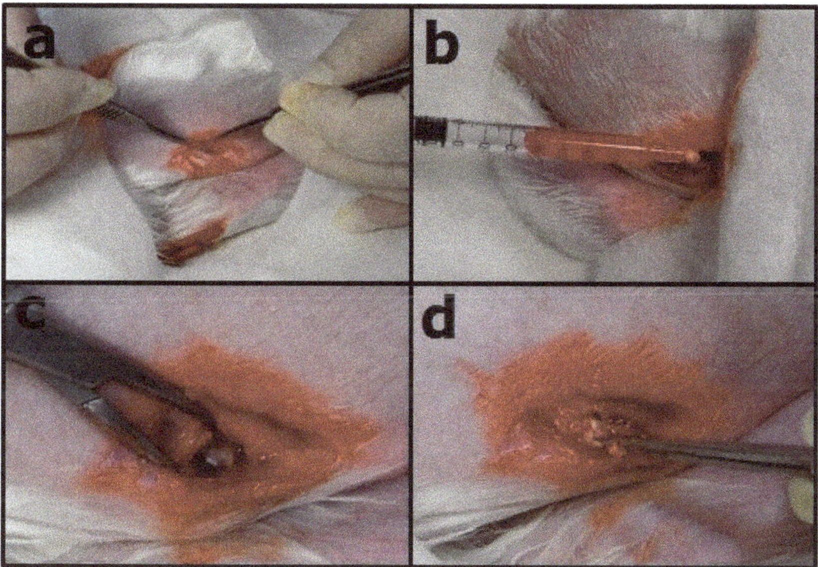

Figure 3. Sampling the specimen from the knee joint of New Zealand white rabbits. (**a**) The knee joint of the rabbit was opened; (**b**) the synovial fluid was sampled using a syringe; (**c**) the wound was opened; (**d**) the residual drug-eluting nano/microparticles could be seen.

3. Results

3.1. Characterization of Electrosprayed Particles

Three different types of solvents (i.e., DCM, CHF, and HFIP) were used to prepare the drug-eluting particles (50:50 PLGAs). Figure 4 shows SEM images of electrosprayed particles prepared with different solvents. Those made with DCM and CHF ranged from 2–5 μm to 5–10 μm, while electrosprayed particles prepared with HFIP were approximately 500 nm. Because the particles prepared with CHF exhibited a more uniform size and geometric distribution compared to those of other solvents, CHF was chosen for use in subsequent in vivo tests.

The influence of polymer types on the particles was also investigated. Figure 5 shows electrosprayed drug-loaded 75:25 PLGA particles (with DCM as the solvent). The size of the particles was approximately 10 μm, which is a little larger than that of the 50:50 PLGA microparticles.

Figure 6 displays the measured spectra of virgin PLGA microparticles and antibiotic/analgesic-loaded PLGA microparticles. The new absorption peak at 3200~3500 cm^{-1} could be attributable to the N–H bonds of vancomycin, ceftazidime, and lidocaine [26–28]. The absorption at 1725 cm^{-1} corresponded to C=O bonds, primarily due to the addition of antibiotics and analgesic; the absorbance peak near 1300 cm^{-1} may be the result of C–O bond enhancement in loaded pharmaceuticals. The FTIR spectra illustrates that the antibiotics and analgesic were successfully incorporated into PLGA microparticles.

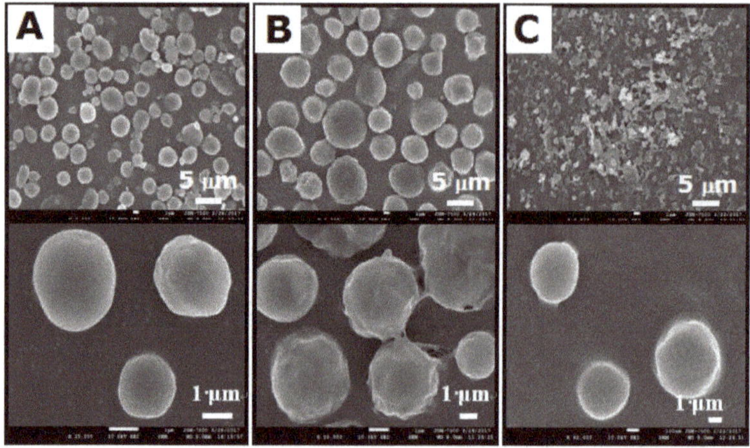

Figure 4. Drug-eluting 50:50 PLGA nano/microparticles fabricated using different solvents: (**A**) dichloromethane (DCM); (**B**) chloroform (CHF); (**C**) 1,1,1,3,3,3-hexafluoro-2-propanol (HFIP).

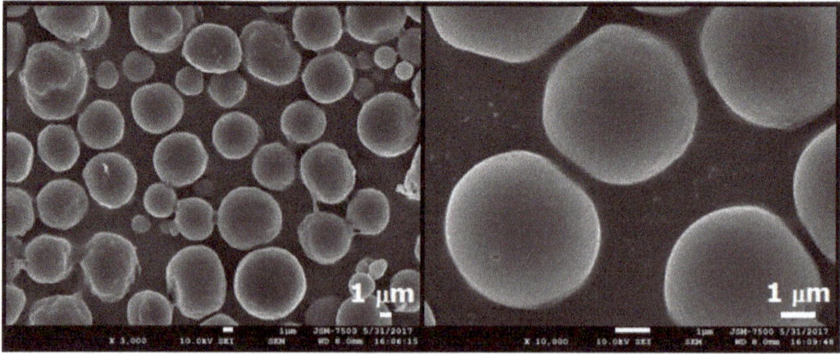

Figure 5. Drug-eluting 75:25 PLGA nano/microparticles (DCM was used as the solvent).

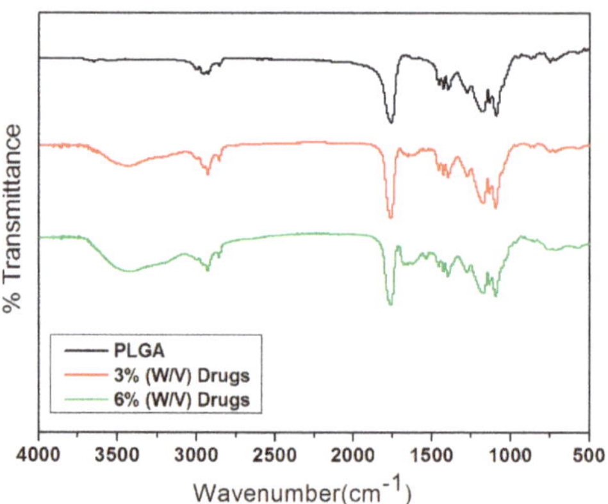

Figure 6. FTIR spectra of electrosprayed nano/microparticles. Abbreviations: FTIR, Fourier transform infrared; PLGA, poly(D,L)–lactide–*co*–glycolide.

3.2. In Vitro Release of Drug-Eluting Nano/Microparticles

Figures 7–10 show the measured daily and accumulated release curves of vancomycin, ceftazidime, and lidocaine from the electrosprayed drug-eluting nano/microparticles of varying particle weight loadings (50, 100, 200, 400 mg) immersed in 1 mL PBS. Triphasic and biphasic curve patterns were observed for antibiotics and lidocaine, respectively. The initial burst release pattern was observed within days for all pharmaceuticals at all particle weight loads. The second peak release of pharmaceuticals was observed on Days 24, 23, 22, and 19 for particle loads of 50, 100, 200, and 400 mg, respectively. The third peak release of antibiotics was noted on Day 28.

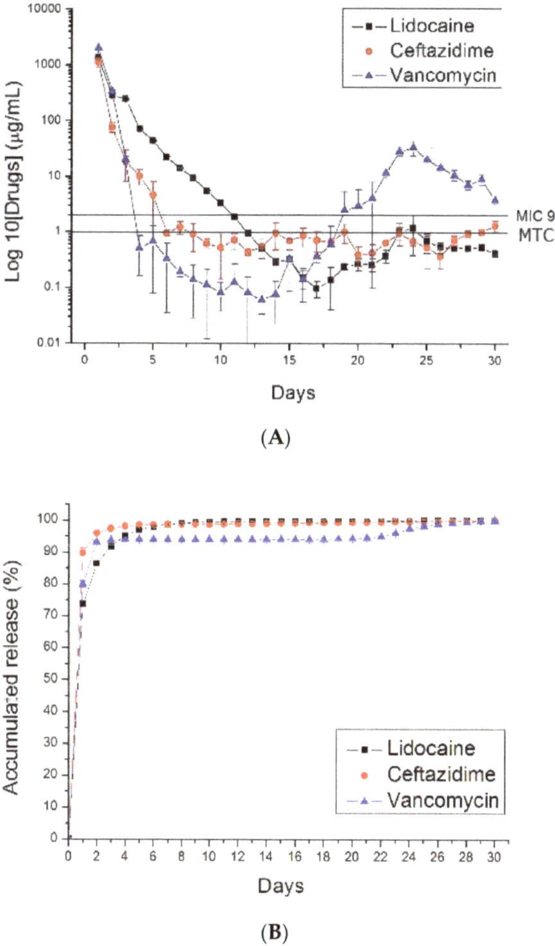

Figure 7. PLGA (50:50)—(**A**) daily and (**B**) accumulated release curves of pharmaceuticals from 50 mg PLGA nano/microparticle-loaded buffered solution.

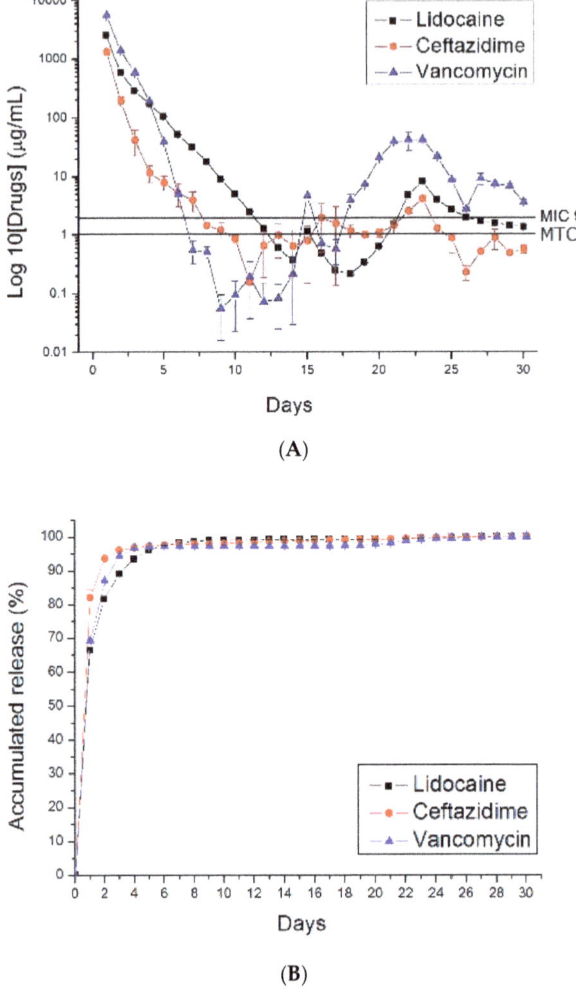

Figure 8. PLGA (50:50)—(**A**) daily and (**B**) accumulated release curves of pharmaceuticals from 100 mg PLGA nano/microparticle-loaded buffered solution.

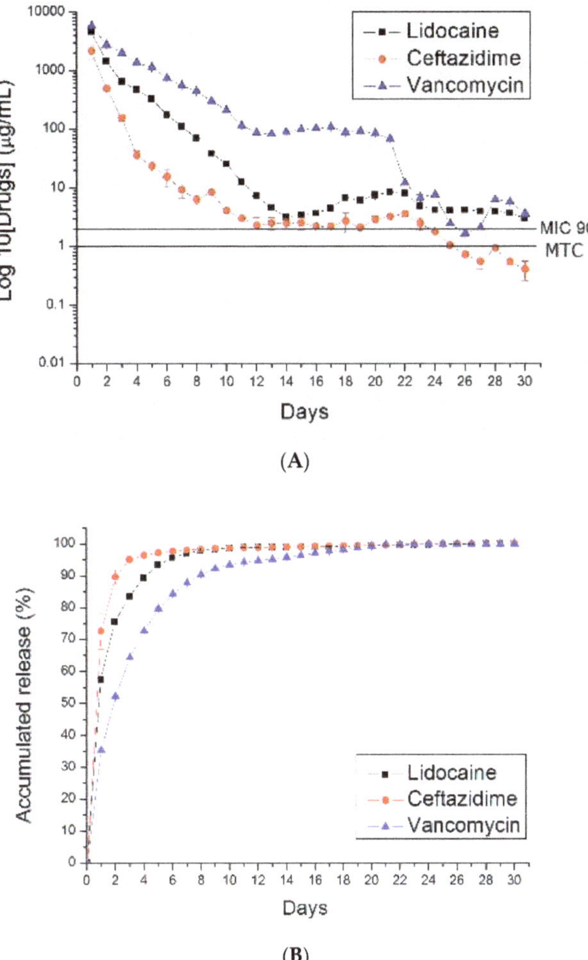

(**A**)

(**B**)

Figure 9. PLGA (50:50)—(**A**) daily and (**B**) accumulated release curves of pharmaceuticals from 200 mg PLGA nano/microparticle-loaded buffered solution.

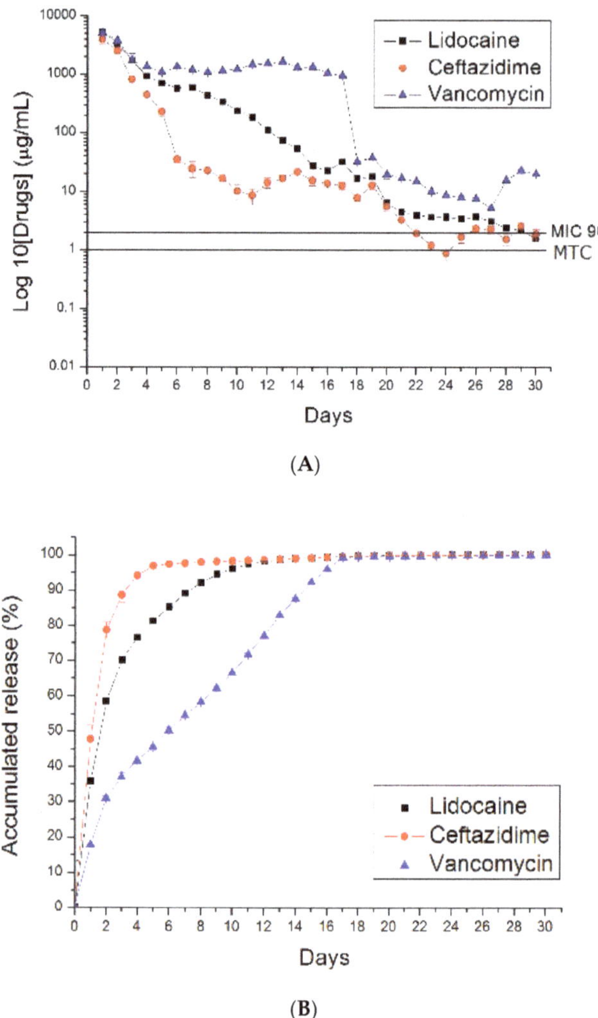

Figure 10. PLGA (50:50)—(**A**) daily and (**B**) accumulated release curves of pharmaceuticals from 400 mg PLGA nano/microparticle-loaded buffered solution.

The 50 and 100 mg nano/microparticle-loaded solutions exhibited similar release patterns, characterized by an initially high release on Day 1 and a second peak release after 20 days, followed by a more gradual and sustained release of the drugs. As drug loads increased to 200 and 400 mg, the second drug release peaks gradually declined and the total period of sustainable drug release increased. For example, the 50 mg drug-loaded nano/microparticle solution released high concentrations of vancomycin and ceftazidime, well above the minimum inhibition concentration 90% (MIC$_{90}$), for only 3 and 5 days, respectively, whereas the 400 mg drug-loaded solution liberated high antibiotic strengths (above MIC$_{90}$) for 30 and 21 days, respectively.

For comparison, drug-loaded 75:25 PLGA nano/microparticles were also prepared. Figure 11 shows the daily and accumulated release curves of vancomycin, ceftazidime, and lidocaine from electrosprayed drug-eluting 75:25 PLGA nano/microparticles (with a particle loading of 200 mg).

Compared to 50:50 PLGA (Figure 9), 75:25 PLGA nano/microparticles exhibited a slower drug release rate. This is probably due to the fact that 75:25 PLGA has a greater molecular weight than 50:50 PLGA. Degradation and drug release were reduced accordingly.

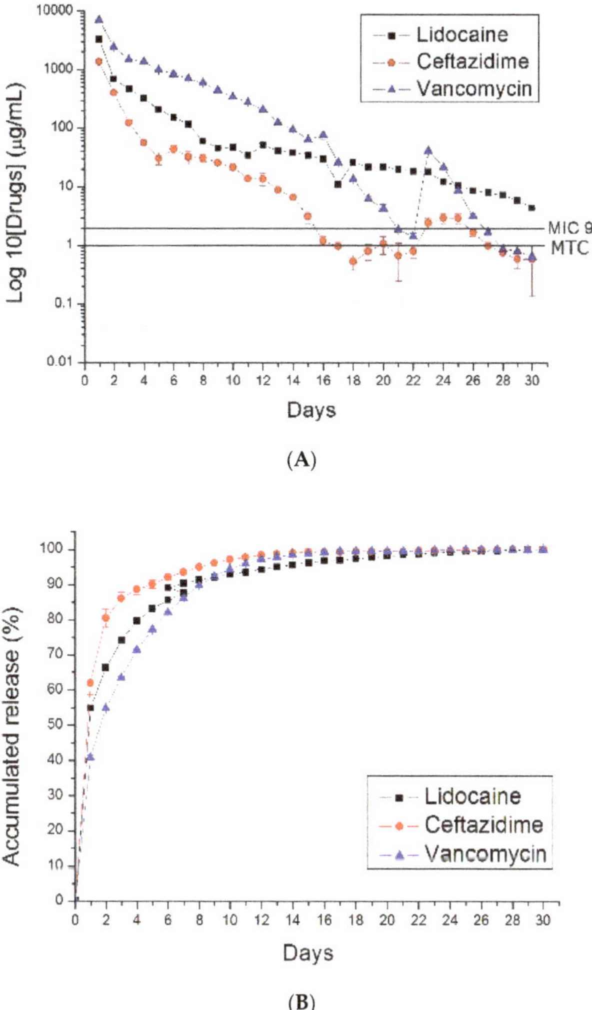

Figure 11. PLGA (75:25)-(**A**) daily and (**B**) accumulated release curves of pharmaceuticals from 200 mg PLGA nano/microparticle-loaded buffered solution.

3.3. *In Vivo Release*

The in vivo release of nano/microparticles in the knee joints of rabbits was characterized. Figure 12 displays the measured drug concentrations in synovial fluid and synovial tissue. The results demonstrate that drug-eluting nano/microparticles can provide a sustained release of high concentrations of vancomycin/ceftazidime (above MIC_{90}) and lidocaine (above the minimum therapeutic concentration, MTC [29]) for more than 14 days and 3 days, respectively.

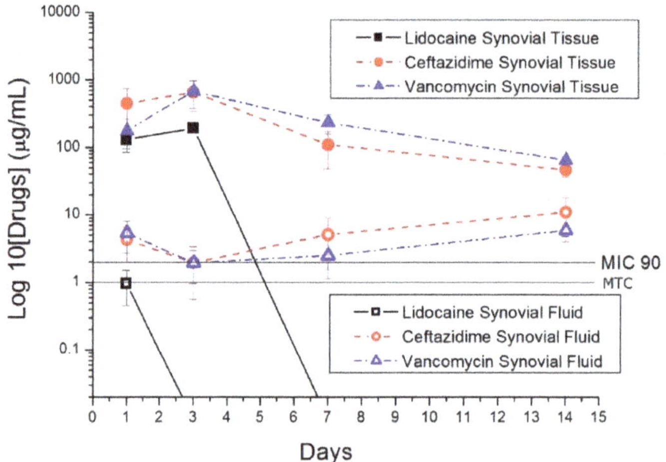

Figure 12. In vivo drug release curves of nano/microparticles in New Zealand white rabbits.

4. Discussion

Although both osteomyelitis and periprosthetic joint infections are commonly treated with vancomycin and ceftazidime combined with bone cement consisting of poly(methyl methacrylate) (PMMA) [30,31], PMMA is nonbiodegradable; it requires surgical insertion and removal. Thus, such treatment is not appropriate for native septic arthritis because the presence of foreign bodies may reduce the motion of knee joints. In contrast, the much small size of biodegradable nano/microparticles allows them to be easily injected into knee joints. They have less influence on the knees' range of motion than do larger particles, and their biodegradable nature eliminates the need for surgical removal.

Debates have continued during the past years over what standard of care is needed to treat native septic arthritis, by medical treatment or surgical intervention. Both surgical intervention with arthroscopic irrigation or open arthrotomy to remove purulent material and serial irrigation have been employed as an effective treatment for native septic knee arthritis [7]. Vinod et al. demonstrated the satisfactory therapeutic outcomes of medical treatment (closed serial needle aspiration and intravenous antibiotics treatment) for most uncomplicated cases of native septic arthritis [8]. However, no strong evidence is available and even no randomized clinical trial has been completed regarding the appropriate duration of antibiotics usage in septic arthritis treatment. The consensus milestone therapy includes the removal of purulent material and the subsequent use of adequate antibiotics for 6 weeks [5]. This study reports the successful development of multidrug-loaded PLGA nano/microparticles using electrospraying technology. The biodegradable drug-eluting nano/microparticles released high concentrations of antibiotics for over 2 weeks in vivo, which is advantageous for infection control at the knee joint. Based on the literature, the current work is the first study that explores the sustained release of high and local antibiotic concentrations in vivo at the native septic arthritis area with the injection of drug-eluting nano/microparticles. Furthermore, the nano/microparticles also released lidocaine at the target site for more than three days, offering the benefit of early-stage pain relief at the joint. The experimental results demonstrated that most of the pharmaceuticals were absorbed by the surrounding tissues. Moreover, serial injections of various doses of nano/microparticles at the knee may offer a theoretically staged delivery of various broad-spectrum antibiotics to the target site, if the initial antibiotics treatment was not sensitive to the identified bacteria.

PLGA is among the most suitable biodegradable polymeric materials used for the fabrication of drug delivery devices and tissue engineering [32]. The material is biocompatible and biodegradable, exhibits a wide range of erosion times, and possesses tunable mechanical properties. Most importantly,

PLGA is an FDA-approved polymer that has been extensively studied for the controlled release of small molecule drugs, proteins, and other macromolecules. Herein, we fabricated biodegradable drug-eluting PLGA nano/microparticles loaded with vancomycin, ceftazidime, and lidocaine.

Drug release from a biodegradable carrier can be generally divided into three stages, including a primary blast, a diffusion-dominated elution, and a degradation-dominated release. Meanwhile, the clustering effect of nano/microparticles also contributes to and affects release behavior.

After the electrospraying process, most drugs are dispersed into the volume of the nano/microparticles, but a few pharmaceutical compounds located on the particles' surface may lead to an initial drug release burst. All pharmaceuticals used in this study were water soluble. However, the octanol-water ratios (log P) (PubChem Compound Database) of vancomycin, lidocaine, and ceftazidime are 4.21, 2.84, and 0.53, respectively. Vancomycin has relatively non-water-soluble characteristics compared to those of lidocaine and ceftazidime; thus, it exhibited the fastest and greatest amount of release during the first stage. Meanwhile, vancomycin had the lowest percentage of accumulated release during the first stage, suggesting that most of the vancomycin was located inside the nano/microparticles.

Following the initial burst, the drug release profiles were not only controlled by diffusion, but also by polymer degradation. A relatively constant decline in the release rate of the antibiotics and analgesic was thus observed. As the number of nano/microparticles increased, clustering occurred, thus reducing the total surface area of the particles. The release rates of vancomycin, ceftazidime, and lidocaine decreased accordingly. Increased nano/microparticle-loading also raised the total amount of incorporated pharmaceuticals, which prolonged the effective drug release period. The third peak release behavior was only noted in the curves of antibiotics (all drug loads) after approximately 28 days. We believe that PLGA nano/microparticles produced biocompatible hydrolysis products (lactic and glycolic acids) upon biodegradation, liberating antibiotics and analgesic from within the nano/microparticulate carriers with drug release patterns similar to those observed on Day 1.

Substances with smaller molecular weights theoretically exhibit more rapid release rates during the diffusion period than those with larger molecular weights. In this study, the molecular weights of lidocaine, ceftazidime, and vancomycin were 323.34, 546.57, and 1449.27, respectively (PubChem Compound Database). Most of the lidocaine leached out of the nano/microparticles during the diffusion period over the first few days.

Vancomycin is notorious for renal toxicity. There is always a safety concern regarding vancomycin blood levels during the therapeutic period when high-dose drugs are administered systemically [31,33]. Our previous study on New Zealand white rabbits demonstrated that blood creatinine levels remain normal during the experiment when vancomycin combined with a biodegradable nanofibrous carrier is administered locally at a fracture site [34]. The experimental results in this study further suggested that the drug absorption rate of the tissues surrounding the drug-eluting nano/microparticles is high. This would provide advantages in terms of transporting high anti-microbial drug levels to the target site with minimized systemic side influence.

It is generally accepted that the sustained release of high local antibiotic concentrations contributes to infection control. Results of the current study demonstrated that biodegradable drug-eluting nano/microparticles could release high concentrations of vancomycin/ceftazidime (well above the MIC$_{90}$) and lidocaine for more than 14 days and 3 days, respectively. FTIR analysis also suggested that the pharmaceuticals loaded into nano/microparticles were not damaged by the electrospraying process. Thus, we believe that electrospray technology will enable the manufacture of multidrug-eluting nano/microparticles that will benefit the treatment of various diseases by providing sustained local drug delivery via particle injection.

5. Conclusions

This study successfully used electrospray technology to develop multidrug-loaded PLGA nano/microparticles. It is the first in vivo study to explore the local injection of drug-eluting nano/microparticles into knee joints, an area that can be affected by native septic arthritis, in order to achieve the sustained release of high concentrations of antibiotics. Indeed, the biodegradable drug-eluting nano/microparticles released high concentrations of antibiotics for over 2 weeks in vivo, and the drug absorption rate of the surrounding tissues was high, which is advantageous for the control of infection in knee joints. The nano/microparticles also released lidocaine at the target sites for more than 3 days, conferring early-stage pain relief.

In addition to integrating infection-control and pain relief strategies, our novel drug-eluting carriers have other practical clinical implications. For instance, serial injection of various doses of nano/microparticles at the knee might theoretically provide staged delivery of various broad-spectrum antibiotics to the target site, which is potentially useful in cases where the identified bacteria is not sensitive to initial antibiotic treatment. For this reason, further research on multidrug-loaded PLGA nano/microparticles is warranted.

Author Contributions: Y.-H.H. and S.-J.L. conceived and designed the experiments; Y.-H.H., Y.-C.C., and Y.-H.Y. performed the experiments; D.W.-C.C. and M.-J.L. analyzed the data; M.-J.L. and S.-J.L. contributed reagents/materials/analysis tools; Y.-H.H., D.W.-C.C. and S.-J.L. wrote the paper.

Funding: Funding was supported by the Ministry of Science and Technology, Taiwan (Contract No. 107-2221-E-182-017); and the Chang Gung Memorial Hospital (Contract No. CMRPD2G0251).

Conflicts of Interest: The authors declare no conflict of interest.

References

1. Mathews, C.J.; Weston, V.C.; Jones, A.; Field, M.; Coakley, G. Bacterial septic arthritis in adults. *Lancet* **2010**, *375*, 846–855. [CrossRef]
2. Kaandorp, C.J.; Dinant, H.J.; van de Laar, M.A.; Moens, H.J.; Prins, A.P.; Dijkmans, B.A. Incidence and sources of native and prosthetic joint infection: A community based prospective survey. *Ann. Rheum. Dis.* **1997**, *56*, 470–475. [CrossRef] [PubMed]
3. Clerc, O.; Prod'hom, G.; Greub, G.; Zanetti, G.; Senn, L. Adult native septic arthritis: A review of 10 years of experience and lessons for empirical antibiotic therapy. *J. Antimicrob. Chemother.* **2011**, *66*, 1168–1173. [CrossRef] [PubMed]
4. Geirsson, A.J.; Statkevicius, S.; Vikingsson, A. Septic arthritis in Iceland 1990–2002: Increasing incidence due to iatrogenic infections. *Ann. Rheum. Dis.* **2008**, *67*, 638–643. [CrossRef] [PubMed]
5. Coakley, G.; Mathews, C.; Field, M.; Jones, A.; Kingsley, G.; Walker, D.; Phillips, M.; Bradish, C.; McLachlan, A.; Mohammed, R.; et al. On behalf of the British Society for Rheumatology Standards, Guidelines and Audit Working Group. BSR & BHPR, BOA, RCGP and BSAC guidelines for management of the hot swollen joint in adults. *Rheumatology* **2006**, *45*, 1039–1041. [PubMed]
6. Dubost, J.J.; Soubrier, M.; De Champs, C.; Ristori, J.M.; Bussiere, J.L.; Sauvezie, B. No changes in the distribution of organisms responsible for septic arthritis over a 20 year period. *Ann. Rheum. Dis.* **2002**, *61*, 267–269. [CrossRef] [PubMed]
7. Johns, B.P.; Loewenthal, M.R.; Dewar, D.C. Open compared with arthroscopic treatment of acute septic arthritis of the aative knee. *J. Bone Joint Surg. Am.* **2017**, *99*, 499–505. [CrossRef] [PubMed]
8. Ravindran, V.; Logan, I.; Brian, E. Bourke Medical vs surgical treatment for the native joint in septic arthritis: A 6-year, single UK academic centre experience. *Rheumatology* **2009**, *48*, 1320–1322. [CrossRef] [PubMed]
9. Gaskell, S.M. Electrospray: Principles and Practice. *J. Mass Spectrom.* **1997**, *32*, 677–688. [CrossRef]
10. Nguyen, D.N.; Clasen, C.; Van den Mooter, G. Pharmaceutical applications of electrospraying. *J. Pharm. Sci.* **2016**, *105*, 2601–2620. [CrossRef] [PubMed]
11. Hao, S.; Wang, Y.; Wang, B.; Deng, J.; Liu, X.; Liu, J. Rapid preparation of pH-sensitive polymeric nanoparticle with high loading capacity using electrospray for oral drug delivery. *Mater. Sci. Eng. C* **2013**, *33*, 4562–4567. [CrossRef] [PubMed]
12. Enayati, M.; Chang, M.-W.; Bragman, F.; Edirisinghe, M.; Stride, E. Electrohydrodynamic preparation of particles, capsules and bubbles for biomedical engineering applications. *Colloids Surf. A Physicochem. Eng. Asp.* **2011**, *382*, 154–164. [CrossRef]

13. Bock, N.; Woodruff, M.A.; Hutmacher, D.W.; Dargaville, T.R. Electrospraying, a reproducible method for production of polymeric microspheres for biomedical applications. *Polymers* **2011**, *3*, 131–149. [CrossRef]

14. Prabhakaran, M.P.; Zamani, M.; Felice, B.; Ramakrishna, S. Electrospraying technique for the fabrication of metronidazole contained PLGA particles and their release profile. *Mater. Sci. Eng. C* **2015**, *56*, 66–73. [CrossRef] [PubMed]

15. Hao, S.; Wang, Y.; Wang, B.; Deng, J.; Zhu, L.; Cao, Y. Formulation of porous poly(lactic-co-glycolic acid) microparticles by electrospray deposition method for controlled drug release. *Mater. Sci. Eng. C* **2014**, *39*, 113–119. [CrossRef] [PubMed]

16. Songsurang, K.; Praphairaksit, N.; Siraleartmukul, K.; Muangsin, N. Electrospray fabrication of doxorubicin-chitosan-tripolyphosphate nanoparticles for delivery of doxorubicin. *Arch. Pharm. Res.* **2011**, *34*, 583–592. [CrossRef] [PubMed]

17. Malik, S.A.; Ng, W.H.; Bowen, J.; Tang, J.; Gomez, A.; Kenyon, A.J.; Richard, M.; Day, R.M. Electrospray synthesis and properties of hierarchically structured PLGA TIPS microspheres for use as controlled release technologies. *J. Colloid Interface Sci.* **2016**, *467*, 220–229. [CrossRef] [PubMed]

18. He, D.; Wang, S.; Lei, L.; Hou, Z.; Shang, P.; He, X.; Nie, H. Core-shell particles for controllable release of drug. *Chem. Eng. Sci.* **2015**, *125*, 108–120. [CrossRef]

19. Davoodi, P.; Feng, F.; Xua, Q.; Yan, W.-C.; Tong, Y.W.; Srinivasan, M.P.; Sharma, V.K.; Wang, C.-H. Coaxial electrohydrodynamic atomization: Microparticles for drug delivery applications. *J. Control. Release* **2015**, *205*, 70–82. [CrossRef] [PubMed]

20. Cao, L.; Luo, J.; Tu, K.; Wang, L.-Q.; Jianga, H. Generation of nano-sized core–shell particles using a coaxialtri-capillary electrospray-template removal method. *Colloids Surf. B Biointerfaces* **2014**, *115*, 212–218. [CrossRef] [PubMed]

21. Gao, Y.; Zhao, D.; Chang, M.-W.; Ahmad, Z.; Li, X.; Suo, H.; Li, J.-S. Morphology control of electrosprayed core-shell particles via collection media variation. *Mater. Lett.* **2015**, *146*, 59–64. [CrossRef]

22. Kim, W.; Kim, S.S. Synthesis of biodegradable triple-layered capsules using a triaxial electrospray method. *Polymer* **2011**, *52*, 3325–3336. [CrossRef]

23. Amass, W.; Amass, A.; Tighe, B. A review of biodegradable polymers: Uses, current developments in the synthesis and characterization of biodegradable polyesters, blends of biodegradable polymers and recent advances in biodegradation studies. *Polym. Int.* **1998**, *47*, 89–144. [CrossRef]

24. Doppalapudi, S.; Jain, A.; Khan, W.; Domb, A.J. Biodegradable polymers-an overview. *Polym. Adv. Technol.* **2014**, *25*, 427–435. [CrossRef]

25. Chou, Y.C.; Cheng, Y.S.; Hsu, Y.H.; Yu, Y.H.; Liu, S.J. A bio-artificial poly([D,L]-lactide-co-glycolide) drug-eluting nanofibrous periosteum for segmental long bone open fractures with significant periosteal stripping injuries: In vitro and in vivo studies. *Int. J. Nanomed.* **2016**, *11*, 941–953. [CrossRef] [PubMed]

26. Fraceto, L.F.; de Matos Alves, P.L.; Franzoni, L.; Braga, A.A.; Spisni, A.; Schreier, S.; de Paula, E. Spectroscopic evidence for a preferential location of lidocaine inside phospholipid bilayers. *Biophys. Chem.* **2002**, *99*, 229–243. [CrossRef]

27. Zarif, M.S.; Afidah, A.R.; Abdullah, J.M.; Shariza, A.R. Physicochemical characterization of vancomycin and its complexes with β-cyclodextrin. *Biomed. Res.* **2012**, *23*, 513–520.

28. Moreno, A.D.H.; Salgado, H.R.N. Development and validation of the quantitative analysis of ceftazidime in powder for injection by infrared spectroscopy. *Phys. Chem.* **2012**, *2*, 6–11. [CrossRef]

29. Schulz, M.; Schmoldt, A. Therapeutic and toxic blood concentrations of more than 800 drugs and other xenobiotics. *Pharmazie* **2003**, *58*, 447–474. [PubMed]

30. Mader, J.T.; Calhoun, J.; Cobos, J. In vitro evaluation of antibiotic diffusion from antibiotic-impregnated biodegradable beads and polymethylmethacrylate beads. *Antimicrob. Agents Chemother.* **1997**, *41*, 415–418. [PubMed]

31. Hsu, Y.H.; Hu, C.C.; Hsieh, P.H.; Shih, H.N.; Ueng, S.W.N.; Chang, Y. Vancomycin and ceftazidime in bone cement as a potentially effective treatment for knee periprosthetic Joint Infection. *J. Bone Joint Surg. Am.* **2017**, *99*, 223–231. [CrossRef] [PubMed]

32. Middleton, J.C.; Tipton, A.J. Synthetic biodegradable polymers as orthopedic devices. *Biomaterials* **2000**, *21*, 2335–2346. [CrossRef]

33. Springer, B.D.; Lee, G.C.; Osmon, D.; Haidukewych, G.J.; Hanssen, A.D.; Jacofsky, D.J. Systemic safety of high-dose antibiotic-loaded cement spacers after resection of an infected total knee arthroplasty. *Clin. Orthop. Relat. Res.* **2004**, *427*, 47–51. [CrossRef]
34. Hsu, Y.H.; Chen, D.W.; Tai, C.D.; Chou, Y.C.; Liu, S.J.; Ueng, S.W.N.; Chan, E.C. Biodegradable drug-eluting nanofiber-enveloped implants for sustained release of high bactericidal concentrations of vancomycin and ceftazidime: In vitro and in vivo studies. *Int. J. Nanomed.* **2014**, *9*, 4347–4355. [CrossRef] [PubMed]

 *polymers*

Article

Biomass Extraction Using Non-Chlorinated Solvents for Biocompatibility Improvement of Polyhydroxyalkanoates

Guozhan Jiang [1,†], Brian Johnston [1], David E. Townrow [1], Iza Radecka [1,*], Martin Koller [2], Paweł Chaber [3], Grażyna Adamus [3] and Marek Kowalczuk [1,3,*]

[1] School of Biology Chemistry and Forensic Science, University of Wolverhampton, Wolverhampton WV11LY, UK; guozhan.jiang@cranfield.ac.uk (G.J.); b.johnston@wlv.ac.uk (B.J.); D.Townrow@wlv.ac.uk (D.E.T.)
[2] C/o Institute of Chemistry, Office of Research Management and Service, University of Graz, NAWI Graz, Heinrichstrasse 28/III, 8010 Graz, Austria; martin.koller@uni-graz.at
[3] Centre of Polymer and Carbon Materials, Polish Academy of Sciences, M. Curie-Skłodowskiej 34, 41-819 Zabrze, Poland; pchaber@cmpw-pan.edu.pl (P.C.); grazyna.adamus@cmpw-pan.edu.pl (G.A.)
* Correspondence: i.radecka@wlv.ac.uk (I.R.); marek.kowalczuk@cmpw-pan.edu.pl (M.K.); Tel.: +48-483-2271-6077 (ext. 225) (M.K.); +44-1902-322-366 (I.R.)
† Current address: Centre for Bioenergy and Resource Management, School of Water, Energy and Environment, Cranfield University, Bedfordshire MK43 0AL, UK.

Received: 15 May 2018; Accepted: 29 June 2018; Published: 3 July 2018

Abstract: An economically viable method to extract polyhydroxyalkanoates (PHAs) from cells is desirable for this biodegradable polymer of potential biomedical applications. In this work, two non-chlorinated solvents, cyclohexanone and γ-butyrolactone, were examined for extracting PHA produced by the bacterial strain *Cupriavidus necator* H16 cultivated on vegetable oil as a sole carbon source. The PHA produced was determined as a poly(3-hydroxybutyrate) (PHB) homopolyester. The extraction kinetics of the two solvents was determined using gel permeation chromatography (GPC). When cyclohexanone was used as the extraction solvent at 120 °C in 3 min, 95% of the PHB was recovered from the cells with a similar purity to that extracted using chloroform. With a decrease in temperature, the recovery yield decreased. At the same temperatures, the recovery yield of γ-butyrolactone was significantly lower. The effect of the two solvents on the quality of the extracted PHB was also examined using GPC and elemental analysis. The molar mass and dispersity of the obtained polymer were similar to that extracted using chloroform, while the nitrogen content of the PHB extracted using the two new solvents was slightly higher. In a nutshell, cyclohexanone in particular was identified as an expedient candidate to efficiently drive novel, sustainable PHA extraction processes.

Keywords: cyclohexanone; γ-butyrolactone; chloroform; extraction; polyhydroxyalkanoates; PHB

---

1. Introduction

The search for biomaterials that are able to provide the best performance with an appropriate host response in medical devices has been based upon the understanding of all of the interactions within the biocompatibility phenomena [1]. Biomass-derived polyhydroxyalkanoates (PHA) constitute biomaterials of potential value for medical applications [2]. The structure of the deposited PHA granules consists of a polyester core surrounded by an organic shell composed of phospholipids, structural proteins, regulating proteins, polyester synthase, and depolymerase [3]. In order to eliminate microbial components (cell debris or metabolites) from crude PHA, the biopolymer has to be carefully purified before it is processed. In particular, bacterial endotoxins (heat-resistant lipopolysaccharides), which are synthesized primarily by Gram-negative microbes, and protein traces need to be efficiently removed [4]. Recently, the facile method for PHA purity evaluation versus protein residues by elemental analysis has been reported by some studies [5].

The existing recovery processes can be divided into two categories: disruption of the non-PHA cellular materials (NPCM) (including nucleic acids, lipids and phospholipids, peptidoglycan, proteinaceous materials including glycoproteins and, in some cases, lipopolysaccharides and other carbohydrates), followed by the release of the intact PHA granules from inside; and direct solvent extraction from within the cells. In the former method, the cell walls can be disintegrated using chemical, enzymatic, or mechanical means [6]. In a chemical digestion of the non-PHA cellular materials, the chemicals used are aqueous sodium hypochlorite solution (NaClO) [7], aqueous surfactant solutions [8], or a combination of surfactants and NaClO [8,9]. Enzymatic digestion of NPCM was first invented by ICI in the early 1990s where the NPCM was digested by a cocktail of proteolytic enzymes (proteases) in one stage or several stages [10]. Both the polymer yield and purity are low for these types of methods. Further purification is usually carried out by solvent extraction of the obtained PHAs. Some green solvents have been examined for disrupting cell walls. 1-ethyl-3-methylimidazolium methylphosphonate, an ionic liquid, has been used to disrupt cyanobacteria cell walls to extract the PHAs inside [11]. Supercritical CO_2 (scCO_2) has also been used to disrupt *Ralstonia eutropha* (today: *Cupriavidus necator*) biomass for the recovery of PHAs at 200 atm and 40 °C for 100 min, and 89% of PHA was recovered from the biomass [12]. This process was only recently optimized by using both scCO_2 and "CO_2 expanded ethanol", the latter allowing the recovery of highly pure PHA at a reduced temperature when compared to the use of scCO_2 alone. However, the ionic liquid per se is currently expensive, and scCO_2, although it does not leave behind any residues as all the other solvents do, is expensive in terms of equipment investment. For these reasons, their application to recover PHAs is not economically acceptable at present. Moreover, the PHAs undergo significant degradation after disrupting non-PHA cellular materials using these types of methods. For example, the molar mass of PHAs has a reduction of approximately 50% after treatment with NaClO to disrupt the cell walls [7]. Only recently, a rather exotic, biological method to generate intact PHA granules without affecting the native structure of the biopolyester was reported, which is based on the digestion of PHA-rich bacterial biomass by the mealworm *Tenebrio molitor*; as a product of digestion, PHA granules of remarkable purity exceeding 90% were obtained, which could be further purified using water and NaOH; of course, this approach is still in its infancy and it is doubtful regarding its suitability for large scale implementation [13].

In solvent extraction, a PHA solvent is used to dissolve PHA inside the cells, which is then precipitated from the solution using a PHA anti-solvent after removing the cell wall debris from the solution. Before extraction, a cell breakage step should be adopted to facilitate the extraction. In most cases, heating at above 100 °C is enough to weaken the cells of the best described PHA production strains such as *C. necator* [14]. Generally, the efforts needed to break the cell walls strongly depend on the type of production strain; whereas some strains like recombinant *Eschericha coli* (no natural PHA producers) easily burst, Gram-positive cells are typically more recalcitrant towards cell disintegration. In order to facilitate separation of the cell debris from the resultant PHA solution, the PHA concentration should be less than 5% by weight as the solution tends to be very viscous, rendering subsequent separation of the cell debris difficult.

There are not many solvents that can dissolve rather hydrophobic PHAs, especially those with a short chain length, e.g., poly(3-hydroxybutyrate) (PHB); this is especially valid for solvents of considerable polarity. Therefore, direct PHA extraction often resorts to chlorinated hydrocarbons such as chloroform, dichloromethane, or 1,2-dichloroethane [14], by which a highly pure PHA and high PHA recovery yield can be achieved. However, they have a severe toxicity and high environmental impact; their use clearly counteracts the sustainability principles of PHA manufacturing [15]. Hence, less toxic non-halogenated solvents have been investigated to develop solvent-based recovery systems. Many non-halogenated solvents have been mentioned in patents [16–18] such as cyclic carbonic esters, methyl ethyl ketone, cyclohexanone (CYC), and γ-butyrolactone (GBL). Among these potential solvents, only a few have been investigated in detail. Koller et al. [19] developed a high pressure unit to use acetone at 120 °C (above the solvent's boiling point) and a pressure of 7 bar to dissolve a short chain length PHA, poly(3-hydroxybutyrate-*co*-21.8%-3-hydroxyvalerate-*co*-5.14%-4-hydroxybutyrate), with a recovery yield of 98.9% and a purity of 99.0%; precipitation of the PHA from acetone simply occurs during cooling down to room temperature. Yang et al. [20] investigated the use of methyl ethyl ketone as a solvent to extract poly(3-hydroxybutyrate-*co*-3-hydroxyvalerate). The extraction was conducted at 100 °C for 5 min to achieve a recovery yield of 93% and a purity of 91%. Since methyl ethyl ketone has a boiling point of only 79.6 °C, this extraction was also conducted using pressurized equipment. Rosengart et al. [21] investigated anisole and CYC as the solvents to extract PHB from *Burkholderia sacchari* cells. Polymer recovery yields of 97% and 93% were obtained, respectively, at 120–130 °C for about 15–30 min using a cell/solvent ratio of 1.5% (*w/v*). Maximum polymer purities using these experimental conditions were 98% for both solvents.

In the above solvent-based recovery systems, kinetics of the recovery was scarcely investigated, which is important for the design of emerging application areas. In this work, the kinetics of the extraction of PHAs from the cells was investigated via the use of gel permeation chromatography (GPC). This method is universal for other solvent systems. For this purpose, two solvents, CYC and GBL, were selected. The two solvents in this work have high boiling points (156 °C and 204 °C, respectively), which will be an advantage for the extraction to be conducted at atmospheric pressure but high temperature. The former was investigated primarily in terms of health and safety for the extraction of PHAs [21], while the latter has not been studied in detail in the open literature.

2. Materials and Methods

2.1. Synthesis

The PHA contained in biomass was produced via a two-stage fermentation process with strain *C. necator* H16 with vegetable oil as the sole carbon source in a 5 L batch fermenter (Electrolab FerMac 310). The starter medium contained 30 g/L of tryptic soy broth (TSB). For production, a basal salt medium (BSM) with the following composition was used: 1 g/L $Na_2HPO_4\cdot 2H_2O$, 1 g/L KH_2PO_4, 1 g/L $(NH_4)_2SO_4$, 0.1 g/L $MgSO_4\cdot 7H_2O$, 0.1 g/L KNO_3, 0.1 g/L NaCl, and 10 mL/L trace element solution. The trace element solution consisted of 2 g/L $FeCl_3$, 2 g/L $CaCl_2$, 2 g/L $CuSO_4\cdot 5H_2O$, 2 g/L $MnSO_4\cdot 5H_2O$, 2 g/L $ZnSO_4\cdot 5H_2O$, and 2 g/L $(NH_4)_6Mo_7O_{24}\cdot 4H_2O$.

The starter culture was prepared by inoculation of 250 mL TSB solution in a 500 mL flask with a single colony of the strain; incubation was done at 35 °C for 24 h with a shaking rate of 150 rpm. The broth was then centrifuged to remove the clear supernatant, then the solid bacterial pellet was resuspended in 500 mL BSM; this culture was transferred into the fermenter as inoculum. The sole carbon source vegetable oil (from a local Asda supermarket, Waterloo Road, Wolverhampton, UK) was first mixed with 500 mL BSM, and then emulsified using sonication (Bandelin Electronic sonicator, Berlin, Germany). The vegetable oil/BSM mixture was then transferred into the fermenter. The total volume of the production medium was 3500 mL (including the inoculum, oil, and BSM) containing 40 g of vegetable oil in the 5 L batch fermenter.

The production was carried out for 72 h at 30 °C. The pH was automatically controlled at 7.0 ± 0.05 by adding either 2 M HCl or 2 M NaOH. Dissolved oxygen was controlled at below 4.5% of the saturation concentration of oxygen by regulating the stirrer speed and sterile air flow rate.

2.2. Downstream Extraction and Determination of Extraction Kinetics

After centrifugation (4500 rpm, 10 min, Sigma 6-16S, Sigma Laborzentrifugen GmbH, Osterode am Harz, Germany) and lyophilization (−40 °C, 5 mbar, 48 h, Edwards freeze dryer, Crawley, UK), the dried biomass was degreased in acetone (AR, Fisher Scientific, Loughborough, UK) with a volume/mass ratio of 20/1 at room temperature overnight under magnetic stirring [22]. The degreased dried biomass was then extracted using chloroform (AR, Fisher), CYC (AR, Fisher Scientific,

Loughborough, UK), and GBL (AR, Fisher Scientific, Loughborough, UK), respectively. Chloroform was used as the control solvent for this work, acting as the "benchmark" to assess the performance of the other two solvents. Ten times the volume of methanol (AR, Fisher Scientific, Loughborough, UK) was used as the PHA anti-solvent to precipitate the dissolved PHAs from their solutions. The precipitation was performed by dropwise addition of the polymer solution into methanol under vigorous magnetic stirring. When chloroform was used as the extracting solvent, a Soxhlet extractor was used. The biomass was placed in a thimble, and the polymer was extracted into the solution in the distillation flask. When CYC or GBL was used as the extracting solvent, the biomass was placed into the solvent with a ratio of 1/50 (g/mL). The extraction was performed at 80–120 °C under vigorous stirring. After half an hour, the solution was hot filtered to remove the cell debris, and the polymer was then precipitated from the solution by adding methanol.

To determine the kinetics of the extraction, 3 mL of the solvent was heated to a predetermined temperature in a 10 mL round bottom flask. In this work, three temperature levels were used: 80, 100, and 120 °C. After the temperature was reached, 0.30 g of dried biomass was added into the hot solvent with magnetic stirring. At different time intervals, 0.1 mL of the mixture was sampled and immediately dissolved in HPLC grade chloroform for GPC analysis as follows. The concentration of PHA extracted in the solvent was determined using the area ratio of the two peaks on the GPC chromatogram, one for PHA and one for the solvent. The area ratio was calibrated using several ratios of PHA and the solvent in chloroform. The recovery yield of PHA was calculated using the following equation:

$$y = V d_s x / (m x_0) \tag{1}$$

where y is the recovery yield; V is the total volume of the solvent; d_s is the density of the solvent; x is the mass fraction of the PHA in the solution measured by GPC as described in the following; m is the mass of the dried biomass added to the solvent; and x_0 is the PHA content in the dried biomass determined by GC-MS.

2.3. Determination of PHA Using GC-MS

The PHA in the dried biomass and the purity of the PHA extracted were determined using GC-MS. For this purpose, the dried biomass or the extracted PHA was subjected to methanolysis in the presence of sulfuric acid according to the well-established procedure [23]. Benzoic acid was used as the internal standard. The samples were then analyzed on an Agilent 7890B GC coupled with a 5977A mass spectrometer (Santa Clara, CA, USA). The column used was DB5-MS UI with a size of 30 m × 0.25 mm × 0.25 mm. The temperatures of the injection port, interface, quadrupole, and ion source were set at 250 °C, 280 °C, 120 °C, and 250 °C, respectively. The oven temperature was programmed at an initial temperature of 40 °C and subsequently raised at a rate of 10 °C/min to 280 °C and held for 5 min. Helium carrier gas was set at a flow rate of 1.2 mL/min. Solvent delay was set at 2.5 min. The MS detector using electron impact (EI) ionization at 70 eV was operated in full scans (mass range of m/z 40 to 600).

2.4. Nuclear Magnetic Resonance (NMR)

The ^{1}H NMR spectra of the extracted PHA were recorded on a JEOL ECX-400 (400 MHz) spectrometer (Hertfordshire, UK). The solvent for the NMR experiments was chloroform-*d*. For the ^{1}H and ^{13}C spectra, 16 and 1024 scans, respectively, were used. ^{1}H spectrum chemical shifts were referred to CHCl$_3$ (δ = 7.25 ppm), ^{13}C spectrum chemical shifts to CDCl$_3$ (δ = 77.00 ppm).

2.5. Gel Permeation Chromatography (GPC)

The kinetics of the extraction was determined using a TOSOH EcoSec HLC/GPC 8320 system GPC system (gel permeation chromatography). The GPC system was equipped with a UV and RI detector, operated at a temperature of 40 °C. The column used was a TSKgel HZM-N calibrated against monodisperse polystyrene standards ranging from 560 to 70,000 Da. The UV detector was set at a wavelength of 254 nm. Chloroform was used as the eluent at a flow rate of 0.25 mL/min. A sample size of 2 µL was injected into the system using an autosampler.

2.6. Elemental Analysis

The elemental composition of the extracted PHA was determined using a CHN analyzer. Elemental analysis as a measure for product purity was performed using a Perkin Elmer Analyzer 2400 as described in [5].

3. Results and Discussion

Biotechnological production of PHAs with *C. necator* is a well-established procedure [24]. In this work, the PHA was produced using vegetable oil as the sole carbon source in a batch fermenter for 72 h. The final dried biomass collected was 30.1 g with a PHA content of 82.3%. The conversion yield of substrate to PHA was about 0.62 g/g. About 2 g of the dried biomass was extracted using boiling chloroform under reflux in a Soxhlet extractor for 24 h. The extracted PHA was analyzed using NMR to determine the type of the PHA.

The ^{1}H NMR spectrum of the PHA is shown in Figure 1. Peak 1 at 1.20 ppm was attributed to the methyl protons (side chain of 3-hydroxybuytrate). Peak 2 at 2.69–2.24 ppm was due to the diasterotopic protons at position 2 of the chemical structure (backbone of 3-hydroxybutyrate). Peak 3 at 5.19 ppm is due to the proton bonded to the carboxyl oxygen. The ^{13}C NMR spectrum of the obtained PHA is shown in Figure 2. There were four peaks (δ = 20.5 ppm, 41.5 ppm, 68.3 ppm, and 170 ppm) in the spectrum that could be attributed to methyl carbon (side chain of 3-hydroxybuytrate), methylene carbon (backbone of 3-hydroxybutyrate), methine carbon (chiral center of 3-hydroxybutyrate), and carbonyl carbon, respectively. From both the ^{1}H and ^{13}C spectra, it can be confirmed that the PHA produced was primarily PHB.

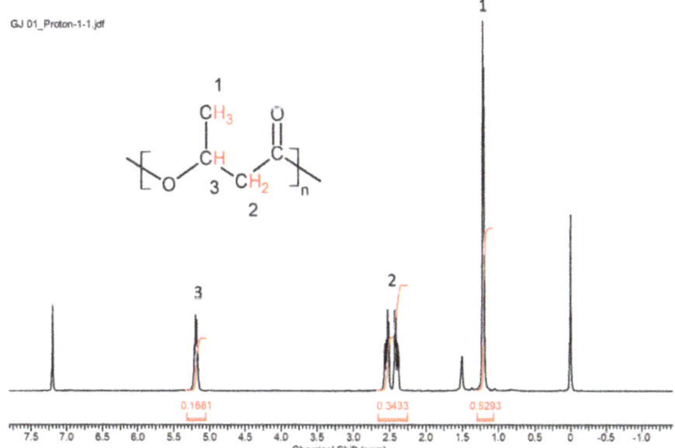

Figure 1. ^{1}H NMR spectrum of the PHA extracted using chloroform. ^{1}H NMR (400 MHz, Chloroform-d) δ 5.19 (d, *J* = 6.7 Hz, 1H), 2.69–2.24 (m, 2H), 1.20 (d, *J* = 6.2 Hz, 3H).

After characterization of the PHA produced, extraction using CYC and GBL was performed. The effects of extraction time and temperature were studied on the recovery yield and molar mass. The PHB in the biomass was first extracted using chloroform, and characterized using GC-MS, but its extraction kinetics was not investigated due to the limitation of the methodology. Mainly the extraction kinetics of CYC and GBL were the focus of our interest. A plot of the percentage of polymer extracted from the cells against extraction time is shown in Figure 3. For each solvent, the extraction was carried out at three temperature levels: 80 °C, 100 °C, and 120 °C. The left panel of Figure 3 shows the kinetics of the extraction using CYC. At 80 °C, only 16% of the polymer was extracted regardless of how long the extraction was conducted. In our experiment, the extraction time was extended to 20 h at 80 °C.

However, the recovery yield was not significantly increased, still maintaining at around 16%. When the temperature was increased to 100 °C, about 90% of the polymer was extracted within 5 min. By further increasing the extraction time, the recovery did not increase significantly. When the temperature was increased to 120 °C, about 99% of the polymer was extracted within 3 min. Further increase in extraction time did not improve the recovery yield. After kinetic measurement, the polymer was precipitated using methanol. To confirm the structure and purity of the extracted PHA, GC-MS analysis after methanolysis was performed (Section 2.3). The presence of methyl 3-hyroxybutyrate was indicated and PHB purity was 99.5%. The right panel of Figure 3 shows the kinetics of extraction using GBL. At 80 °C, only 18% of the polymer was extracted no matter how long the extraction time was extended, which is in accordance with the results obtained using CYC. In contrast to CYC, the recovery yield of PHA was only 45% at 100 °C and 120 °C; no significant difference was observed at these two temperature levels. After 20 min, the recovery yield did not increase further with the extraction time. After kinetic measurement, the polymer was precipitated using methanol. The purity of the PHB was 97.2% as measured by GC-MS.

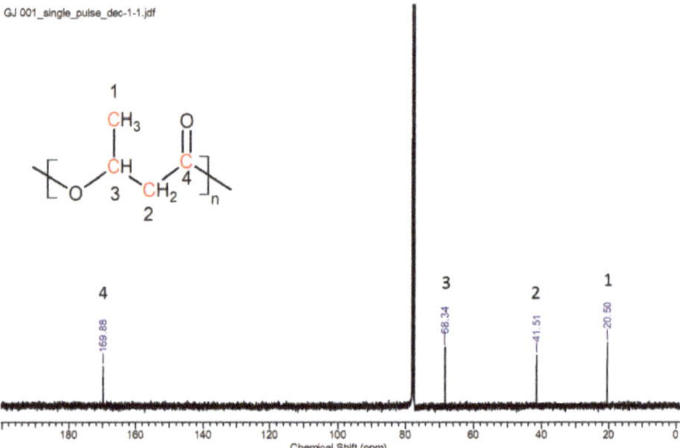

Figure 2. ^{13}C NMR spectrum of the PHA extracted using chloroform.

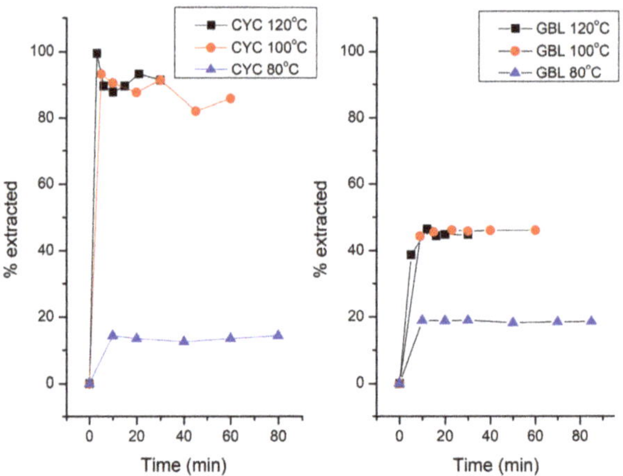

Figure 3. Extraction kinetics of PHA from cells using cyclohexanone (CYC, **left**) and γ-butyrolactone (GBL, **right**) at various temperatures.

From the above results shown in Figure 3, it is interesting to note that the extraction of PHA from the cells was quite fast, especially at higher temperatures (100 °C and 120 °C). The recovery yield depends on both temperature and solvent type, but is less dependent on extraction time. This indicates that the solubility of the solvents restricts the recovery yield, while the solubility of the solvent is again a function of temperature. It is known that the polymer occurs as an intracellular product ("carbonosomes") [25]. The rapid dissolution of the polymer suggests that the synergistic action of the drying process and elevated temperature sufficiently weakens the bacterial cells to enable the PHB to be extracted without the necessity for any previous cell breakage step [14]. As the crystal melting point of PHB is about 120 °C [26], the rapid dissolution of the polymer may be due to the melting of the PHA crystals before dissolution in the solvents. There was a big difference in the recovery yield for the two solvents. This was probably due to the difference in the solubility of PHB in the two solvents. The difference can be illustrated by using the Hansen solubility parameters of PHB and those of the two solvents. The Hansen solubility parameters (δd, δp, δh) for PHB, CYC, and GBL were 15.5, 9.0, 8.6 [27], 17.8, 6.3, 5.1 [28]; and 19.0, 16.6, 7.4 [28], respectively. The solubility radius of PHB in CYC and GBL could then be calculated using the Hansen equation

$$^{ij}R = \left[4\left({}^{i}\delta d - {}^{j}\delta d \right)^2 + \left({}^{i}\delta p - {}^{j}\delta p \right)^2 + \left({}^{i}\delta h - {}^{j}\delta h \right)^2 \right]^{1/2} \tag{2}$$

where i is the solvent and j is the polymer. The calculated results were 6.4 for cyclohexanone and 10.4 for GBL. The solubility radius for PHB was 8.5 [27], indicating that CYC was a better solvent and GBL was a poor solvent for PHB.

The effect of the two extraction solvents on molar mass and dispersity of the extracted polymers is shown in Figure 4. The number average molar mass (Mn) and dispersity (Mw/Mn) of the polymer extracted using chloroform for 24 h was 23 kDa and 2.2, respectively. For both solvents, the dispersity of the extracted polymers had a value of around 2, and the molar mass had an average value of about 23 kDa, regardless of the type of solvent and the temperature and extraction time. Moreover, the molecular weights of the extracted polymers were not significantly affected by extraction time and temperature; this is a considerable benefit when compared to the long term extraction using halogenated solvents at high temperature where losses of molar mass due to random and chain-end scission of PHA macromolecules have been reported [22]. A comparison of the GPC traces of the PHA extracted using CYC, GBL, and chloroform is shown in Figure 5. There was a longer tail in the GPC trace for the PHA extracted using chloroform, which may indicate a slight degradation of the polymer during extraction using chloroform, while the PHAs extracted using CYC and GBL did not have such tails in their GPC traces.

It should be noted that the amount of polymer extracted from the cells was different. At higher temperatures, more polymers were extracted and less polymer remained inside the cells. Therefore, the molar mass and dispersity of the polymer extracted at higher temperatures represented a larger amount of the polymers originally present in the biomass. Since the present study demonstrated that the molar mass and dispersity did not depend on temperature and solvent, it can be assumed that the extraction occurs uniformly without selectivity, i.e., non-selective extraction of both low and high molar mass PHA chains. Most remarkably, the two ecologically benign, non-halogenated solvents displayed extraction behavior similar to the precarious solvent chloroform.

The ^{1}H NMR spectra of the polymers extracted using CYC and GBL, respectively, are shown in Figure 6 in comparison with the polymer extracted using chloroform. It can be seen that the spectra were quite similar for different solvent extractions, indicating that both solvents did not affect the structure of the extracted PHB. The PHB extracted using GBL had a little solvent residue at $\delta = 2.35$ ppm and 4.40 ppm.

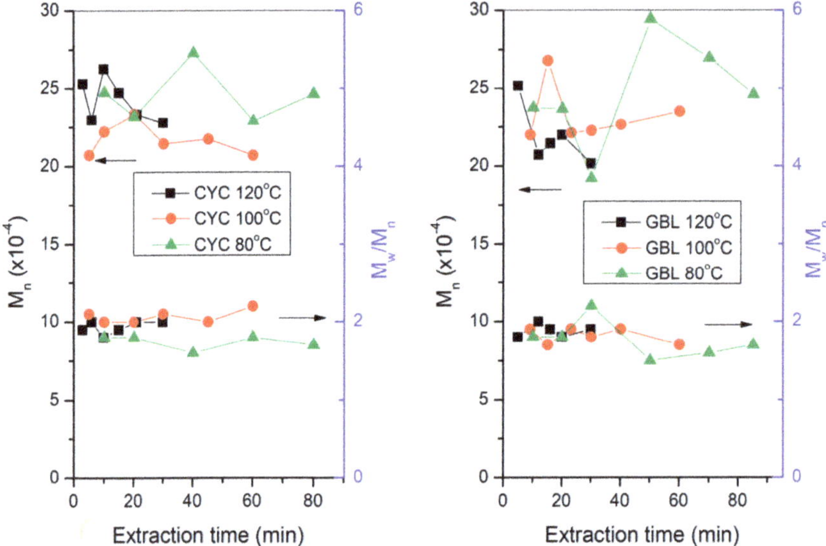

Figure 4. Effect of the extraction solvent cyclohexanone (CYC, **left**) and γ-butyrolactone (GBL, **right**) on molar mass (M_n) and dispersity (M_w/M_n) of the extracted polymers at various temperatures versus extraction time.

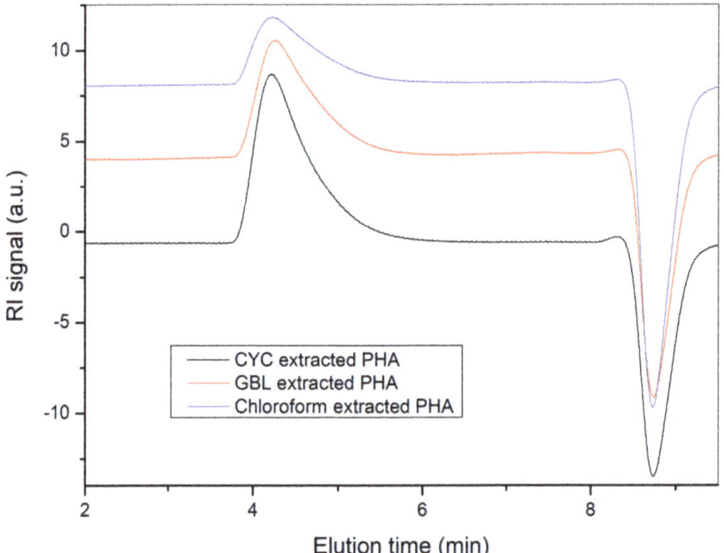

Figure 5. The GPC traces of the PHA extracted using cyclohexanone (CYC), γ-butyrolactone (GBL) at 120 °C for 30 min and chloroform at 60 °C for 24 h. The number average molar mass (M_n) and dispersity (M_w/M_n) of the extracted polymers are 211 kD/2.36, 213 kD/2.23, and 185 kD/2.63, respectively.

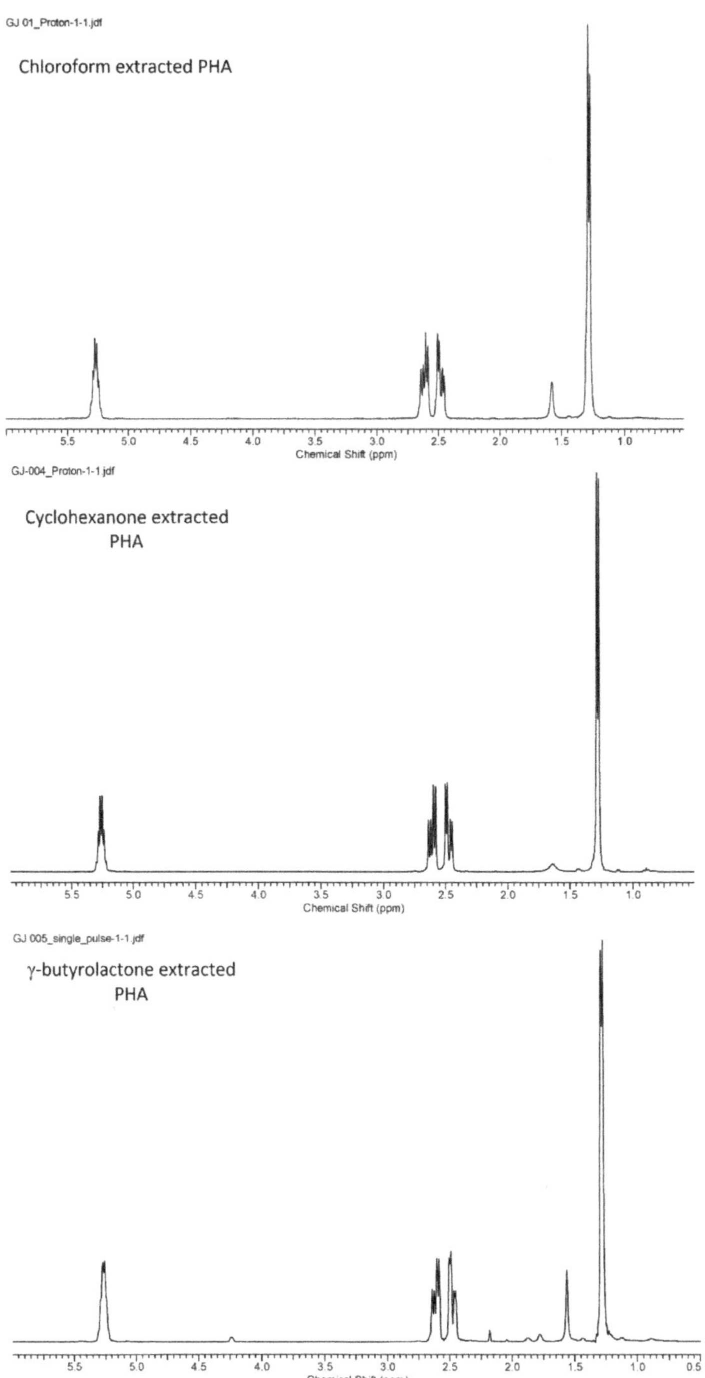

Figure 6. ^{1}H NMR spectra for the PHA extracted from the biomass using chloroform, CYC, and GBL, respectively.

Solvent extracted PHAs may contain residual protein and, in the case of Gram-negative cell factories like *C. necator*, high levels of endotoxin, both of them being potent pyrogens that affect the biocompatibility of the produced PHAs by causing inflammatory reactions [29]. The residual protein can be determined using elemental analysis of the nitrogen content in the polymer. Table 1 shows the elemental analysis of the PHA extracted using various solvents.

Table 1. Elemental analysis of the PHA extracted using non-chlorinated solvents. Two parallel measurements were conducted for each sample.

Extraction Conditions	C (wt %)	H (wt %)	N (wt %)
CYC, 80 °C, 1 h	57.05	6.802	0.113
	56.72	7.099	0.009
CYC, 100 °C, 30 min	57.54	7.160	0.008
	55.60	6.920	0.006
CYC, 120 °C, 10 min	56.05	7.088	0
	55.13	7.096	0.002
GBL, 120 °C, 10 min	56.18	6.894	0.005
	55.29	7.060	0
Chloroform, 61 °C, 24 h	55.28	7.044	0
	55.64	7.033	0

As can be seen from Table 1, the element nitrogen was not detected for the chloroform extracted polymers. For the polymers extracted using CYC, the average nitrogen content decreased with an increase in extraction temperature, from 0.111 wt % at 80 °C to 0.001 wt % at 120 °C. At the same temperature and extraction time, the nitrogen content of the GBL extracted polymer was 0.003 wt %, which was higher than that extracted using CYC. The nitrogen content in the extracted polymers may be related to the solubility of proteins in the solvents and the variation of solubility with temperature. The nitrogen content of the polymer extracted using CYC at 120 °C was quite similar to the *mcl*-PHA extracted using ethanol [5]. It is interesting to note that the polymer could, in addition to the NMR studies described above, be further verified as PHB according to the carbon to hydrogen ratio based on the elemental analysis listed in Table 1.

For industrial application, several factors have to be considered. Rosengart et al. [21] examined 10 non-halogenated solvents for PHB extraction in terms of waste problems, environmental impact, health, flammability and explosion, reactivity and stability, life cycle score, legislation flag, EHS (environmental, health and safety) flag, boiling point and melting point according to the GSK Solvent Selection Guide. In this study, CYC was demonstrated to be a "good" solvent. Beyond that, we demonstrated its suitability in terms of kinetics of extraction and the effect of extraction time on the molar mass and dispersity of the extracted polymer in the present study. In order to reduce the cost of separation of CYC and the anti-solvent used to precipitate the dissolved polymer, the polymer can be precipitated out of the solution by cooling down the solution to room temperature, at which the solubility of PHA is very low; a similar approach has been previously reported in the case of using acetone as the PHA-solvent [18]. In the CYC case, this can be seen from Figure 3, where the solubility of PHA had a significant drop from 120 to 80 °C.

4. Conclusions

Currently, solvent-based recovery of PHAs is still the most effective technique in terms of polymer yield and purity for medical applications. However, the most often used solvents are halogenated, eco-toxic compounds. The harmful character of halogenated solvents is a major drawback of this method when compared with non-solvent based recovery. In this work, both CYC and GBL were examined for their performance in the extraction of PHAs from cells. At 120 °C, the extraction was completed in 3 min with a recovery yield of 95% in the case of CYC, while GBL could only recover 50%. Moreover, both solvents had no effect on the molar mass and dispersity of the extracted polymer.

Furthermore, the solubility of PHB in CYC decreased rapidly at low temperatures. This can be used to precipitate the dissolved PHB by cooling down the solution, which can improve the biocompatibility of this biopolyester for medical or cosmetic purposes. Therefore, it can be concluded that the easily recyclable ketone CYC is a potential substitute for halogenated solvents that can provide fast PHA extraction with high purity and recovery yield without compromising the molecular structure and molar mass distribution. Moreover, the described methodology is of potential use for the purification of commercially available PHA (which could contain higher levels of nitrogen), and further studies in this area are currently under way at our laboratories.

It is worth noting that among the commercially produced PHA, aqueous extraction is currently being used [30–33]. This particularly concerns the promising biodegradable copolymer PHBHHx, whose structure at the molecular level has been already established [34]. Thus, the biocompatibility improvement achieved by the methodology described herein is of potential use for the purification of commercially available PHA (which could contain higher levels of nitrogen), and further studies in this direction are currently under way at our laboratories.

Author Contributions: I.R. and M.K. (Marek Kowalczuk) conceived and designed the experiments; M.K. (Martin Koller), P.C. and G.A. provided advice and technical support; G.J., B.J., and D.E.T. performed the experiments and analyzed the data; G.J. wrote the paper

Funding: This work was funded by the Research Investment Fund, University of Wolverhampton (Wolverhampton, UK). Partial financial support from the European Regional Development Fund Project EnTRESS No. 01R16P00718 and UM0-2016/22/Z/STS/00692 PELARGODONT Project financed under the M-ERA.NET 2 Program of Horizon 2020 is gratefully acknowledged.

Acknowledgments: The authors are indebted to Mariola Siwy for performing the elemental analysis measurements.

Conflicts of Interest: The authors declare no conflict of interest.

References

1. Williams, D.F. On the mechanisms of biocompatibility. *Biomaterials* **2008**, *29*, 2941–2953. [CrossRef] [PubMed]
2. Peptu, C.; Kowalczuk, M. Biomass-derived polyhydroxyalkanoates: Biomedical applications. In *Biomass as Renewable Raw Material to Obtain Bioproducts of High-Tech Value*; Popa, V., Volf, I., Eds.; Elsevier: New York, NY, USA, 2018; pp. 271–313.
3. Wampfler, B.; Ramsauer, T.; Rezzonico, S.; Hischier, R.; Rohling, R.; Thony-Meyer, L.; Zinn, M. Isolation and purification of medium chain length poly (3-hydroxyalkanoates) (mcl-PHA) for medical applications using nonchlorinated solvents. *Biomacromolecules* **2010**, *11*, 2716–2723. [CrossRef] [PubMed]
4. Koller, M. Biodegradable and biocompatible polyhydroxy-alkanoates (PHA): Auspicious microbial macromolecules for pharmaceutical and therapeutic applications. *Molecules* **2018**, *23*, 363. [CrossRef] [PubMed]
5. Muhr, A.; Rechberger, E.M.; Salerno, A.; Reiterer, A.; Schiller, M.; Kwiecien, M.; Adamus, G.; Kowalczuk, M.; Strohmeier, K.; Schober, S.; et al. Biodegradable latexes from animal-derived waste: Biosynthesis and characterization of mcl-PHA accumulated by Ps. citronellolis. *React. Funct. Polym.* **2013**, *73*, 1391–1398. [CrossRef]
6. Koller, M.; Neibelschutz, H.; Braunegg, G. Strategies for recovery and purification of poly [(R)-3-hydroxyalkanoates] (PHA) biopolyesters from surrounding biomass. *Eng. Life Sci.* **2013**, *13*, 549–562. [CrossRef]
7. Berger, E.; Ramsay, B.A.; Ramsay, J.A.; Chavarie, C. PHB recovery by hypochlorite digestion of non-PHB biomass. *Biotechnol. Tech.* **1989**, *3*, 227–232. [CrossRef]
8. Ramsay, J.A.; Berger, E.; Ramsay, B.A.; Chavarie, C. Recovery of poly-3-hydroxyalkanoic acid granules by a surfactant-hypochlorite treatment. *Biotechnol. Tech.* **1990**, *4*, 221–226. [CrossRef]
9. Yang, Y.; Brigham, C.J.; Willis, L.; Rha, C.; Sinskey, A.J. Improved detergent-based recovery of polyhydroxyalkanoates (PHAs). *Biotechnol. Lett.* **2011**, *33*, 937–942. [CrossRef] [PubMed]
10. Holmes, P.A.; Lim, G.B. *Separation Process*; Metabolix Inc.: Woburn, MA, USA, 1990.

11. Kobayashi, D.; Fujita, K.; Nakamura, N.; Ohno, H. A simple recovery process for biodegradable plastics accumlated in cyanobacteria treated with ionic liquids. *Appl. Microbiol. Biotechnol.* **2015**, *99*, 1647–1653. [CrossRef] [PubMed]

12. Hejazi, P.; Vasheghani-Farahani, E.; Yamini, Y. Supercritical fluid disruption of Ralstonia eutropha for poly (b-hydroxybutyrate) recovery. *Biotechnol. Prog.* **2003**, *19*, 1519–1523. [CrossRef] [PubMed]

13. Ong, S.Y.; Kho, H.P.; Riedel, S.L.; Kim, S.W.; Gan, C.Y.; Taylor, T.D.; Sudesh, K. An integrative study on biologically recovered polyhydroxyalkanoates (PHAs) and simultaneous assessment of gut microbiome in yellow mealworm. *J. Biotechnol.* **2018**, *265*, 31–39. [CrossRef] [PubMed]

14. Holmes, P.A.; Wright, L.F.; Alderson, B.; Senior, P.A. A Process for the Extraction of Poly-3-Hydroxy-Butyric Acid from Microbial Cells. Eur. Patent EP0015123A1, 3 September 1980.

15. Koller, M.; Maršálek, L.; Miranda de Sousa Dias, M.; Braunegg, G. Producing microbial polyhydroxyalkanoate (PHA) biopolyesters in a sustainable manner. *New Biotechnol.* **2017**, *37*, 24–38. [CrossRef] [PubMed]

16. Noda, I.; Schechtman, L.A. Solvent Extraction of Polyhydroxyalkanoates from Biomass. U.S. Patent 5,942,597, 24 August 1999.

17. Narasimhan, K.; Severson, R.G.; Buescher, S.E.; Cearley, A.C.; Gibson, M.S.; Welling, S.J. Process for the Extraction of Polyhydroxyalkanoates from Biomass. U.S. Patent 7,118,897, 10 October 2006.

18. Lafferty, R.M.; Heinzle, E. Cyclic Caronic acid Esters as Solvents for Poly (b-Hydroxybutyric Acid). U.S. Patent 4,101,533, 18 July 1978.

19. Koller, M.; Bona, R.; Chiellini, E.; Braunegg, G. Extraction of short-chain-length poly [(R)-hydroxyalkanoates] (scl-PHA) by the "anti-solvent" acetone under elevated temperature and pressure. *Biotechnol. Lett.* **2013**, *35*, 1023–1028. [CrossRef] [PubMed]

20. Yang, Y.; Jeon, J.; Yi, D.H.; Kim, J.; Seo, H.; Sinskey, A.J.; Bigham, C.J. Application of a non-halogenated solvent, methyl ethyl ketone (MEK) for recovery of poly (3-hydroxybutyrate-co-3-hydroxyvalerate) [P (HB-co-HV)] from bacterial cells. *Biotechnol. Bioprocess Eng.* **2015**, *20*, 291–297. [CrossRef]

21. Rosengart, A.; Cesario, M.T.; de Almeida, M.D.; Raposo, R.S.; Espert, A.; de Apodaca, E.D.; da Fonseca, M.M.R. Efficient P (3HB) extraction from Burkholderia sacchari cells using non-chlorinated solvents. *Biochem. Eng. J.* **2015**, *103*, 39–46. [CrossRef]

22. Ramsay, J.A.; Berger, E.; Voyer, R.; Chavarie, C.; Ramsay, B.A. Extraction of poly-3-hydroxybutyrate using chlorinated solvents. *Biotechnol. Tech.* **1994**, *8*, 589–594. [CrossRef]

23. Braunegg, G.; Sonnleitner, B.; Lafferty, R.M. A rapid gas chromatographic method for the determination of poly-b-hydroxybutyric acid in microbial biomass. *Eur. J. Appl. Microbiol.* **1978**, *6*, 29–37. [CrossRef]

24. Rodriguez-Contreras, E.; Koller, M.; de Sousa Dias, M.M.; Calafell-Monfort, M.; Braunegg, G.; Marques, M.S. Influence of glycerol on poly (3-hydroxybutyrate) production by *Cupriavidus necator* and Burkholderia sacchari. *Biochem. Eng. J.* **2015**, *94*, 50–57. [CrossRef]

25. Jendrossek, D.; Pfeiffer, D. New insights in the formation of polyhydroxyalkanoate granules (carbonosomes) and novel functions of poly (3-hydroxybutyrate). *Environ. Microbiol.* **2014**, *16*, 2357–2373. [CrossRef] [PubMed]

26. Sato, H.; Ando, Y.; Dybal, J.; Iwata, T.; Noda, I.; Ozaki, Y. Crystal structures, thermal behaviours, and C-H … O=C hydrogen bondings of poly (3-hydroxyvalerate) and poly (3-hydroxybutyrate) studied by infrared spectroscopy and X-ray diffraction. *Macromolecules* **2008**, *41*, 4305–4312. [CrossRef]

27. Terada, M.; Marchessault, R.H. Determination of solubility parameters for poly (3-hydroxyalkanoates). *Int. J. Biol. Macromol.* **1999**, *25*, 207–215. [CrossRef]

28. Hansen, C.M. *Hansen Solubility Parameters A User's Handbook*, 2nd ed.; CRC Press: Boca Raton, FL, USA, 2007.

29. Williams, S.F.; Martin, D.P.; Horowitz, D.M.; Peoples, O.P. PHA applications: Addressing the price performance issue: I. Tissue engineering. *Int. J. Biol. Macromol.* **1999**, *25*, 111–121. [CrossRef]

30. Kunasundari, B.; Sudesh, K. Isolation and recovery of microbial polyhydroxyalkanoates. *Express Polym. Lett.* **2011**, *5*, 620–634. [CrossRef]

31. Anis, S.N.S.; Iqbal, N.M.; Kumar, K.; Al-Ashraf, A. Increased recovery and improved purity of PHA from recombinant *Cupriavidus necator*. *Bioengineered* **2013**, *4*, 115–118. [CrossRef] [PubMed]

32. Murugan, P.; Han, L.; Gan, C.Y.; Maurer, F.H.; Sudesh, K. A new biological recovery approach for PHA using mealworm, *Tenebrio molitor*. *J. Biotechnol.* **2016**, *239*, 98–105. [CrossRef] [PubMed]

33. Senda, K. Aqueous Dispersion of Biodegradable Polyester and Method for Production Thereof. EP1 566 409 B1, 24 August 2005.
34. Adamus, G.; Sikorska, W.; Kowalczuk, M.; Noda, I.; Satkowski, M.M. Electrospray ion-trap multistage mass spectrometry for characterisation of co-monomer compositional distribution of bacterial poly (3-hydroxybutyrate-*co*-3-hydroxyhexanoate) at the molecular level. *Rapid Commun. Mass Spectrom.* **2003**, *17*, 2260–2266. [CrossRef] [PubMed]

Review

Hyaluronic Acid in the Third Millennium

Arianna Fallacara, Erika Baldini, Stefano Manfredini * and Silvia Vertuani

Department of Life Sciences and Biotechnology, Master Course in Cosmetic Science and
Technology (COSMAST), University of Ferrara, Via L. Borsari 46, 44121 Ferrara, Italy;
arianna.fallacara@student.unife.it (A.F.); erika.baldini@student.unife.it (E.B.); vrs@unife.it (S.V.)
* Correspondence: smanfred@unife.it; Tel.: +39-0532-455294; Fax: +39-0532-455378

Received: 28 May 2018; Accepted: 20 June 2018; Published: 25 June 2018

Abstract: Since its first isolation in 1934, hyaluronic acid (HA) has been studied across a variety
of research areas. This unbranched glycosaminoglycan consisting of repeating disaccharide
units of *N*-acetyl-D-glucosamine and D-glucuronic acid is almost ubiquitous in humans and
in other vertebrates. HA is involved in many key processes, including cell signaling, wound
reparation, tissue regeneration, morphogenesis, matrix organization and pathobiology, and has
unique physico-chemical properties, such as biocompatibility, biodegradability, mucoadhesivity,
hygroscopicity and viscoelasticity. For these reasons, exogenous HA has been investigated as a drug
delivery system and treatment in cancer, ophthalmology, arthrology, pneumology, rhinology, urology,
aesthetic medicine and cosmetics. To improve and customize its properties and applications, HA can
be subjected to chemical modifications: conjugation and crosslinking. The present review gives an
overview regarding HA, describing its history, physico-chemical, structural and hydrodynamic
properties and biology (occurrence, biosynthesis (by hyaluronan synthases), degradation (by
hyaluronidases and oxidative stress), roles, mechanisms of action and receptors). Furthermore,
both conventional and recently emerging methods developed for the industrial production of HA
and its chemical derivatization are presented. Finally, the medical, pharmaceutical and cosmetic
applications of HA and its derivatives are reviewed, reporting examples of HA-based products that
currently are on the market or are undergoing further investigations.

Keywords: biological activity; crosslinking; drug delivery; cosmetic; food-supplement;
functionalization; hyaluronan applications; hyaluronan derivatives; hyaluronan synthases;
hyaluronic acid; hyaluronidases; physico-chemical properties

---

1. Introduction and Historical Background of HA

Research on hyaluronic acid (HA) has expanded over more than one century.

The first study that can be referred to regarding HA dates from 1880: the French scientist Portes
observed that mucin from vitreous body was different from other mucoids in cornea and cartilage
and called it "hyalomucine" [1]. Nevertheless, only in 1934, Meyer and Palmer isolated from bovine
vitreous humor a new polysaccharide containing an amino sugar and a uronic acid and named it HA,
from "hyaloid" (vitreous) and "uronic acid" [2]. During the 1930s and 1950s, HA was isolated also
from human umbilical cord, rooster comb and streptococci [3,4].

The physico-chemical properties of HA were widely studied from the 1940s [5–9], and its chemical
structure was solved in 1954 by Meyer and Weissmann [10]. During the second half of the Twentieth
Century, the progressive understanding of HA's biological roles [11–13] determined an increasing
interest in its production and development as a medical product for a number of clinical applications.
Hence, the extraction processes from animal tissues were progressively optimized, but still carried
several problems of purification from unwanted contaminants (i.e., microorganisms, proteins). The first

studies on HA production through bacterial fermentation and chemical synthesis were carried out before the 1970s [1].

The first pharmaceutical-grade HA was produced in 1979 by Balazs, who developed an efficient method to extract and purify the polymer from rooster combs and human umbilical cords [14]. Balazs' procedure set the basis for the industrial production of HA [14]. Since the early 1980s, HA has been widely investigated as a raw material to develop intraocular lenses for implantation, becoming a major product in ophthalmology for its safety and protective effect on corneal endothelium [15–22]. Additionally, HA was found to be beneficial also for the treatment of joint [23–27] and skin diseases [28,29], for wound healing [30–33] and for soft tissue augmentation [34,35]. Since the late 1980s, HA has also been used to formulate drug delivery systems [36–41], and efforts continue still to today to develop HA-based vehicles to improve therapeutic efficacy [42–45]. During the 1990s and 2000s, particular attention was paid to identifying and characterizing the enzymes involved in HA metabolism, as well as developing bacterial fermentation techniques to produce HA with controlled size and polydispersity [1]. Nowadays, HA represents a key molecule in a variety of medical, pharmaceutical, nutritional and cosmetic applications. For this reason, HA is still widely studied to elucidate its biosynthetic pathways and molecular biology, to optimize its biotechnological production, to synthesize derivatives with improved properties and to optimize and implement its therapeutic and aesthetic uses [1,42–44,46–58].

Considering the great interest in HA from different fields, the fast growing number of studies and our interest in this topic, we decided to provide a comprehensive overview regarding HA and its potentialities, giving a concise update on the latest progress. As an example, a search on the most common public databases (i.e., Pubmed, Scopus, Isi Web of Science, ScienceDirect, Google Scholar, ResearchGate and Patent Data Base Questel) with the keyword "hyaluron*", gave a total of 161,863 hits: 142,575 papers and 19,288 patents. This huge amount of data are continuously growing. Thus, with the aim to give a clearer picture about where researches and applications in the field are going, the present work starts with an update of HA's physico-chemical, structural and hydrodynamic properties and proceeds with the discussion of HA biology: occurrence, biosynthesis (by hyaluronan synthases), degradation (by hyaluronidases and oxidative stress), roles, mechanisms of action and receptors. Furthermore, both conventional and recently-emerging methods developed for the industrial production of HA and its chemical derivatization are described. Finally, the medical, pharmaceutical, cosmetic and dietary applications of HA and its derivatives are reviewed, reporting examples of HA-based products that currently are on the market or are undergoing further investigations.

Literature search: we searched the Cochrane Controlled Trials Register (Central), Medline, EMBase and Cinahl from inception to November 2006 using truncated variations of preparation names including brand names combined with truncated variations of terms related to osteoarthritis, all as text. No methodologic filter for controlled clinical trials was applied (the exact search strategy is available from the authors). We entered relevant articles into the Science Citation Index to retrieve reports that have cited these articles, manually searched conference proceedings and textbooks, screened reference lists of all obtained articles and checked the proceedings of the U.S. Food and Drug Administration advisory panel related to relevant approval applications. Finally, we asked authors and content experts for relevant references and contacted manufacturers known to have conducted trials on viscosupplementation.

2. Physico-Chemical, Structural and Hydrodynamic Properties of HA

HA is a natural and unbranched polymer belonging to a group of heteropolysaccharides named glycosaminoglycans (GAGs), which are diffused in the epithelial, connective and nervous tissues of vertebrates [46,59,60]. All the GAGs (i.e., HA, chondroitin sulfate, dermatan sulfate, keratin sulfate, heparin sulfate and heparin) are characterized by the same basic structure consisting of disaccharide units of an amino sugar (*N*-acetyl-galactosamine or *N*-acetyl-glucosamine) and a uronic sugar (glucuronic acid, iduronic acid or galactose). However, HA differs as it is not sulfated and it is not synthesized by Golgi enzymes in association with proteins [46,59,60]. Indeed, HA is produced at the inner face of the plasma membrane without any covalent bond to a protein core. Additionally, HA can reach a very high molecular weight (HMW, 10^8 Da), while the other GAGs are relatively smaller in size (<5 × 10^4 Da, usually 1.5–2 × 10^4 Da) [46,59,60].

The primary structure of HA is a linear chain containing repeating disaccharide units linked by ß-1,4-glycosidic bonds. Each disaccharide consists of *N*-acetyl-D-glucosamine and D-glucuronic acid connected by ß 1,3-glycosidic bonds (Figure 1) [10,61]. When both the monosaccharides are in the ß configuration, a very energetically-stable structure is formed, as each bulky functional group (hydroxyl, carboxyl, acetamido, anomeric carbon) is in the sterically-favorable equatorial position, while each small hydrogen atom occupies the less energetically-favorable axial position [62]. Thus, the free rotation around the glycosidic bonds of HA backbone is limited, resulting in a rigid conformation where hydrophobic patches (CH groups) are alternated with polar groups [63,64], which are linked by intra- and inter-molecular hydrogen bonds (H-bonds) (Figure 1) [65]. At physiological pH, each carboxyl group has an anionic charge, which can be balanced with a mobile cation such as Na^+, K^+, Ca^{2+} and Mg^{2+}. Hence, in aqueous solution, HA is negatively charged and forms salts generally referred to as hyaluronan or hyaluronate [66,67], which are highly hydrophilic and, consequently, surrounded by water molecules. More precisely, as displayed in Figure 1, water molecules link HA carboxyl and acetamido groups with H-bonds that stabilize the secondary structure of the biopolymer, described as a single-strand left-handed helix with two disaccharide residues per turn (two-fold helix) [68]. In aqueous solution, HA two-fold helices form duplexes, i.e., a ß-sheet tertiary structure, due to hydrophobic interactions and inter-molecular H-bonds, which enable the aggregation of polymeric chains with the formation of an extended meshwork [64,65].

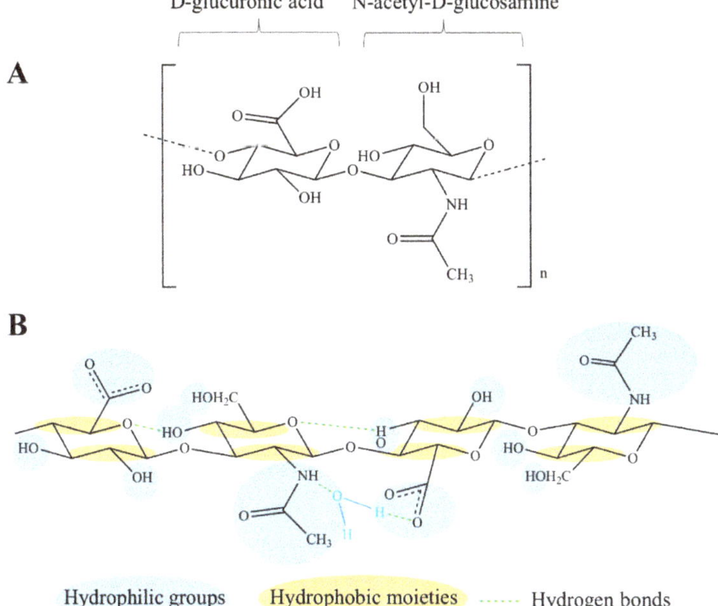

Figure 1. Chemical structures of HA disaccharide unit (**A**) and HA tetrasaccharide unit where the hydrophilic functional groups and the hydrophobic moieties are respectively evidenced in blue and yellow, while the hydrogen bonds are represented by green dashed lines (**B**).

The establishment of this network depends on HA molecular weight (MW) and concentration; for example, HMW native HA (>10^6 Da) forms an extended network even at a very low concentration of 1 µg/mL [64,69]. With increasing MW and concentration, HA networks are strengthened, and consequently, HA solutions display progressively increased viscosity and viscoelasticity [70]. Since hyaluronan is a polyelectrolyte [71], its rheological properties in aqueous solutions are influenced also by ionic strength, pH and temperature [46,70,72]: as these factors increase, HA viscosity declines markedly, suggesting a weakening of the interactions among the polymer chains [73]. In particular, HA is highly sensitive to pH alterations: in acidic and alkaline environments, a critical balance between repulsive and attractive forces occurs [74], and when the pH is lower than four or higher than 11, HA is degraded by hydrolysis [75]. In alkaline conditions, this effect is more pronounced, due to the disruption of H bonds, which take part in the structural organization of HA chains [74,76,77]. Therefore, both the structural properties and the polyelectrolyte character of HA determine its rheological profile [65,73,78,79]. HA solutions are characterized by a non-Newtonian, shear-thinning and viscoelastic behavior. The shear-thinning (or pseudoplastic) profile of HA is due to the breakdown of the inter-molecular hydrogen bonds and hydrophobic interactions under increasing shear rates: HA chains deform and align in the streamlines of flow, and this results in a viscosity decrease [74,78] (Figure 2). Additionally, HA solutions are non-thixotropic: as the shear rate decreases and ends, they recover their original structure and viscosity proceeding through the same intermediate states of the breakdown process [73] (Figure 2). Hence, the breakdown of the polymeric network is transient and reversible. This unique rheological behavior is peculiar and extremely important, as it determines many physiological roles and pharmaceutical, medical, food and cosmetic applications of hyaluronan.

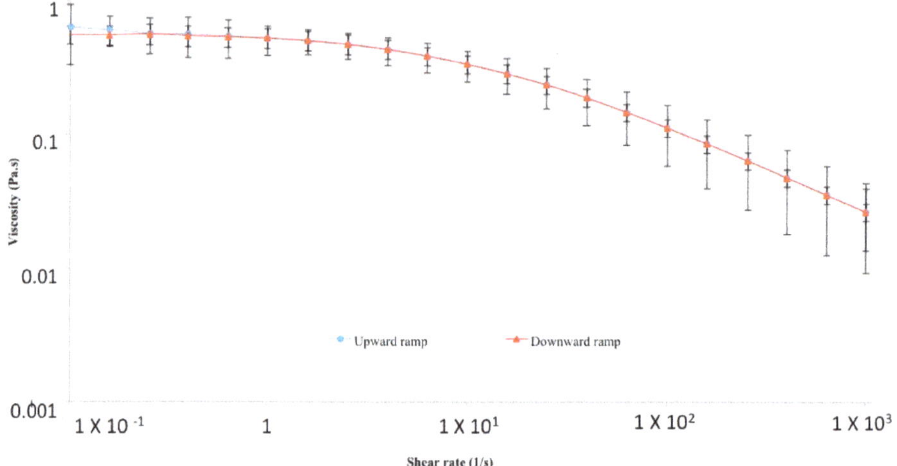

Figure 2. Shear-thinning and non-thixotropic behavior of 0.5% HA solution (2 MDa) analyzed using the rotational rheometer AR2000 (TA instruments, New Castle, DE, USA), connected to the Rheology Advantage software (Version V7.20) and equipped with an aluminum cone/plate geometry (diameter 40 mm, angle 2°, 64-µm truncation). The viscosity decreases in response to gradual increases of the shear rate over time (upward ramp), and then, the viscosity increases in response to gradual decreases of the shear rate over time (downward ramp). The initial viscosity is recovered through the same intermediate states of the breakdown process: the breakdown of the polymeric network is transient and reversible, and therefore, the original structure of HA is recovered.

3. Biology of HA

3.1. HA Occurrence in Living Organism and Diffusion in the Human Body

Hyaluronan is widely diffused in nature: it is present in humans, animals, such as, rabbits, bovines, roosters, bacteria, such as *Streptococcus equi*, *Streptococcus zooepidermicus*, *Streptococcus equisimilis*, *Streptococcus pyogenes*, *Streptococcus uberis*, *Pasteurella multocida* [49,80–82], algae, such as the green algae *Chlorella* sp. infected by the *Chlorovirus* [49,83], yeasts, such as *Cryptococcus neoformans* [49], and mollusks [84]. However, it is not found in fungi, plants and insects [85].

In the human body, the total content of HA is about 15 g for a 70-kg adult [86]. HA is prevalently distributed around cells, where it forms a pericellular coating, and in the extracellular matrix (ECM) of connective tissues [61,82]. Approximately 50% of the total HA resides in the skin, both in the dermis and the epidermis [82]. Synovial joint fluid and eye vitreous body, being mainly composed of ECM, contain important amounts of hyaluronan: 3–4 mg/mL and 0.1 mg/mL (wet weight), respectively [61,82]. Moreover, HA is also abundant in the umbilical cord (4 mg/mL), where it represents the major component of Wharton's jelly together with chondroitin sulfate [87,88]. The turnover of HA is fast (5 g/day) and is finely regulated through enzymatic synthesis and degradation [86].

3.2. HA Synthesis in the Human Body

In the human body, HA is synthesized as a free linear polymer by three transmembrane glycosyltransferase isoenzymes named hyaluronan synthases, HAS: HAS1, HAS2 and HAS3, whose catalytic sites are located on the inner face of the plasma membrane. HA growing chains are extruded onto the cell surface or into the ECM through the plasma membrane and HAS protein complexes [89,90] (Figure 3). The three HAS isoforms share the 50–71% of their amino acid sequences (55% HAS1/HAS2, 57% HAS1/HAS3, 71% HAS2/HAS3), and indeed, they are all characterized by seven membrane-spanning regions and a central cytoplasmic domain [50,86,89]. However, HAS gene sequences are located on different chromosomes (hCh19-HAS1, hCh8-HAS2 and hCh16-HAS3) [91,92], and the expression and the activity of HAS isoforms are controlled by growth factors, cytokines and other proteins such as kinases in different fashions, which appear cell and tissue specific [50,90,93,94]. Hence, the three HAS genes may respond differently to transcriptional signals: for example, in human fibroblasts like synoviocytes, transforming growth factor ß upregulates HAS1 expression, but downregulates HAS3 expression [95]. Moreover, HAS biochemical and synthetic properties are different: HAS1 is the least active isoenzyme and produces HMW hyaluronan (from 2×10^5 to 2×10^6 Da). HAS2 is more active and synthesizes HA chains greater than 2×10^6 Da. It represents the main hyaluronan synthetic enzyme in normal adult cells, and its activity is finely regulated [96]. HAS2 also regulates the developmental and reparation processes of tissue growth, and it may be involved in inflammation, cancer, pulmonary fibrosis and keloid scarring [55,86,97–99]. HAS3 is the most active isoenzyme and produces HA molecules with MW lower than 3×10^5 Da [60].

Dysregulation and misregulation of HAS genes' expression result in abnormal production of HA and, therefore, in increased risk of pathological events, altered cell responses to injury and aberrant biological processes such as malignant transformation and metastasis [47,48,50,100].

Even if the exact regulation mechanisms and functions of each HAS isoenzyme have not been fully elucidated yet [96], all the aforementioned studies suggest that HAS are critical mediators of physiological and pathological processes, as they are involved in development, injury and disease.

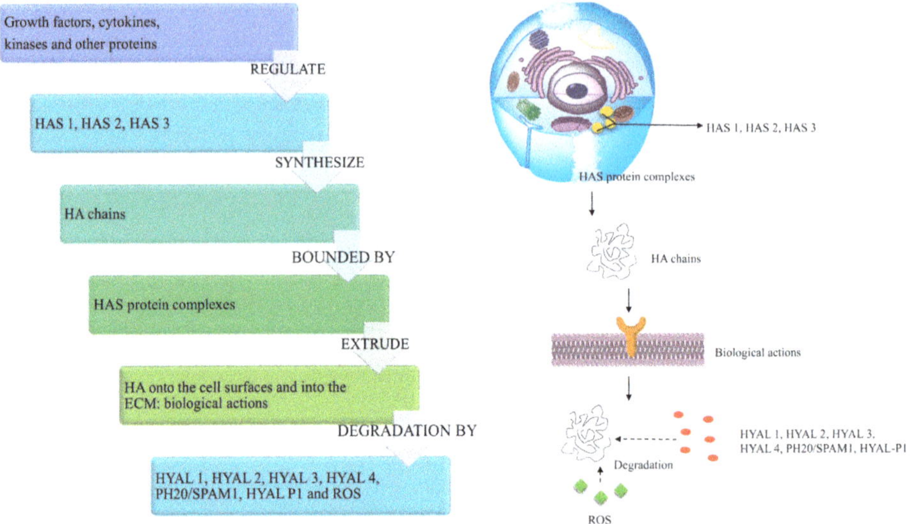

Figure 3. Schematic diagram showing HA key steps from its synthesis to its degradation.

3.3. HA Degradation in the Human Body

HA degradation in the human body is accomplished by two different mechanisms: one is specific, mediated by enzymes (hyaluronidases (HYAL)), while the other is nonspecific, determined by oxidative damage due to reactive oxygen species (ROS) (Figure 3). Together, HYAL and ROS locally degrade roughly 30% of the 15 g HA present in the human body. The remaining 70% is catabolized systemically: hyaluronan is mostly transported by the lymph to the lymph nodes, where it is internalized and catabolized by the endothelial cells of the lymphatic vessels. Additionally, a small part of HA is carried to the bloodstream and degraded by liver endothelial cells [50].

HYAL have a pivotal regulatory function in the metabolism of hyaluronan. These enzymes predominantly degrade HA, even if they are able to catabolize also chondroitin sulfate and chondroitin [101]. Randomly cleaving the β-*N*-acetyl-D-glucosaminidic linkages (β-1,4 glycosidic bonds) of HA chains, HYAL are classified as endoglycosidases. In the human genome, six HYAL gene sequences have been identified in two linked triplets: HYAL 1, HYAL 2, HYAL 3 genes, clustered on chromosome 3p21.3; HYAL-4 and PH20/SPAM1 genes, similarly located on chromosome 7p31.3, together with HYAL-P1 pseudogene [102]. HYAL have a consistent amino acid sequence in common: in particular, HYAL 1, HYAL 2, HYAL 3, HYAL 4 and PH20/SPAM1 share about 40% of their identity [101]. The expression of HYAL appears tissue specific. Nowadays, much is still unknown about HYAL activity, functions and posttranslational processing. HYAL-1, HYAL 2 and PH20/SPAM1 are the most characterized human HYAL. Both HYAL-1 and HYAL 2 have an optimal activity at acidic pH (≤4) [103,104] and are highly expressed in human somatic tissues [102]. HYAL 1 was the first human HYAL to be isolated: it was purified from serum (60 ng/mL) [105] and, successively, from urine [106]. HYAL 1 was found to regulate cell cycle progression and apoptosis: it is the main HYAL expressed in cancers, and therefore, it may regulate tumor growth and angiogenesis [107]. HYAL 1 works together with HYAL 2 to degrade HA, possibly according to the following mechanism, which is still the object of study. HYAL 2 is anchored on the external side of the cell surface: here, it cleaves into oligosaccharides (approximately 25 disaccharide units, 2×10^4 Da) and the extracellular HMW HA ($\geq 10^6$ Da), which is linked to its receptor cluster of differentiation-44 (CD44). These intermediate fragments are internalized, transported first to endosomes and then to lysosomes, where they are degraded into tetrasaccharide units (800 Da) by HYAL-1 [51]. Differently from HYAL-1 and HYAL-2, PH20/SPAM1 shows not only endoglycosidase activity both at acidic and neutral pH, but also a role in fertilization [108]. Hence, PH20/SPAM1 is unique among HYAL, as it behaves as a multifunctional enzyme.

HMW hyaluronan can also be naturally degraded in the organism by ROS, including superoxide, hydrogen peroxide, nitric oxide, peroxynitrite and hypohalous acids, which are massively produced during inflammatory responses, tissue injury and tumorigenesis [60,109]. The depolymerization of HA occurs through mechanisms of the reaction that are dependent on the ROS species, but always involve the scission of the glycosidic linkages [86,110]. Studies have shown that oxidation-related inflammatory processes, determining HA fragmentation, can increase the risk of injury in the airways and determine loss of viscosity in synovial fluid, with consequent cartilage degeneration, joint stiffness and pain [111–113]. ROS-induced degradation of HA might suggest why its antioxidant activity is one of its possible roles in reducing inflammation; however, so far, this biological function of HA has only been hypothesized, as it is not sufficiently supported by experimental data.

Due to these degradation mechanisms, which continuously occur in vivo, it has been estimated that the half-life of HA in the skin is about 24 h, in the eye 24–36 h, in the cartilage 1–3 weeks and in the vitreous humor 70 days [82].

3.4. Biological Roles of HA in Relation to Its MW

The equilibrium between HA synthesis and degradation plays a pivotal regulatory function in the human body, as it determines not only the amount, but also the MW of hyaluronan. Molecular mass and circumstances of synthesis/degradation are the key factors defining HA's biological actions [50,51,100]. Indeed, high molecular weight (HMW) and low molecular weight (LMW) hyaluronan can even display opposite effects [51,60], and when they are simultaneously present in a specific tissue, they can exert actions different from the simple sum of those of their separate size-related effects [51].

Extracellular HMW HA ($\geq 10^6$ Da) is anti-angiogenic, as it is able to inhibit endothelial cell growth [51,60,114]. Additionally, due its viscoelasticity, it acts as a lubricating agent in the synovial joint fluid, thus protecting the articular cartilage [115]. HMW HA has also important and beneficial roles in inflammation, tissue injury and repair, wound healing and immunosuppression: it binds fibrinogen and controls the recruitment of inflammatory cells, the levels of inflammatory cytokines and the migration of stem cells [60,93,114].

During some environmental and pathological conditions, such as asthma, pulmonary fibrosis and hypertension, chronic obstructive pulmonary disease and rheumatoid arthritis, HMW HA is cleaved into LMW HA (2×10^4–10^6 Da), which has been shown to possess pro-inflammatory and pro-angiogenic activities [51,100]. Indeed, LMW hyaluronan is able to stimulate the production of proinflammatory cytokines, chemokines and growth factors [51] and to promote ECM remodeling [50]. Moreover, LMW HA can also induce tumor progression, exerting its influence on cells [51,116] and provoking ECM remodeling.

Both anti- and pro-inflammatory properties have been displayed by oHA and HA fragments ($\leq 2 \times 10^4$ Da), depending on cell type and disease. Certain studies have shown that oHA are able to reduce Toll-like receptors (TLRs)-mediated inflammation [117], inhibit HA-CD44 activation of kinases [118] and retard the growth of tumors [119]. However, oHA have been also found to promote inflammation in synovial fibroblasts [120], stimulate cell adhesion [121] and enhance angiogenesis during wound healing [53].

Therefore, HA is clearly a key molecule involved in a number of physiological and pathological processes. However, despite the intensive studies carried out so far, still little is known about HA's biological roles, the factors determining HA accumulation in transformed connective tissues and the consequent cancer progression, and much less is known about their dependence on hyaluronan molecular size and localization (intra- or extra-cellular). Further researches focusing on HA molecular biology and mechanisms of action are necessary to clarify all these aspects and may facilitate the development of novel HA-based therapies.

3.5. Mechanisms of Action of HA

HA performs its biological actions (Section 3.4.) according to two basic mechanisms: it can act as a passive structural molecule and as a signaling molecule. Both of these mechanisms of action have been shown to be size-dependent [51,86].

The passive mechanism is related to the physico-chemical properties of HMW HA. Due to its macromolecular size, marked hygroscopicity and viscoelasticity, HA is able to modulate tissue hydration, osmotic balance and the physical properties of ECM, structuring a hydrated and stable extracellular space where cells, collagen, elastin fibers and other ECM components are firmly maintained [59,86,88].

HA also acts as a signaling molecule by interacting with its binding proteins. Depending on HA MW, location and on cell-specific factors (receptor expression, signaling pathways and cell cycle), the binding between HA and its proteins determines opposite actions: pro- and anti-inflammatory activities, promotion and inhibition of cell migration, activation and blockage of cell division and differentiation. All the factors that determine HA activities as a signaling molecule could be related: MW may influence HA uptake by cells and may affect receptor affinity. Additionally, receptor complexes may cluster differently depending on HA MW [51].

HA binding proteins can be distinguished into HA-binding proteoglycans (extracellular or matrix hyaloadherins) and HA cell surface receptors (cellular hyaloadherins) [51]. HA has shown two different molecular mechanisms of interaction with its hyaloadherins. First, HA can interact in an autocrine fashion with its receptors on the same cell [60]. Second, it can behave as a paracrine substance, which binds its receptors on neighboring cells and thus activates different intracellular signal cascades. If HA has an HMW, a single chain can interact simultaneously with several cell surface receptors and can bind multiple proteoglycans. These structures, in turn, can aggregate with additional ECM proteins to form complexes, which can be linked to the cell surface through HA receptors [60,100]. Hence, HA acts as a scaffold that stabilizes the ECM structure not only through its passive structural action, but also through its active interaction with several extracellular hyaloadherins, such as aggrecan (prominent in the cartilage), neurocan and brevican (prominent in the central nervous system) and versican (present in different soft tissues) [60]. For these reasons, pericellular HA is involved in the preservation of the structure and functionality of connective tissues, as well as in their protection from environmental factors [88].

HA Cell Surface Receptors

HA interactions with its cell surface receptors mediate three biological processes: signal transduction, formation of pericellular coats and receptor-mediated internalization [60]. The present subsection describes HA cell surface receptors and the biological actions that they control when linked by HA (Figure 4).

The principal receptor for HA is CD44, which is a multifunctional transmembrane glycoprotein. It is expressed in many isoforms diffused in almost all human cell types. CD44 can interact not only with HA, but also with different growth factors, cytokines and extracellular matrix proteins as fibronectin [96]. CD44 intracellular domain interacts with cytoskeleton; hence, when its extracellular domain binds ECM hyaluronan, a link between the cytoskeletal structures and the biopolymer is created [46]. HA-CD44 interaction is involved in a variety of intracellular signaling pathways that control cell biological processes: receptor-mediated hyaluronan internalization/degradation, angiogenesis, cell migration, proliferation, aggregation and adhesion to ECM components [46,51,60,100,122]. Hence, CD44 plays a critical role in inflammation and wound healing [46,96]. However, abnormal activation of HA-CD44 signaling cascades, as well as overexpression and upregulation of CD44 (due to pro-inflammatory cytokines such as interleukin-1, and growth factors such as epidermal growth factors) can result into development of pathological lesions and malignant transformation [60,100]. Indeed, CD44 is overexpressed in many solid tumors, such as pancreatic, breast and lung cancer [54].

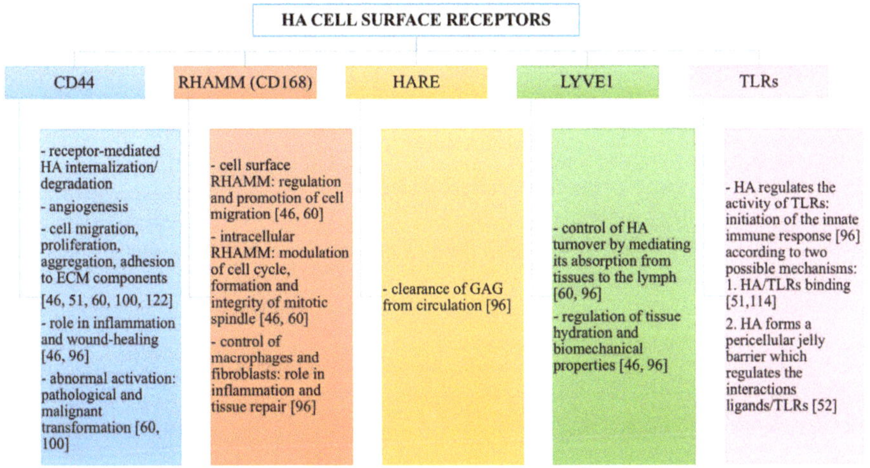

Figure 4. Summary of HA cell surface receptors and of the actions that they control when linked by HA.

The receptor for HA-mediated cell motility (RHAMM) is also known as CD168, and it was the first isolated cellular hyaloadherin. It exists in several isoforms, which can be present not only in the cell membrane, but also in the cytoplasm and in the nucleus [96]. When liked by HA, cell surface RHAMM mediates and promotes cell migration, while intracellular RHAMM modulates the cell cycle, the formation and the integrity of mitotic spindle [46,60]. Interactions of HA with RHAMM play important roles in inflammation and tissue repair, by triggering a variety of signaling pathways and thus controlling cells such as macrophages and fibroblasts [96].

Hyaluronan receptor for endocytosis (HARE) was initially isolated from endothelial cells in the liver, lymph nodes and spleen and successively found also in endothelial cells of eye, brain, kidney and heart [96]. It is able to bind not only HA, but also other GAGs, with the exception of keratin sulfate, heparin sulfate and heparin. It is involved in the clearance of GAGs from circulation [96].

Furthermore, lymphatic vessel endothelial hyaluronan receptor 1 (LYVE1, a HA-binding protein expressed in lymph vascular endothelium and macrophages) controls HA turnover by mediating its adsorption from tissues to the lymph [60,96]. In this way, LYVE1 is involved in the regulation of tissue hydration and their biomechanical properties [46,96]. Additionally, LYVE1 forms complexes with growth factors, prostaglandins and other tissue mediators, which are implicated in the regulation of lymphangiogenesis and intercellular adhesion [46,96].

Finally, HA is involved in the regulation of the activity of TLRs that, recognizing bacterial lipopolysaccharides and lipopeptides, are able to initiate the innate immune response [96]. Two possible mechanisms have been proposed to explain how HA can influence TLRs. According to the first theory, LMW hyaluronan acts as an agonist for TLR2 and TLR4, thus provoking an inflammatory reaction [51,114]. On the contrary, according to the second theory, hyaluronan does not bind to TLRs, but it is able to regulate TLRs interactions with their ligands through the pericellular jelly barrier that it forms [52]. Indeed, in physiological conditions, HMW HA creates a dense and viscous protective coat around the cells, thus covering surface receptors such as TLRs and limiting their interactions with ligands. During inflammation, an imbalance between HA synthesis and degradation occurs, and this alters the thickness and the viscosity of HA pericellular barrier [52]. More precisely, HA is rapidly degraded due to pH reduction, ROS increase and the possible presence of pathogens producing HYAL [46,109]. Hence, HA MW decreases, reducing the polymer water binding ability and the thickness and the viscosity of its pericellular shield [52]. This results in an increased accessibility of the cell receptors to their ligands, in the initiation of the innate immune response and in the enhancement of the inflammatory reaction [52]. For this reason, HA can also be involved in the pathogenesis of diseases sustained by immunological processes [46].

4. Industrial Production of HA

The plethora of activities of a food-contained molecule has raised important interest for public health: the global market of HA was USD 7.2 billion in 2016, and it is expected to reach a value of USD 15.4 billion by 2025 [123]. Indeed, hyaluronan is gaining an exponentially growing interest for many pharmaceutical, medical, food and cosmetic applications, due to its important activities—anti-inflammatory, wound healing and immunosuppressive—and its numerous and incomparable biological and physico-chemical properties, such as biocompatibility, biodegradability, non-immunogenicity, mucoadhesivity, hygroscopicity, viscoelasticity and lubricity. Hence, there is a strong interest in optimizing HA production processes to obtain products that fulfill high quality standards and are characterized by great yield and accessible costs. Both the source and the purification process co-occur to determine the characteristics of the produced HA in terms of purity, MW, yield and cost [124,125]. Therefore, producing high quality HA with high yield and less costly methods represents one of the biggest challenges in the field of hyaluronan applied research.

The first production process applied at an industrial scale consisted of HA extraction from animal sources, such as bovine vitreous and rooster combs [46,49,124]. Despite the extraction protocols being improved over the years, this methodology was always hampered by several technical limitations, which led to the production of highly polydispersed HA (MW $\geq 10^6$ Da) with a low yield [1,46]. This was due to the polymer intrinsic polydispersity, its low concentration in tissues and its uncontrolled degradation caused by the endogenous HYAL and the harsh isolation conditions [46,49]. Additional disadvantages of animal-derived HA were represented by the risk of biological contamination—the presence of proteins, nucleic acids and viruses—and by the high purification costs [46,49,124]. Therefore, alternative methodologies for the industrial production of HA have been developed.

Currently, commercial hyaluronan is principally produced with biotechnology (microbial fermentation). Microorganism-derived HA is biocompatible with the human body because the HA structure is highly conserved among the different species [1,49]. Streptococci strains A and C were the first bacteria used for HA production, and nowadays, many commercial products are derived from *Streptococcus equi* (such as Restylane® by Q-med AB and Juvederm® by Allergan). Optimum bacterial culture conditions to obtain HMW HA (3.5–3.9 × 10^6 Da) have been determined at 37 °C, pH 7, in the presence of lactose or sucrose [125,126]. Hyaluronan yield has been optimized up to 6–7 g/L, which is the upper technical limit of the process due to mass transfer limitation caused by the high viscosity of the fermentation broth [1]. As streptococci genera include several human pathogens, an accurate and expensive purification of the produced HA is necessary [46,49]. Hence, other microorganisms have been and are currently investigated to synthesize HA. An ideal microorganism for HA biosynthesis should be generally regarded as safe (GRAS), not secrete any toxins and be able to produce at least 10^6 Da HA, as the polymer quality and market value increase with its purity and MW, which affect rheological and biological properties and define suitable applications [49,127]. Since the natural hyaluronan-producing organisms are mostly pathogenic, metabolic engineering currently represents an interesting opportunity to obtain HA from non-pathogenic, GRAS microorganisms. Endotoxin-free HA has already been synthesized by recombinant hosts including *Lactococcus lactis* [128], *Bacillus subtilis* [129], *Escherichia coli* [130] and *Corynebacterium glutamicum* [131]. However, up to now, there has been no heterologous bacterial host producing as much HA as the natural ones. Hence, there is an increasing effort to find an ideal bioreactor for HA production: in addition to bacteria, also eukaryotic organisms such as yeasts, like *Saccharomyces cerevisiae* [132] and *Pichia pastoris* [133], and plant cell cultures, like transformed tobacco-cultured cells [134], have been explored in the last few years.

Finally, to obtain HA of defined MW and narrow polydispersity, other approaches have been used. For example, to produce monodisperse oHA, chemoenzymatic synthesis has been performed [135]. This technique has successfully led to a product commercialized under the name Select HA™ (Hyalose LLC), characterized by a low polydispersity index value. Moreover, other studies have shown the possibility to prepare HA monodisperse fragments by controlling the degradation of HMW hyaluronan using different methods, including acidic, alkaline, ultrasonic and thermal degradation [110].

5. Synthetic Modifications of HA

HA has several interesting medical, pharmaceutical, food and cosmetic uses in its naturally-occurring linear form. However, chemical modifications of the HA structure represent a strategy to extend the possible applications of the polymer, obtaining better performing products that can satisfy specific demands and can be characterized by a longer half-life. During the design of novel synthetic derivatives, particular attention is paid to avoiding the loss of native HA properties such as biocompatibility, biodegradability and mucoadhesivity [46].

5.1. General Introduction of the Chemical Approaches to Modify HA

HA chemical modifications mainly involve two functional sites of the biopolymer: the hydroxyl (probably the primary alcoholic function of the *N*-acetyl D glucosamine) and the carboxyl groups. Furthermore, synthetic modifications can be performed after the deacetylation of HA *N*-acetyl groups, a strategy that allows one to recover amino functionalities [136]. All these functional groups of HA can be modified through two techniques, which are based on the same chemical reactions, but lead to different products: conjugation and crosslinking (Figure 5). Conjugation consists of grafting a monofunctional molecule onto one HA chain by a single covalent bond, while crosslinking employs polyfunctional compounds to link together different chains of native or conjugated HA by two or more covalent bonds [136]. Crosslinked hyaluronan can be prepared from native HA (direct crosslinking) [56,58,137] or from its conjugates (see below). Conjugation and crosslinking are generally performed for different purposes. Conjugation permits crosslinking with a variety of molecules; to obtain carrier systems with improved drug delivery properties with respect to native HA; to develop pro-drugs by covalently linking active molecules to HA [136]. On the other hand, crosslinking is normally intended to improve the mechanical, rheological and swelling properties of HA and to reduce its degradation rate, in order to develop derivatives with a longer residence time in the site of application and greater release properties [58,138,139]. A recent trend is to conjugate and crosslink HA chains using bioactive molecules in order to develop derivatives with improved and customized activities [58] for a variety of applications in medicine, aesthetics and bioengineering, including cell and molecule delivery, tissue engineering and the development of scaffolds [46,56,58,140–144].

A number of synthetic approaches have been developed to produce conjugated or crosslinked hyaluronan [136].

Generally, HA is chemically modified in the liquid phase. Since it is hydrophilic, several reactions are performed in aqueous media also from its conjugates [145–147]: however, they are pH dependent and, therefore, require acidic or alkaline conditions, which if too strong, can determine HA degradation [75,145]. Other synthetic methods, involving the use of reagents sensitive to hydrolysis, are performed in anhydrous organic solvents such as dimethylsulfoxide (DMSO) [146] or dimethylformamide (DMF) [148]. These approaches necessarily introduce a preparation step to convert native HA into tetrabutylammonium (TBA) salt, soluble in organic ambient [148]: this increases the reaction time and cost, as well as the chances of HA chain fragmentation due to physico-chemical treatments. Additionally, when HA modifications take place in organic solvents, longer final purification processes are necessary [136]. The basic and classic chemistry that underlies the possible modifications of HA functional groups in the liquid phase is overviewed in the following Sections 5.2–5.4.

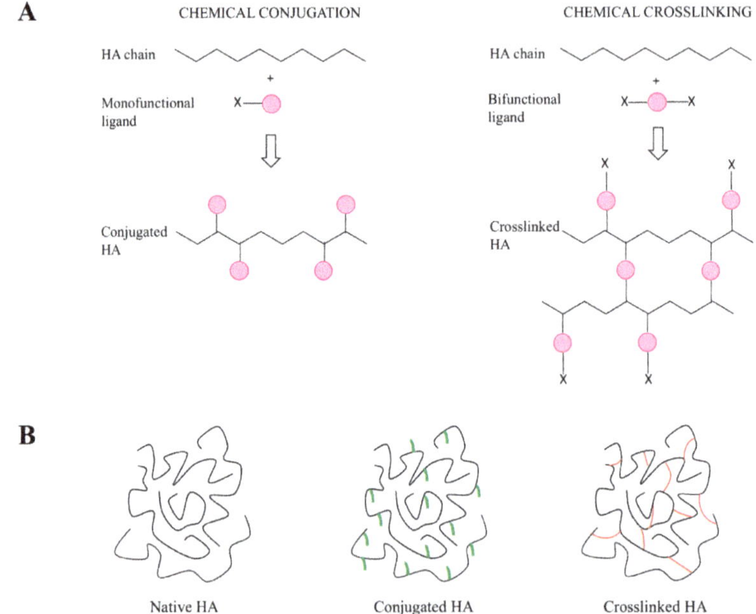

Figure 5. Chemical modifications of HA: conjugation and crosslinking (**A**). HA forms used for pharmaceutical, medical, food and cosmetic applications: native, conjugated and crosslinked (**B**).

Generally, HA is chemically modified in the liquid phase. Since it is hydrophilic, several reactions are performed in aqueous media also from its conjugates [145–147]: however, they are pH dependent and, therefore, require acidic or alkaline conditions, which if too strong, can determine HA degradation [75,145]. Other synthetic methods, involving the use of reagents sensitive to hydrolysis, are performed in anhydrous organic solvents such as dimethylsulfoxide (DMSO) [146] or dimethylformamide (DMF) [148]. These approaches necessarily introduce a preparation step to convert native HA into tetrabutylammonium (TBA) salt, soluble in organic ambient [148]: this increases the reaction time and cost, as well as the chances of HA chain fragmentation due to physico-chemical treatments. Additionally, when HA modifications take place in organic solvents, longer final purification processes are necessary [136]. The basic and classic chemistry that underlies the possible modifications of HA functional groups in the liquid phase is overviewed in the following Sections 5.2–5.4.

Since HA derivatives of high quality and purity are necessary to develop injectable products, implantable scaffolds, drug delivery systems and 3D hydrogel matrices encapsulating living cells, techniques for efficient, low-cost and safe modification of HA are continuously being explored [46,147,149]. Hence, in the last few years, several efforts have been made to introduce one-pot reactions that preferably proceed in an aqueous environment, under mild and, possibly, environmentally-friendly conditions, without the use of toxic catalysts and reagents [147,149]. Additionally, alternative approaches to efficiently modify HA have been introduced: solvent-free methods, i.e., reactions in solid phase [57], "click chemistry" syntheses, which are simple and chemoselective, proceeding with fast kinetics in an aqueous environment, under mild conditions, leading to quantitative yields, without appreciable amounts of side products, i.e., the thiol-ene reaction [150], the Dies–Alder cycloaddition [151] and the azide-alkyne cycloaddition [152]; in situ crosslinking of functionalized HA through air oxidation [153]; photo-crosslinking of functionalized HA in the presence of photosensitizers [154,155].

5.2. Modification of HA Hydroxyl Groups

By modifying HA's hydroxyl groups, the carboxyl groups remain unchanged, thus preserving HA's natural recognition by its degradative enzymes [136]. Over the years, different derivatives of HA (ethers, hemiacetals, esters and carbamates) have been produced through reactions that occur between the polymeric hydroxyl groups and mono- or bi-functional agents.

Epoxides and bisepoxides like butanediol-diglycidyl ether (BDDE) [137], ethylene glycol-diglycidyl ether, polyglycerol polyglycidyl ether [156], epichlorohydrin and 1,2,7,8 diepoxyoctane [157] have been widely used to synthesize ether derivatives of hyaluronan in alkaline aqueous solution. Currently, HA-BDDE ether represents one of the most marketed HA derivative: it can be obtained through simple synthetic procedures in an aqueous environment, and it is degraded into non-cytotoxic fragments [136]. Other efficient methods to form ether derivatives of HA involve the use of divinyl sulfone (DVS) [158] or ethylene sulfide [159] in basic water.

Many studies showed that hemiacetal bonds can be formed between the hydroxyl groups of HA and glutaraldehyde in an acetone-water medium. Since glutaraldehyde is toxic, particular handling is required during the reaction and purification of the final product [160,161].

The hydroxyl groups of HA can be also esterified by reacting with octenyl succinic anhydride [162] or methacrylic anhydride [163] under alkaline conditions. Alternatively, HA can be converted into a DMSO-soluble salt, which can undergo esterification with activated compounds such as acyl-chloride carboxylates [164].

Finally, the activation of HA hydroxyl groups to cyanate esters, and the subsequent reaction in basic water with amines, allows one to synthesize carbamate derivatives with high degrees of substitution, in a reaction time of only 1 h [165].

5.3. Modification of HA Carboxyl Groups

Strategies for the derivatization of HA also involve esterification and amidation, which can be performed after the activation of the polymeric carboxyl groups using different reagents. By modifying HA's carboxyl groups, derivatives more stable to HYAL degradation can be synthesized: hence, if a drug is conjugated on the carboxyl groups of HA, a slow drug release may occur [136].

Esterification can be performed by alkylation of HA carboxyl groups using alkyl halides [166] or tosylate activation [167]. Moreover, HA esters can be synthesized using diazomethane as the activator of the carboxyl groups [168]. All these reactions proceed in DMSO from the TBA salt of HA. Alternatively, HA can undergo esterification also in water using epoxides such as glycidyl methacrylate and excess trimethylamine as a catalyst [169] The conversion of HA carboxyl groups into less hydrophilic esters represents a strategy to decrease the water solubility of HA, with the aim to reduce its susceptibility to HYAL degradation and enhance its in situ permanence time [46]. A well-known biopolymer synthesized to this end is HA benzyl ester (HYAFF 11), the properties of which are finely regulated by its degree of functionalization [36,170].

Amidation represents a further approach to modify HA: over the years, several synthetic procedures have been developed. However, some of these present important drawbacks: for example, Ugi condensation (useful to crosslink HA chains through diamide linkages) requires a strongly acidic pH (3), the use of formaldehyde, which is carcinogenic, and cyclohexyl isocyanide, which determines a pending undesired cyclohexyl group in the final product [145,160]. HA amidation with 1,1'-carbonyldiimidazole [171] or 2-chloro-1-methylpyridinium iodide [148] as activating agents are performed in DMSO and DMF, respectively: hence, HA conversion into TBA salt and longer purification steps are needed. On the contrary, other methods are based on reaction conditions that meet the modification requirements for HA. Particularly efficient is the activation of HA carboxyl groups by carbodiimide (i.e., N-(3-dimethylaminopropyl)-N'-ethylcarbodiimide hydrochloride) (EDC) and co-activators such as N-hydroxysuccinimide (NHS) or 1-hydroxybenzotriazole in water: proceeding under mild conditions, this reaction does not lead to HA chains' cleavage, and it is suitable also for the derivatization with biopolymers easily susceptible to denaturation, such as protein or peptides [146,172,173]. Another promising method to synthesize HA derivatives with high grafting yields, in mild conditions, is based on triazine-activated amidation, typically performed with 2-chloro-dimethoxy-1,3,5-triazine [174] or (4-(4,6-dimethoxy-1,3,5-triazin-2-yl)-4-methylmorpholinium (DMTMM) [175]. A recent study made a systematic comparison of EDC/NHS and DMTMM activation chemistry for modifying HA via amide formation in water [175]. The results showed that DMTMM is more efficient than EDC/NHS for ligation of amines to HA and does not require accurate pH control during the reaction to be effective [175]. Using these mild conditions of amidation, it is possible to synthesize highly hydrophilic and biocompatible derivatives, such as urea-crosslinked HA, which has already shown interesting applications in the ophthalmic and aesthetics field [56,58].

5.4. Modification of HA N-Acetyl Groups

The deacetylation of the *N*-acetyl groups of HA recovers amino functionalities, which can then react with activated acids using the same amidation methods described above. However, this approach is not frequently used to synthesize HA derivatives for two main reasons: first of all, even the mildest deacetylation techniques have been shown to induce chain fragmentation [146,171,176]. Moreover, the deacetylation is a strong structural modification, which could importantly change the unique biological properties typical of native HA: indeed, it has been recently found that it reduces the interactions with the receptor CD44 [177].

6. Applications of HA and Its Derivatives

Due to their unique biological and physico-chemical properties and to their safety profile, native HA and many of its derivatives represent interesting biomaterials for a variety of medical, pharmaceutical, food and cosmetic applications (Figure 6). Some HA-based products are already on the market and/or have already a consolidated clinical practice, while others are currently undergoing further investigations to confirm their effectiveness. Since the literature concerning HA derivatives and their applications is very extensive, only some examples are reported hereafter.

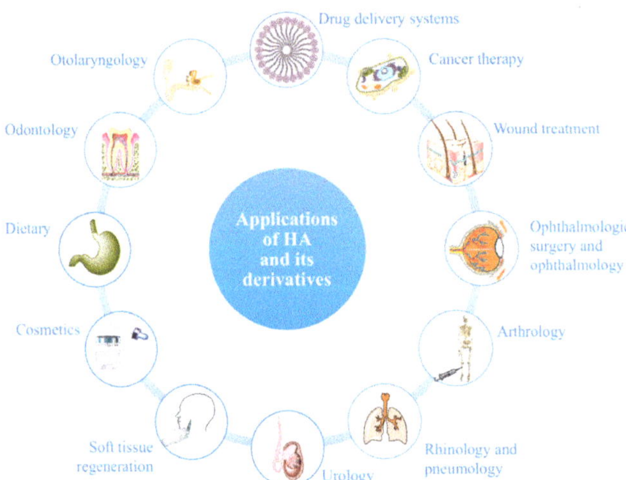

Figure 6. Medical, pharmaceutical, cosmetic and dietary applications of HA and its derivatives.

6.1. Drug Delivery Systems

HA and its derivatives of synthesis represent useful and emerging tools to improve drug delivery. They have been used alone or in combination with other substances to develop pro-drugs, surface-modified liposomes, nanoparticles, microparticles, hydrogel and other drug carriers. All these delivery systems are undergoing a continuous optimization as they are the object of intensive research. However, the industrialization and extensive clinical application of HA and its derivatives as drug carriers still have a long way to go, as many scientific studies are only at an in vitro experimental stage.

The conjugation of active ingredients to HA is intended to develop pro-drugs with improved physico-chemical properties, stability and therapeutic efficacy compared to free drugs. Considering that hyaluronan has several biological functions, HA-drug conjugates can exert their activities as such. Alternatively, their therapeutic actions are accomplished when the drugs are released, i.e., when the covalent bonds, which link drugs and HA, are broken down in the organism, ideally at the specific target sites. A variety of active ingredients can be conjugated to HA for topical or systemic uses. For example, HA can be conjugated with antidiabetic peptides such as exendin 4: this derivative shows a prolonged half-life, a protracted hypoglycemic effect and improved insulinotropic activity compared to the free exendin 4 in type 2 diabetic mice [178]. However, since HA has a short half-life in the blood, the majority of HA-drug conjugates have been developed for local, i.e., intraarticular, intratumoral, subcutaneous, intravesical and intraperitoneal, rather than systemic administration [179]. For instance, a variety of anti-inflammatory drugs, including hydrocortisone, prednisone, prednisolone and dexamethasone, have been conjugated to HA and investigated for intraarticular therapy of arthritis [166]. Furthermore, a recent study has displayed the potential of an emulsion containing a novel HA-P40 conjugate to treat a mouse model of dermatitis induced by oxazolone [180]. P40 is a particulate fragment isolated from *Corynebacterium granulosum* (actually known as *Propionibacterium acnes*), which has immunomodulatory, antibacterial, antiviral and antitumor properties. The conjugation of P40 to HA successfully prevents its systemic absorption and, therefore, improves its topical therapeutic effect [180]. Other research has shown that conjugation of curcumin with HA represents a strategy to enhance the water solubility and stability of curcumin [181]. Moreover, the HA-curcumin conjugate displays improved healing properties compared to free curcumin and free HA, both in vitro (wound model of human keratinocytes) and in vivo (wound model of diabetic mouse). Hence, HA-conjugated curcumin may be useful to treat diabetic wounds [182]. Finally, the most promising and thoroughly studied conjugations of active ingredients to HA involve antitumoral agents, which can be strategically carried to malignant cells by hyaluronan, as explained in Section 6.2 [179,183].

HA can also be conjugated to phospholipids in order to develop surface-modified liposomes: the chemical modification can be performed prior to liposome formulation [184] or after, on the outside shell [185,186]. Moreover, HA can also be non-covalently linked on the liposome surface: indeed, liposomes can be covered by HA through ionic interaction mechanism [187,188] or the simple lipid film hydration technique [189]. HA-modified liposomes appear to be promising carriers, as they have been shown to enhance the stability of drugs in the bloodstream, prolong their half-life, reduce their systemic toxicity [186], enhance their tissue permeability, sustain their prolonged release [189] and ameliorate their therapeutic effects through synergistic actions [190]. HA-coated liposomes could improve the safety and the efficacy of antitumoral therapies: they appear proficient in mediating site-specific delivery of siRNA [191] and anticancer drugs such as doxorubicin [185], gemcitabine [186], imatinib mesylate [188] and docetaxel [184], via CD44 cell receptors. Additionally, HA-surface-modified liposomes have been investigated as delivery systems also in ophthalmology [189], pneumology [190] and topical treatment of wounds and burns [40].

Another promising type of drug delivery system, which can be formulated with HA and its derivatives of synthesis, is represented by nanoparticles. Hyaluronan can be a constituent element of nanoparticles [192,193], but can also be used to cover nanoparticles, in order to improve the targeting efficiency and the therapeutic action of the encapsulated drugs [194]. HA-nanoparticles are being investigated for a number of administration routes and customized applications: for example, HA-Flt1 peptide conjugate nanoparticles might represent a next-generation pulmonary delivery carrier for dexamethasone in the management of asthma [193], while chitosan nanoparticles coated with HA and containing betamethasone valerate have displayed a great potential for the topical treatment of atopic dermatitis [194]. Hyaluronan nanoparticles included in polymeric films have shown potential as innovative therapeutic system for the prolonged release of vitamin E for the management of skin wounds [195]. Moreover, the HA-poly(*N*-isopropylacrylamide) conjugate appears to be a promising candidate to treat osteoarthritis: once injected subcutaneously or intra-articularly, it spontaneously forms biocompatible nanoparticles able to control inflammation with a long-lasting action [192]. Finally, other HA surface-modified nanoparticles have been investigated for cancer-targeted therapies [196–198].

Additionally, over the years, HA microspheres and microparticles have been explored as formulations to improve the biomucoadhesive property and the drug release profile and to ameliorate the texturing feeling in the case of dermal formulations. For example, it has been shown that spray-dried HA microspheres allow favorable ofloxacin delivery to the lung via inhalation, determining a superior pharmacological effect compared to free ofloxacin and to other routes of ofloxacin administration [199]. Similarly, inhaled HA microparticles have displayed prolonged pulmonary retention of salbutamol sulfate and reduced systemic exposure and side effects in a rat model [200]. HA microspheres have been evaluated also as possible materials for bone supplementation: indeed, they could be introduced in mineral bone cements to extend the release of active compounds [201]. A recent work has shown the potential of caffeine-loaded HA microparticles dispersed in a lecithin organogel as a dermal formulation for the long-term treatment of cellulite: this drug delivery system is not only effective at repairing cellulite tissue damage, but has also an intrinsic moisturizing action [202]. Besides the microparticles prepared from native HA, the scientific literature also describes microspheres formulated from HA derivatives of synthesis such as hyaluronan benzyl esters [203] and DVS-crosslinked hyaluronan [138] as topical drug delivery systems.

Finally, hydrogels prepared from linear HA and its chemical derivatives are 3D polymeric networks, which can be well-suited systems for topical delivery of cells [204] and many active ingredients, such as anti-inflammatories [44], anti-bacterials [205], antibodies and proteins in general [42]. To implement the mechanical and release properties, HA hydrogels can incorporate thermoresponsive polymers [42] or other drug carriers such as liposomes [43,44]. Up to now, HA hydrogels have shown a great potential for intraocular [42,204], intratympanic [44], intraarticular [206] and dermal delivery [207]. A topical 2.5% HA hydrogel containing 3% diclofenac has been commercialized for the treatment of actinic keratosis under the tradename of Solaraze® (Pharmaderm) in Europe, USA and Canada [207].

6.2. Cancer Therapy

It has been shown that the receptor CD44 is overexpressed in a variety of tumor cells, which consequently, display an increase of HA binding and internalization [46,54,208]. Hence, the receptor CD44 has been identified as a potential target in cancer therapy, and hyaluronan, its primary ligand, has been recognized as a powerful tool to develop targeted therapies [54,198]. Many research works have shown that HA can act as a drug carrier and targeting agent at the same time, under the form of polymer-antitumoral conjugates or delivery systems encapsulating anticancer drugs [179]. Additionally, hyaluronan can be employed to design surface-modified nanoparticles [196–198] or liposomes [184–188]. After CD44 receptor-mediated cell internalization, all these HA derivatives are hydrolyzed by intracellular enzymes, and therefore, drugs are released inside the cancer target cells [179]. This should improve the pharmacokinetic profile and the delivery process of many anticancer drugs, overcoming the limitations that reduce their clinical potential, such as low solubility, short in vivo half-life, lack of discrimination between healthy and malignant tissues, consequent off-target accumulation and side effects [54,179]. Up to now, there are no HA-antitumoral conjugates and anticancer loaded carriers of HA on the market; however, the promising results of many research works and clinical trials outline their potential and encourage further studies [54,179]. For example, it has been shown that HA-modified polycaprolactone nanoparticles encapsulating naringenin enhance drug uptake by cancer cells in vitro and inhibit tumor growth in rat with urethane-induced lung cancer [198]. Additionally, HA-coated chitosan nanoparticles have been found to promote 5-fluorouracil delivery into tumor cells that overexpress the CD44 receptor [196]. Further studies have displayed that a novel unsaturated derivative of HA [209] and different types of HA-paclitaxel conjugates [179,183] have a great potential as anticancer therapies.

6.3. Wound Treatment

As previously explained (Section 3.4.), endogenous HA sustains wound healing and re-epithelialization processes thanks to several actions including the promotion of fibroblast proliferation, migration and adhesion to the wound site, as well as the stimulation of collagen production [210]. For this reason, HA is used in topical formulations (such as Connettivina® by Fidia) to treat skin irritations and wounds such as abrasions, post-surgical incisions, metabolic and vascular ulcers and burns [82,210–213]. Currently, HA derivatives [173,182,214,215] and HA-based wound dressings, films or hydrogels enriched with other therapeutic agents [216,217] are being evaluated in order to understand if the cicatrization process could be further enhanced. The wound healing properties of HA and its derivatives are being explored not only in dermatology, but also in other medical fields such as ophthalmology [56,218], otolaryngology [142], rhinology [219] and odontology [220].

6.4. Ophthalmologic Surgery and Ophthalmology

HA is a natural component of the human eye: it has been found in vitreous body, lacrimal gland, corneal epithelium and conjunctiva and tear fluid [56]. Therefore, ophthalmic products based on HA are fully biocompatible and do not trigger foreign body reactions [136].

HA solutions are the most used viscosurgical devices to protect and lubricate the delicate eye tissues, replace lost vitreous fluid and provide space for manipulation during ophthalmic interventions [221]. Indeed, the viscosity of HA permits keeping the tissues in place, reducing the risk of displacement, which can potentially compromise both the surgery and the repairing process [46]. The first ophthalmic viscosurgical device containing HA was approved by the FDA in 1980 and is still marketed under the trademark Healon® (Abbott).

Moreover, HA is the active ingredient of many eye drops, such as DropStar® by Bracco and Lubristil® by Eyelab, which, hydrating the ocular surface and improving the quality of vision, are the mainstay to treat diseases such as dry eye syndrome and are useful at increasing the comfortability of contact lenses [56,222,223]. Many studies have proven the safety and the efficacy of native HA solutions as artificial tears [222,224–226]. More recently, also novel derivatives of hyaluronan with improved mechanical and biological properties are being investigated to formulate eye drops with enhanced ocular residence times. For example, promising preliminary results have been obtained with solutions of HA-cysteine ethyl ester [227] and urea-crosslinked HA (HA-CL) [56].

6.5. Arthrology

HA is one of the major lubricating agents of the ECM of synovial joint fluid: due to its viscoelasticity, it absorbs mechanical impacts and avoids friction between the bone-ends [61,82,115]. When the synovial fluid is reduced or inflamed, and the HA level decreases, disorders such as rheumatoid arthritis and osteoarthritis occur. Viscosupplementation represents an approach to treat and slow down the progression of these conditions: intraarticular injections of HMW HA, such as Supartz FX® by Bioventus and Hyalgan® by Fidia, allow maximizing the topical effect and the reduction of pain, as well as minimizing systemic adverse effects [140]. Locally-injected HA has been shown to provide long-term clinical benefits, suggesting that it acts with more than one mechanism [140,228], as the restoration of synovial fluid viscoelasticity is only temporary because HA is degraded within 24 h [229]. Hence, the therapeutic effect of HA intra-articular injection appears prevalently due to biological activities: induction of the synthesis of new HA in synovial cells, stimulation of chondrocyte proliferation and resulting reduction of cartilage degradation [140,228,230].

In order to increase the half-life after injection and, consequently, the therapeutic efficacy, crosslinked HA derivatives have been investigated and introduced on the market (such as Synvisc® by Genzyme) as viscosupplementation agents [140,231].

6.6. Rhinology and Pneumology

Endogenous HMW HA plays a pivotal role in the homeostasis of the upper and the lower airways: it is an important component of the normal airway secretions, exerts anti-inflammatory and anti-angiogenic actions, promotes cell survival and mucociliary clearance, organizes extracellular matrix, stabilizes connective tissues, sustains healing processes and regulates tissues hydration [144,232–234]. Hence, exogenous HMW HA represents a promising therapeutic agent for the treatment of nasal and lung diseases that involve inflammation, oxidative stress and epithelial remodeling, such as allergic and non-allergic rhinitis, asthma, chronic obstructive pulmonary disease and cystic fibrosis [143,233,235–238]. Examples of marketed formulations containing HA to treat respiratory diseases are Ialoclean® (Farma-Derma), a nasal spray to treat nasal dryness and rhinitis and to promote nasal wound healing, Hyaneb® (Chiesi Farmaceutici), a hypertonic saline solution containing HA to hydrate and reduce mucus viscosity in cystic fibrosis patients [239], and Yabro® (Ibsa Farmaceutici), a high viscosity nebulizer solution of HA to treat bronchial hyper-reactivity [143].

6.7. Urology

Recently, the possible therapeutic uses of HA in urology have been explored. Preliminary evidence has shown that intravesical HA, administered alone or in combination with chondroitin sulfate or alpha blockers, could be able to reduce the recurrence of urinary tract infections such as bacterial cystitis, to alleviate the symptoms of these diseases and to protect the mucosa of urinary bladder [240,241]. However, further clinical studies are necessary to confirm the effectiveness of HA treatment in urology.

6.8. Soft Tissue Regeneration

HA skin content decreases with aging, and the most visible effects are the loss of facial skin hydration, elasticity and volume, which are responsible for wrinkles [88]. Over the last few years, HA has been widely used as a biomaterial to develop dermal fillers (DFs), which are class III medical devices that, injected into or under the skin, restore lost volumes and correct facial imperfections such as wrinkles or scars [58]. Being characterized by most of the properties that an ideal DF should have—biocompatibility, biodegradability, viscoelasticity, safety, versatility—HA DFs have become the most popular agents for viscoaugmentation, i.e., for soft tissue contouring and volumizing [58]. Indeed, according to data from the American Society of Plastic Surgeons (ASPS), in 2017, out of a total of 2,691,265 treatments with soft tissue fillers, 2,091,476 were performed with HA DFs [242]. One of the reasons for this success resides in the reversibility of the HA DF effect: they correct wrinkles in a reversible manner, as a hypothetical medical error or complication can be remedied through the injection of HYAL (Vitrase®, ISTA Pharmaceuticals; Hylenex®, Halozyme Therapeutics) [58]. The duration of the corrective effect of HA DFs varies between three and 24 months, depending prevalently on HA concentration, crosslinking (degree and type), the treated area and the individual [58,243]. For example, Hylaform® (Genzyme Biosurgery) contains 4.5–6 mg/mL HA crosslinked with DVS (20% degree), and its effect lasts 3–4 months, while the Juvederm® DFs family (Allergan) contains 18–30 mg/mL HA crosslinked with 9–11% BDDE, and its effect lasts 6–24 months [58]. Finally, DFs that combine BTX-A and HA have been developed to ensure an optimal correction even in patients with extremely deep wrinkles [58].

6.9. Cosmetics

HA represents a moisturizing active ingredient widely used in cosmetic formulations (gels, emulsions or serums) to restore the physiological microenvironment typical of youthful skin. HA-based cosmetics such as Fillerina® (Labo Cosprophar Suisse) claims to restore skin hydration and elasticity:

this is reported to exert an anti-wrinkle effect, although no rigorous scientific proof is able to fully substantiate this claim [82,141,244]. It has to be considered that HA's hydrating effect largely depends on its MW, and its longevity depends on HA stability to hyaluronidases. Indeed, HMW HA mainly works as a film-forming polymer: it reduces water evaporation, with an occlusive-like action. On the other hand, medium MW and LMW HA mainly work by binding moisture from the environment, do to their high hygroscopicity [141,244]. In some cases, this capacity may reverse HA's expected hydrating activity as at a high concentration, HA may even extract humidity from the skin. Furthermore, also sunscreens containing hyaluronan may contribute to maintaining a youthful skin, protecting it against the harmful effects of ultraviolet irradiation, due to the possible free radical scavenging properties of HA [245,246]. The same capability has been demonstrated by dietary intake of HA (Section 6.10).

6.10. Dietary

HA can also represent an interesting ingredient in enriched food and food supplements: it has gained the unofficial designation as a nutri-cosmetic because of its capability to improve skin appearance [247]. For a long time, the fact that HA can cross in its "intact" form the intestinal barrier has been debated; recently, a few studies have appeared in the literature to highlight this question. Kimura et al. have evaluated the degradation and absorption of HA (300 KDa and 2 KDa) after oral ingestion in rats, demonstrating intestinal degradation to oligosaccharides, which are subsequently absorbed in the large intestine, translocated into the blood and distributed in the skin [248]. Orally-ingested LMW HA has shown the opposite effects: some studies have reported inflammatory properties with the activation of the immune response [249], while other research has highlighted the efficacy in reducing knee joint pain without inflammation [250]. Nowadays, the mechanisms at the base of these different actions of LMW ingested HA remain still unclear. In another study, the absorption, distribution and excretion of HMW-labeled (1 MDa) HA were evaluated after oral administration in rats and dogs: for the first time, it was shown that dietary HMW HA can be distributed to connective tissues [251]. In particular, these reproducible results suggest that orally-administered HMW HA may reach joints, bones and skin, even if in small amounts [251], thus highlighting that a rationale may exist in the use of HA-based food supplements designed for joint and skin health.

In these regards, several studies have reported on the safety of HA as a food supplement, confirming its possible use as a food ingredient itself. HA is marketed as a food supplement in the USA, Canada, Europe and Asia (particularly in Korea and Japan) with some difference in the suggested use: to treat joint pain in the USA and in Europe; to treat wrinkles and to moisturize in Japan, even if the involvement of this polymer in the skin moisture retention effect needs to be further elucidated in the future [252]. In the present review, we focused our attention prevalently on studies that exclude complex mixtures, in order to evidence HA's effects alone; however, several research works describe the use of HA as a food ingredient in enriched extracts or mixtures with collagen and other supplements. For example, in a preliminary double-blind, controlled, randomized, parallel trial over 12 weeks, rooster comb extract was added to low fat yoghurt, which was given to mild knee pain patients (n: 40), resulting in significative improvement in muscle strength in men [253]. In another recent clinical study, an oral HA preparation diluted in a cascade fermented organic whole food concentrate supplemented with biotin, vitamin C, copper and zinc (Regulatpro® Hyaluron, Dr. Niedermaier®) led to a significant cutaneous antiaging effect in twenty female subjects after 40 days of daily consumption [254]. From such mixtures, it is difficult to elucidate HA's specific contribution.

Regarding supplementation with HA alone, some studies have demonstrated a direct correlation between ingestion and body effects. In a recent review, Kawada et al. underlined a number of studies in support of the contribution of ingested HA to hydrate skin, thus improving the quality of life for people who suffer from skin dryness induced by UV, smoking and pollutants, responsible for cutaneous reduction of HA [255]. Indeed, the review of Kawada and coworkers reported that, in different randomized, double-blind, placebo-controlled trials, the amounts of HA ranging from 37.52–240 mg per day, ingested in a period comprised between four and six weeks, significantly improved cutaneous moistness [255]. These results, as always happens with biopolymers as food supplements, are difficult to correlate with quantitative effects, as polymer sources and MW always vary. However, qualitatively, the correlation between HA consumption and the decrease of skin dryness is evident. The authors suggest that partially-digested HA, regardless of its MW, is adsorbed in the gut, while intact HA is absorbed by the lymphatic system; both are distributed to the skin, where they can work as inducers of fibroblast proliferation and endogenous HA synthesis [255]. In a more recent study, the same authors have investigated, in a double blind, placebo controlled, randomized study of 61 subjects with dry skin, the effects of HA (120 mg/day) of two different MW (800 KDa and 300 KDa) over six weeks [256]. Both the HAs were effective, but the 300-KDa group showed the best improvements in skin dryness and moisture content [256]. Again, Kawada et al., in an experiment on hairless mice, demonstrated that the oral administration of 200 mg/kg body weight per day of two different MW HA (300 KDa and less than 10 KDa) for six weeks reduced epidermal thickness and improved skin hydration upon UV irradiation [257]. The effect was stronger for the less than 10-KDa HA [257]. The authors also demonstrated that the less than 10 KDa orally-administered HA also stimulated HAS2 expression, thus highlighting an overall role of LMW HA in the prevention of skin photoageing with different mechanisms [257]. Thus, combining HA oral treatment with HA topical and injective administration could be very successful in the control of skin ageing. A further study suggesting that orally-administered HA can migrate into the skin of rats, thus possibly reducing skin dryness, was conducted by Oe et al., who demonstrated that about 90% of the ingested HA was absorbed from the digestive tract and was used as an energy source or a structural component [252].

Dietary HA can be beneficial not only for skin, but also for joints, as evidenced by a number of randomized, double-blinded, placebo-controlled studies relative to the treatment of knee pain, relief of synovial effusion or inflammation and improvement of muscular knee strength [258].

6.11. 3D Cell Culture Models

HA and its synthetic derivatives can be used as 3D scaffold structures, which represent physical support systems for in vitro cell culture [259]. Indeed, 3D tissue models can be obtained by culturing cells on pre-fabricated polymeric scaffolds or matrices, designed to simulate the in vivo ECM [259]. Cells attach, migrate and fill the interstices within the scaffold to form 3D cultures [259]. 3D scaffolds can be promising also for in vivo tissue regeneration, reproducing the natural physical and structural environment of living tissue [259]. An example of a cell substrate suitable for a 3D environment, for both in vitro and in vivo research, is represented by HyStem® Hydrogels (ESI·BIO™).

For example, HA derivatives characterized by aldehyde and hydrazide groups have been used to develop a biomimetic, 3D culture system for poorly-adherent metastatic prostate cancer cells, employed as an in vitro platform to test the efficacy of anticancer drugs [260]. The hyaluronan-3D cell culture system provided a useful interesting alternative to study antineoplastic drugs, with results superior compared to those from conventional 2D monolayers [260]. Another work showed that methacrylated HA is useful to develop in vitro 3D culture models to assess glial scarring in a robust and repeatable way, in order to evaluate, for example, the foreign body response to implants such as electrodes in the central nervous system [261]. Additionally, a recent study highlighted the importance of ink formulation and crosslinking on the printing of stable structures: a dual-crosslinking HA system was evaluated as printable hydrogel ink in biomedicine [262]. It encompassed shear-thinning and self-healing properties through guest-host bonding, and showed an improved cell adhesion after further functionalization (i.e., peptides) [262].

7. Conclusions, Future Trends and Perspectives

Figure 7 summarize some of the most common medical, pharmaceutical, cosmetic and dietary applications of HA and its derivatives. The present review underlines the interest of academic and industrial research on HA: a comprehensive overview of this polymer is provided through the description of its structural, physico-chemical and hydrodynamic properties, occurrence, metabolism, biological roles, mechanisms of action, methods of production and derivatization, pharmaceutical, biomedical, food supplement and cosmetic applications.

Application	Examples	Beneficial actions and key features	State of the art
DRUG DELIVERY SYSTEMS	- Pro-drugs: HA-exedin 4 [178], HA-hydrocortisone, HA-prednisone, HA-prednisolone, HA-dexamethasone [166], HA-P40 [180], HA-curcumin [181] conjugates. - Surface modified liposomes: for the delivery of siRNA [191], doxorubicin [185], gembicitabine [186], imatinib masylate [188], docetaxel [184]; for ophthalmology [189], pneumology [190] and topical treatments of wounds and burns [40]. - Nanoparticles: consisting of HA [192, 193] or covered with HA [194]. Pulmonary delivery of dexamethasone to treat asthma[193], skin delivery of betametasone valerate for atopic dermatitis [194], skin delivery of vitamin E for the management of wounds [195], treatment of osteoarthritis [192], cancer targeted therapies [196-198]. Microspheres and microparticles: pulmonary delivery of ofloxacin [199] and salbutamol sulphate [200], bone supplementation [201], skin delivery of caffeine [202]. Microspheres can be formulated from native HA or from its synthetic derivatives such as HA benzyl esters [203] and DVS-crosslinked HA [138]. - Hydrogel: topycal delivery of anti-inflammatories [44], anti-bacterials [205], antibodies and proteins [42]. Great potential for intraocular [42, 204], intratympanic [44], intraarticular [206] and dermal delivery [207].	- Development of pro-drugs and derivatives with improved physico-chemical properties, stability, half-life time, and therapeutic efficacy compared to free drugs. Reduction of toxicity. - Development of biocompatible and biodegradable drug delivery systems with intrinsic moisturizing activity.	Except for gels, the industrialization and extensive clinical application are still a long way to go; many scientific studies are only at an *in vitro* experimental stage. Example of commercialized hydrogel: Solaraze® (Pharmaderm).
CANCER THERAPY	Polymer-antitumoral conjugates and delivery systems encapsulating anticancer drugs: - HA-modified polycaprolactone nanoparticles encapsulating naringenin [198]. - HA-coated chitosan nanoparticles encapsulating 5-fluorouracil [196]. - novel unsaturated derivative of HA [209]. HA-paclitaxel conjugates [179, 183].	HA can act as drug carrier and targeting agent (for CD44 receptor) at the same time [54, 179, 198].	Promising results of many researches and clinical trials; further studies are strongly encouraged to develop marketable products.
WOUND TREATMENT	- HA based wound dressings, films and hydrogels, eventually enriched with other therapeutic agents [216, 217]. - Applications in: dermatology, ophtalmology [56, 218], otolaryngology [142], rhinology [219], odontology [220].	- Promotion of fibroblast proliferation, migration and adhesion to the wound site. - Stimulation of collagen production. - Improvement and acceleration of wound healing and re-epithelialization processes: treatment of skin irritations and wounds such as abrasion, post-surgical incisions, metabolic and vascular ulcers, burns [86, 210-213].	Several products already marketed, such as Connettivina® (Fidia).
OPHTHALMOLOGIC SURGERY AND OPHTHALMOLOGY	Solutions as viscosurgical devices and eye drops, consisting of native HA or its derivatives, such as HA-cysteine ethyl ester [227] and urea-crosslinked HA [56].	- Protection and lubrication of eye tissues, replacement of lost vitreous fluid and creation of space for manipulation during ophthalmic interventions [46, 221]. - Hydration of the ocular surface, treatment of dry eye syndrome [56, 222, 223].	Several products already marketed, such as the viscosurgical device Healon® (Abbott) and the eye drops DropStar® (Bracco) and Lubristil® (Eyelab). Many studies show the safety and the efficacy of these formulations [222, 224-226].
ARTHROLOGY	Intrarticular injections of native HMW HA and its crosslinked derivatives as viscosupplementation agents [140, 229, 231].	- Absorption of mechanical impacts and avoidance of frictions between the bone-ends [61, 82, 115]. - Restoration of synovial fluid viscoelasticity [229]. - Induction of HA synthesis in synovial cells stimulation of chondrocyte proliferation and resulting reduction of cartilage degradation [140, 228, 230]. - Treatment of rheumatoid arthritis and osteoarthritis, with reduction of pain and adverse effects [140].	Many studies display the safety and the efficacy of HA intraarticular injections. Marketed products: Supartz FX® (Bioventus), Hyalgan® (Fidia) and Synvisc® (Genzyme).
RHINOLOGY AND PNEUMOLOGY	Nasal sprays, hypertonic saline solution, high viscosity nebulizer of HA [143, 239].	HMW HA: promising therapeutic agent for the treatment of nasal and lung diseases that involve inflammation, oxidative stress and epithelial remodeling, such as allergic and non-allergic rhinitis, asthma, chronic obstructive pulmonary disease and cystic fibrosis [143, 233, 235-238].	Many studies display the safety and the efficacy of HMW HA for nasal and lung applications. Marketed products: Ialoclean® (Farma-Derma), Hyaneb® (Chiesi Farmaceutici) and Yabro® (Ibsa Farmaceutici).
UROLOGY	Intravescical HA, alone or in combination with chondroitin sulfate or alpha blockers [240, 241].	Reduction of recurrence and sympthoms of urinary tract infections such as bacterial cystitis, protection of the mucosa of urinary bladder [240, 241].	Preliminary evidences; further clinical studies are necessary to confirm the effectiveness of HA treatment in urology.
SOFT TISSUE REGENERATION	Injectable dermal fillers (DFs, class III medical devices) consisting of gels of native HA or HA derivatives -such as DVS-crosslinked HA and BDDE crosslinked HA [58].	Injected into or under the skin, restore lost volumes and correct facial imperfections such as wrinkle or scars [58]. Biocompatibility, biodegradability, viscoelasticity, safety, versatility, reversibility: HA DFs are the most popular DFs [58, 242].	Many studies show the safety and the efficacy of HMW HA as DFs. Marketed products: Hylaform® (Genzyme-Biosurgery) and Juvederm® (Allergan).
COSMETICS	Several cosmetic formulations: gels, emulsions and serums.	- Restoration of the physiological microenviroment typical of youthful skin. Hydration depending on HA MW: HMW HA forms films with an occlusive-like action; medium MW and LMW HA bind moisture from the enviroment. - Possible antiwrinkle effect [82, 141, 144]. - Possible protection from UV irradiation due to the possible free-radical scavenging properties [245, 246].	Several marketed products, such as Fillerina® (Labo Cosprophar Suisse).
DIETARY	- HA contained in enriched food and food supplements. - HA as *nutri-cosmetic*. HA as food ingredient itself.	- Dietary HMW HA can be distributed to connective tissues: it can reach joints, bones and skin [251]. - Some studies have underlined that also LMW HA may reduce knee joint pain [250]. - Relief of synovial effusion and inflammation, and improvement of muscolar knee strenght [258].	HA is marketed as food supplement in Canada, USA, Europe (joint pain), and in Japan (treatment of wrinkles) [252]. Example of marketed product: Regulatpro® Hyaluron, Dr. Niedermaier®. Several researches are being performed: randomized, double-blinded, placebo-controlled studies [248-258].
3D cell culture	HA and its synthetic derivatives as 3D scaffold structures for *in vitro* cel culture and for *in vivo* tissue regeneration [259-262].	- Simulation of the *in vivo* ECM and reproduction of the natural physical and structural enviroment of living tissues [259]. - *In vitro* platform to test the efficacy of drugs with results superior compared to those from conventional 2D monolayers [260].	Example of marketed product: HyStem® Hydrogels (ESI · BIO™).

Figure 7. Summary of the medical, pharmaceutical, cosmetic and dietary applications of HA and its derivatives, reporting some examples, the beneficial actions, the key features and the state of the art.

During the last few decades, HA has shown great success due to its numerous and unique properties, such as biodegradability, biocompatibility, mucoadhesivity, hygroscopicity and viscoelasticity, and to the broad spectrum of chemical modifications that it can undergo, allowing the development of derivatives with specific targeting and long-lasting drug delivery. Several in vitro and in vivo studies have shown the beneficial actions of HA treatment, with anti-inflammatory, wound healing, chondroprotective, antiangiogenic, anti-ageing and immunosuppressive effects, among others [51,56,60,115]. This supported the development of a great number of HA-based commercial products: from native HA for ophthalmic and arthritic therapies, to food supplement, esthetic and cosmetic formulations. More recently, also some chemical derivatives of HA have received FDA approval and have been successfully introduced on the market, especially as DFs. As a proof of the great potential of this molecule in terms of real benefits for health, we also conducted (on 20 April 2018) a search on the patent database Questel (Paris, France), which resulted in the following table (Table 1) that illustrates the interest toward hyaluronan.

Table 1. Data on hyaluronan patents resulting from the database Questel (Paris, France).

Total	13,684
Alive	8749
Dead	4935
1st application year	**unknown**
After 2015	2717
2011–2015	4568
2006–2010	2647
2001–2005	1694
Before 2001	2058

In our opinion, although hyaluronan displays a great number of potential applications, further investigations and technological improvements are required, as there are still some questions to be answered and some issues to be addressed. First of all, many aspects of HA metabolism, receptor clustering and affinity still need to be explored to understand the different biological actions that hyaluronan has through changes in MW fully. Additional insight needs to be gained in understanding whether there is a relation between HA size and localization and how concomitant use of different HA sizes may modulate signaling. The comprehension of all these mechanisms could provide opportunities to extend and improve hyaluronan pharmaceutical, biomedical, cosmetics and food supplements applications, obtaining more targeted effects. Towards this aim, the key mechanisms that control MW during HA biotechnological synthesis should be clarified to develop methods to produce more uniform size-defined HA. Additionally, progresses in metabolic engineering are necessary to improve HA yield and find biosynthetic strategies with good sustainability and acceptable production cost. Furthermore, the preparation of hyaluronan chemical derivatives needs to be optimized, using strategies such as one-pot reactions, chemo-selective synthesis, solvent-free methods and "click chemistry" approaches. Furthermore, the reproducibility of HA derivatives during scale-up, their pharmacokinetic and pharmacodynamic properties must be improved to allow their successful commercialization. Finally, all the HA-based next generation products, such as innovative crosslinked derivatives, polymer-drug conjugates and delivery systems, should be developed, enabling high biocompatibility, prolonged half-life and improved in situ permanence: hence, in vivo and clinical studies are required to characterize their safety and efficacy fully. Nevertheless, to date, recent in vitro research works have shown promising results, which open encouraging perspectives for safe and health uses of these novel derivatives: for example, HA-CL has displayed high biocompatibility towards human corneal and lung epithelial cell, as well as interesting anti-inflammatory, antioxidant and wound healing properties [56,263].

Acknowledgments: This work was supported by a PhD grant (to Arianna Fallacara) from I.R.A. Srl (Istituto Ricerche Applicate, Usmate-Velate, Monza-Brianza, Italy) and by Ambrosialab Srl (Ferrara, Italy, Grant 2017 to Stefano Manfredini and Silvia Vertuani).

Conflicts of Interest: The authors declare no conflict of interest.

Abbreviations

BDDE	butanediol-diglycidyl ether
CD44	cluster of differentiation-44
CDMT	2-chloro-dimethoxy-1,3,5-triazine
DFs	dermal fillers
DMF	dimethylformamide
DMTMM	4-(4,6-dimethoxy-1,3,5-triazin-2-yl)-4-methylmorpholinium
DMSO	dimethylsulfoxide
DVS	divinyl sulfone
ECM	extracellular matrix
EDC	N-(3-dimethylaminopropyl)-N'-ethylcarbodiimide hydrochloride
HA	hyaluronic acid; hyaluronate; hyaluronan
HA-CL	urea-crosslinked hyaluronic acid
HARE	hyaluronan receptor for endocytosis
HAS	hyaluronan synthases
H-bonds	hydrogen bonds
HYAL	hyaluronidases
HMW	high molecular weight
GAGs	glycosaminoglycans
GRAS	generally regarded as safe
LYVE1	lymphatic vessel endothelial hyaluronan receptor 1
LMW	low molecular weight
MW	molecular weight
NHS	N-hydroxysuccinimide
oHA	oligosaccharides of hyaluronic acid
RHAMM	receptor for HA-mediated cell motility
ROS	reactive oxygen species
TBA	tetrabutylammonium
TLRs	Toll-like receptors

References

1. Boeriu, C.G.; Springer, J.; Kooy, F.K.; van den Broek, L.A.M.; Eggink, G. Production methods for hyaluronan. *Int. J. Carbohydr. Chem.* **2013**, *2013*, 14. [CrossRef]
2. Meyer, K.; Palmer, J.W. The polysaccharide of the vitrous humor. *J. Biol. Chem.* **1934**, *107*, 629–634.
3. Kendall, F.E.; Heidelberger, M.; Dawson, M.H. A serologically inactive polysaccharide elaborated by mucoid strains of group a hemolytic streptococcus. *J. Biol. Chem.* **1937**, *118*, 61–69.
4. Boas, N.F. Isolation of hyaluronic acid from the cock's comb. *J. Biol. Chem.* **1949**, *181*, 573–575. [PubMed]
5. Kaye, M.A.; Stacey, M. Observations on the chemistry of hyaluronic acid. *Biochem. J.* **1950**, *2*, 13. [PubMed]
6. Meyer, K. Highly viscous sodium hyaluronate. *J. Biol. Chem.* **1948**, *176*, 993–994. [PubMed]
7. Varga, L. Studies on hyaluronic acid prepared from the vitreous body. *J. Biol. Chem.* **1955**, *217*, 651–658. [PubMed]
8. Fletcher, E.; Jacobs, J.H.; Markham, R.L. Viscosity studies on hyaluronic acid of synovial fluid in rheumatoid arthritis and osteoarthritis. *Clin. Sci.* **1955**, *14*, 653–660. [PubMed]
9. Blumberg, B.S.; Ogston, A.G.; Lowther, D.A.; Rogers, H.J. Physicochemical properties of hyaluronic acid formed by Streptococcus haemolyticus. *Biochem. J.* **1958**, *70*, 1–4. [CrossRef] [PubMed]
10. Weissmann, B.; Meyer, K. The structure of hyalobiuronic acid and of hyaluronic acid from umbilical cord. *J. Am. Chem. Soc.* **1954**, *76*, 1753–1757. [CrossRef]

11. Meyer, K. The biological significance of hyaluronic acid and hyaluronidase. *Physiol. Rev.* **1947**, *27*, 335–359. [CrossRef] [PubMed]

12. Ogston, A.G.; Stanier, J.E. The physiological function of hyaluronic acid in synovial fluid; viscous, elastic and lubricant properties. *J. Physiol.* **1953**, *119*, 244–252. [CrossRef]

13. Pinkus, H.; Perry, E.T. The influence of hyaluronic acid and other substances on tensile strength of healing wounds. *J. Investig. Dermatol.* **1953**, *21*, 365–374. [CrossRef] [PubMed]

14. Balazs, E.A. Ultrapure Hyaluronic Acid and the Use Thereof. U.S. Patent 4,141,973, 27 February 1979.

15. Miller, D.; Stegmann, R. Use of Na-hyaluronate in anterior segment eye surgery. *J. Am. Intraocul. Implant Soc.* **1980**, *6*, 13–15. [CrossRef]

16. Binkhorst, C.D. Advantages and disadvantages of intracamerular Na-hyaluronate (Healon) in intraocular lens surgery. *Doc. Ophthalmol.* **1981**, *50*, 233–235. [CrossRef] [PubMed]

17. Binkhorst, C.D. Inflammation and intraocular pressure after the use of Healon in intraocular lens surgery. *J. Am. Intraocul. Implant Soc.* **1980**, *6*, 340–341. [CrossRef]

18. Percival, P. Results of a clinical trial of sodium hyaluronate in lens implantation surgery. *J. Am. Intraocul. Implant Soc.* **1985**, *11*, 257–259. [CrossRef]

19. Graue, E.L.; Polack, F.M.; Balazs, E.A. The protective effect of Na-hyaluronate to corneal endothelium. *Exp. Eye Res.* **1980**, *31*, 119–127. [CrossRef]

20. Percival, P. Protective role of Healon during lens implantation. *Trans. Ophthalmol. Soc. UK* **1981**, *101*, 77–78. [PubMed]

21. Regnault, F.; Bregeat, P. Treatment of severe cases of retinal detachment with highly viscous hyaluronic acid. *Mod. Probl. Ophthalmol.* **1974**, *12*, 378–383. [PubMed]

22. Kanski, J.J. Intravitreal hyaluronic acid injection. A long-term clinical evaluation. *Br. J. Ophthalmol.* **1975**, *59*, 255–256. [CrossRef] [PubMed]

23. Auer, J.A.; Fackelman, G.E.; Gingerich, D.A.; Fetter, A.W. Effect of hyaluronic acid in naturally occurring and experimentally induced osteoarthritis. *Am. J. Vet. Res.* **1980**, *41*, 568–574. [PubMed]

24. Namiki, O.; Toyoshima, H.; Morisaki, N. Therapeutic effect of intra-articular injection of high molecular weight hyaluronic acid on osteoarthritis of the knee. *Int. J. Clin. Pharmacol. Ther. Toxicol.* **1982**, *20*, 501–507. [PubMed]

25. Leardini, G.; Perbellini, A.; Franceschini, M.; Mattara, L. Intra-articular injections of hyaluronic acid in the treatment of painful shoulder. *Clin. Ther.* **1988**, *10*, 521–526. [PubMed]

26. Dougados, M.; Nguyen, M.; Listrat, V.; Amor, B. High molecular weight sodium hyaluronate (hyalectin) in osteoarthritis of the knee: A 1year placebo-controlled trial. *Osteoarthr. Cartil.* **1993**, *1*, 97–103. [CrossRef]

27. Jones, A.C.; Pattrick, M.; Doherty, S.; Doherty, M. Intra-articular hyaluronic acid compared to intra-articular triamcinolone hexacetonide in inflammatory knee osteoarthritis. *Osteoarthr. Cartil.* **1995**, *3*, 269–273. [CrossRef]

28. Juhlin, L. Hyaluronan in skin. *J. Intern. Med.* **1997**, *242*, 61–66. [CrossRef] [PubMed]

29. Pavicic, T.; Gauglitz, G.G.; Lersch, P.; Schwach-Abdellaoui, K.; Malle, B.; Korting, H.C.; Farwick, M. Efficacy of cream-based novel formulations of hyaluronic acid of different molecular weights in anti-wrinkle treatment. *J. Drugs Dermatol.* **2011**, *10*, 990–1000. [PubMed]

30. Abatangelo, G.; Martelli, M.; Vecchia, P. Healing of hyaluronic acid-enriched wounds: Histological observations. *J. Surg. Res.* **1983**, *35*, 410–416. [CrossRef]

31. Doillon, C.J.; Silver, F.H. Collagen-based wound dressing: Effects of hyaluronic acid and fibronectin on wound healing. *Biomaterials* **1986**, *7*, 3–8. [CrossRef]

32. Hellström, S.; Laurent, C. Hyaluronan and healing of tympanic membrane perforations. An experimental study. *Acta Otolaryngol. Suppl.* **1987**, *442*, 54–61. [CrossRef] [PubMed]

33. King, S.R.; Hickerson, W.L.; Proctor, K.G. Beneficial actions of exogenous hyaluronic acid on wound healing. *Surgery* **1991**, *109*, 76–84. [PubMed]

34. Duranti, F.; Salti, G.; Bovani, B.; Calandra, M.; Rosati, M.L. Injectable hyaluronic acid gel for soft tissue augmentation. A clinical and histological study. *Dermatol. Surg.* **1998**, *24*, 1317–1325. [CrossRef] [PubMed]

35. Lin, K.; Bartlett, S.P.; Matsuo, K.; LiVolsi, V.A.; Parry, C.; Hass, B.; Whitaker, L.A. Hyaluronic acid-filled mammary implants: An experimental study. *Plast. Reconstr. Surg.* **1994**, *94*, 306–315. [CrossRef] [PubMed]

36. Benedetti, L.M.; Topp, E.M.; Stella, V.J. Microspheres of hyaluronic acid esters—Fabrication methods and in vitro hydrocortisone release. *J. Control. Release* **1990**, *13*, 33–41. [CrossRef]

37. Lim, S.T.; Martin, G.P.; Berry, D.J.; Brown, M.B. Preparation and evaluation of the in vitro drug release properties and mucoadhesion of novel microspheres of hyaluronic acid and chitosan. *J. Control. Release* **2000**, *66*, 281–292. [CrossRef]

38. Moreira, C.A.J.; Armstrong, D.K.; Jelliffe, R.W.; Moreira, A.T.; Woodford, C.C.; Liggett, P.E.; Trousdale, M.D. Sodium hyaluronate as a carrier for intravitreal gentamicin. An experimental study. *Acta Ophthalmol.* **1991**, *69*, 45–49. [CrossRef]

39. Morimoto, K.; Yamaguchi, H.; Iwakura, Y.; Morisaka, K.; Ohashi, Y.; Nakai, Y. Effects of viscous hyaluronate-sodium solutions on the nasal absorption of vasopressin and an analogue. *Pharm. Res.* **1991**, *8*, 471–474. [CrossRef] [PubMed]

40. Yerushalmi, N.; Arad, A.; Margalit, R. Molecular and cellular studies of hyaluronic acid-modified liposomes as bioadhesive carriers for topical drug delivery in wound healing. *Arch. Biochem. Biophys.* **1994**, *313*, 267–273. [CrossRef] [PubMed]

41. Camber, O.; Edman, P. Sodium hyaluronate as an ophthalmic vehicle: Some factors governing its effect on the ocular absorption of pilocarpine. *Curr. Eye Res.* **1989**, *8*, 563–567. [CrossRef] [PubMed]

42. Egbu, R.; Brocchini, S.; Khaw, P.T.; Awwad, S. Antibody loaded collapsible hyaluronic acid hydrogels for intraocular delivery. *Eur. J. Pharm. Biopharm.* **2018**, *124*, 95–103. [CrossRef] [PubMed]

43. El Kechai, N.; Geiger, S.; Fallacara, A.; Cañero Infante, I.; Nicolas, V.; Ferrary, E.; Huang, N.; Bochot, A.; Agnely, F. Mixtures of hyaluronic acid and liposomes for drug delivery: Phase behavior, microstructure and mobility of liposomes. *Int. J. Pharm.* **2017**, *523*, 246–259. [CrossRef] [PubMed]

44. El Kechai, N.; Mamelle, E.; Nguyen, Y.; Huang, N.; Nicolas, V.; Chaminade, P.; Yen-Nicolaÿ, S.; Gueutin, C.; Granger, B.; Ferrary, E.; et al. Hyaluronic acid liposomal gel sustains delivery of a corticoid to the inner ear. *J. Control. Release* **2016**, *226*, 248–257. [CrossRef] [PubMed]

45. Xie, Y.; Upton, Z.; Richards, S.; Rizzi, S.C.; Leavesley, D.I. Hyaluronic acid: Evaluation as a potential delivery vehicle for vitronectin: Growth factor complexes in wound healing applications. *J. Control. Release* **2011**, *153*, 225–232. [CrossRef] [PubMed]

46. Knopf-Marques, H.; Pravda, M.; Wolfova, L.; Velebny, V.; Schaaf, P.; Vrana, N.E.; Lavalle, P. Hyaluronic Acid and Its Derivatives in Coating and Delivery Systems: Applications in Tissue Engineering, Regenerative Medicine and Immunomodulation. *Adv. Healthc. Mater.* **2016**, *5*, 2841–2855. [CrossRef] [PubMed]

47. Adamia, S.; Pilarski, P.M.; Belch, A.R.; Pilarski, L.M. Aberrant splicing, hyaluronan synthases and intracellular hyaluronan as drivers of oncogenesis and potential drug targets. *Curr. Cancer Drug Targets* **2013**, *13*, 347–361. [CrossRef] [PubMed]

48. Adamia, S.; Maxwell, C.A.; Pilarski, L.M. Hyaluronan and hyaluronan synthases: Potential therapeutic targets in cancer. *Curr. Drug Targets Cardiovasc. Haematol. Disord.* **2005**, *5*, 3–14. [CrossRef] [PubMed]

49. De Oliveira, J.D.; Carvalho, L.S.; Gomes, A.M.; Queiroz, L.R.; Magalhães, B.S.; Parachin, N.S. Genetic basis for hyper production of hyaluronic acid in natural and engineered microorganisms. *Microb. Cell Fact.* **2016**, *15*, 119. [CrossRef] [PubMed]

50. Heldin, P.; Lin, C.Y.; Kolliopoulos, K.; Chen, Y.H.; Skandalis, S.S. Regulation of hyaluronan biosynthesis and clinical impact of excessive hyaluronan production. *Matrix Biol.* **2018**. [CrossRef] [PubMed]

51. Cyphert, J.M.; Trempus, C.S.; Garantziotis, S. Size matters: Molecular weight specificity of hyaluronan effects in cell biology. *Int. J. Cell Biol.* **2015**, *2015*, 563818. [CrossRef] [PubMed]

52. Ebid, R.; Lichtnekert, J.; Anders, H.J. Hyaluronan is not a ligand but a regulator of toll-like receptor signaling in mesangial cells: Role of extracellular matrix in innate immunity. *ISRN Nephrol.* **2014**, *2014*. [CrossRef] [PubMed]

53. Gao, F.; Yang, C.X.; Mo, W.; Liu, Y.W.; He, Y.Q. Hyaluronan oligosaccharides are potential stimulators to angiogenesis via RHAMM mediated signal pathway in wound healing. *Clin. Investig. Med.* **2008**, *31*, E106–E116. [CrossRef]

54. Mattheolabakis, G.; Milane, L.; Singh, A.; Amiji, M.M. Hyaluronic acid targeting of CD44 for cancer therapy: From receptor biology to nanomedicine. *J. Drug Target.* **2015**, *23*, 605–618. [CrossRef] [PubMed]

55. Supp, D.M.; Hahn, J.M.; McFarland, K.L.; Glaser, K. Inhibition of hyaluronan synthase 2 reduces the abnormal migration rate of keloid keratinocytes. *J. Burn Care Res.* **2014**, *35*, 84–92. [CrossRef] [PubMed]

56. Fallacara, A.; Vertuani, S.; Panozzo, G.; Pecorelli, A.; Valacchi, G.; Manfredini, S. Novel Artificial Tears Containing Cross-Linked Hyaluronic Acid: An In Vitro Re-Epithelialization Study. *Molecules* **2017**, *22*, 2104. [CrossRef] [PubMed]

57. Larrañeta, E.; Henry, M.; Irwin, N.J.; Trotter, J.; Perminova, A.A.; Donnelly, R.F. Synthesis and characterization of hyaluronic acid hydrogels crosslinked using a solvent-free process for potential biomedical applications. *Carbohydr. Polym.* **2018**, *181*, 1194–1205. [CrossRef] [PubMed]

58. Fallacara, A.; Manfredini, S.; Durini, E.; Vertuani, S. Hyaluronic acid fillers in soft tissue regeneration. *Facial Plast. Surg.* **2017**, *33*, 87–96. [CrossRef] [PubMed]

59. Fraser, J.R.; Laurent, T.C.; Laurent, U.B. Hyaluronan: Its nature, distribution, functions and turnover. *J. Intern. Med.* **1997**, *242*, 27–33. [CrossRef] [PubMed]

60. Girish, K.S.; Kemparaju, K. The magic glue hyaluronan and its eraser hyaluronidase: A biological overview. *Life Sci.* **2007**, *80*, 1921–1943. [CrossRef] [PubMed]

61. Laurent, T.C.; Fraser, J.R. Hyaluronan. *FASEB J.* **1992**, *6*, 2397–2404. [CrossRef] [PubMed]

62. Hascall, V.C.; Laurent, T.C. Hyaluronan: Structure and Physical Properties. GlycoForum—Hyaluronan Today Website. 1997. Available online: http://glycoforum.gr.jp/science/hyaluronan/HA01/HA01E.html (accessed on 22 February 2018).

63. Scott, J.E. Secondary structures in hyaluronan solutions: Chemical and biological implications. *Ciba Found. Symp.* **1989**, *143*, 6–15. [PubMed]

64. Scott, J.E.; Cummings, C.; Brass, A.; Chen, Y. Secondary and tertiary structures of hyaluronan in aqueous solution, investigated by rotary shadowing-electron microscopy and computer simulation. Hyaluronan is a very efficient network-forming polymer. *Biochem. J.* **1991**, *274*, 699–705. [CrossRef] [PubMed]

65. Scott, J.E.; Heatley, F. Hyaluronan forms specific stable tertiary structures in aqueous solution: A 13C NMR study. *Proc. Natl. Acad. Sci. USA* **1999**, *96*, 4850–4855. [CrossRef] [PubMed]

66. Laurent, T. The biology of hyaluronan. Introduction. *Ciba Found. Symp.* **1989**, *143*, 1–20. [PubMed]

67. Balazs, E.A.; Laurent, T.C.; Jeanloz, R.W. Nomenclature of hyaluronic acid. *Biochem. J.* **1986**, *235*, 903. [CrossRef] [PubMed]

68. Heatley, F.; Scott, J.E. A water molecule participates in the secondary structure of hyaluronan. *Biochem. J.* **1988**, *254*, 489–493. [CrossRef] [PubMed]

69. Scott, J.E. Supramolecular organization of extracellular matrix glycosaminoglycans, in vitro and in the tissues. *FASEB J.* **1992**, *6*, 2639–2645. [CrossRef] [PubMed]

70. Kobayashi, Y.; Okamoto, A.; Nishinari, K. Viscoelasticity of hyaluronic acid with different molecular weights. *Biorheology* **1994**, *31*, 235–244. [CrossRef] [PubMed]

71. Cleland, R.L. Ionic polysaccharides. II. Comparison of polyelectrolyte behavior of hyaluronate with that of carboxymethyl cellulose. *Biopolymers* **1968**, *6*, 1519–1529. [CrossRef] [PubMed]

72. Balazs, E.A. The physical properties of synovial fluid and the special role of hyaluronic acid. In *Disorders of the Knee*; Helfet, A.J., Ed.; J.B. Lippincott Company: Philadelphia, PA, USA, 1974; pp. 63–75.

73. Rwei, S.P.; Chen, S.W.; Mao, C.F.; Fang, H.W. Viscoelasticity and wearability of hyaluronate solutions. *Biochem. Eng. J.* **2008**, *40*, 211–217. [CrossRef]

74. Lapcík, L.J.; Lapcík, L.; De Smedt, S.; Demeester, J.; Chabrecek, P. Hyaluronan: Preparation, Structure, Properties, and Applications. *Chem. Rev.* **1998**, *98*, 2663–2684. [CrossRef]

75. Maleki, A.; Kjøniksen, A.L.; Nystrom, B. Effect of pH on the behavior of hyaluronic acid in dilute and semidilute aqueous solutions. *Macromol. Symp.* **2008**, *274*, 131–140. [CrossRef]

76. Ghosh, S.; Kobal, I.; Zanette, D.; Reed, W.F. Conformational contraction and hydrolysis of hyaluronate in sodium hydroxide solutions. *Macromolecules* **1993**, *26*, 4685–4693. [CrossRef]

77. Morris, E.R.; Rees, D.A.; Welsh, E.J. Conformation and dynamic interactions in hyaluronate solutions. *J. Mol. Biol.* **1980**, *138*, 383–400. [CrossRef]

78. Pisárčik, M.; Bakoš, D.; Čeppan, M. Non-Newtonian properties of hyaluronic acid aqueous solution. Colloids Surf. A Physicochem. *Eng. Asp.* **1995**, *97*, 197–202. [CrossRef]

79. Gura, E.; Hückel, M.; Müller, P.J. Specific degradation of hyaluronic acid and its rheological properties. *Polym. Degrad. Stab.* **1998**, *59*, 297–302. [CrossRef]

80. DeAngelis, P.L.; Jing, W.; Drake, R.R.; Achyuthan, A.M. Identification and molecular cloning of a unique hyaluronan synthase from Pasteurella multocida. *J. Biol. Chem.* **1998**, *273*, 8454–8458. [CrossRef] [PubMed]

81. Balazs, E.A.; Leshchiner, E.; Larsen, N.E.; Band, P. Applications of hyaluronan and its derivatives. In *Biotechnological Polymers*; Gebelein, C.G., Ed.; Technomic: Lancaster, UK, 1993; pp. 41–65.

82. Schiraldi, C.; La Gatta, A.; De Rosa, M. Biotechnological Production and Application of Hyaluronan. In *Biopolymers*; Elnashar, M., Ed.; IntechOpen: London, UK, 2010; Available online: https://www.intechopen.

com/books/biopolymers/biotechnological-production-characterization-and-application-of-hyaluronan (accessed on 20 June 2018).

83. DeAngelis, P.L. Hyaluronan synthases: Fascinating glycosyltransferases from vertebrates, bacterial pathogens, and algal viruses. *Cell. Mol. Life Sci.* **1999**, *56*, 670–682. [CrossRef] [PubMed]

84. Volpi, N.; Maccari, F. Purification and characterization of hyaluronic acid from the mollusc bivalve Mytilus galloprovincialis. *Biochimie* **2003**, *85*, 619–625. [CrossRef]

85. Kogan, G.; Soltés, L.; Stern, R.; Gemeiner, P. Hyaluronic acid: A natural biopolymer with a broad range of biomedical and industrial applications. *Biotechnol. Lett.* **2007**, *29*, 17–25. [CrossRef] [PubMed]

86. Volpi, N.; Schiller, J.; Stern, R.; Soltés, L. Role, metabolism, chemical modifications and applications of hyaluronan. *Curr. Med. Chem.* **2009**, *16*, 1718–1745. [CrossRef] [PubMed]

87. Sobolewski, K.; Bańkowski, E.; Chyczewski, L.; Jaworski, S. Collagen and glycosaminoglycans of Wharton's jelly. *Biol. Neonate* **1997**, *71*, 11–21. [CrossRef] [PubMed]

88. Robert, L.; Robert, A.M.; Renard, G. Biological effects of hyaluronan in connective tissues, eye, skin, venous wall. Role in aging. *Pathol. Biol.* **2010**, *58*, 187–198. [CrossRef] [PubMed]

89. Weigel, P.H.; Hascall, V.C.; Tammi, M. Hyaluronan synthases. *J. Biol. Chem.* **1997**, *272*, 13997–14000. [CrossRef] [PubMed]

90. Itano, N.; Kimata, K. Mammalian hyaluronan synthases. *IUBMB Life* **2002**, *54*, 195–199. [CrossRef] [PubMed]

91. Spicer, A.P.; McDonald, J.A. Characterization and molecular evolution of a vertebrate hyaluronan synthase gene family. *J. Biol. Chem.* **1998**, *273*, 1923–1932. [CrossRef] [PubMed]

92. Itano, N.; Sawai, T.; Yoshida, M.; Lenas, P.; Yamada, Y.; Imagawa, M.; Shinomura, T.; Hamaguchi, M.; Yoshida, Y.; Ohnuki, Y.; et al. Three isoforms of mammalian hyaluronan synthases have distinct enzymatic properties. *J. Biol. Chem.* **1999**, *274*, 25085–25092. [CrossRef] [PubMed]

93. Jiang, D.; Liang, J.; Noble, P.W. Hyaluronan in tissue injury and repair. *Annu. Rev. Cell Dev. Biol.* **2007**, *23*, 435–461. [CrossRef] [PubMed]

94. Vigetti, D.; Viola, M.; Karousou, E.; De Luca, G.; Passi, A. Metabolic control of hyaluronan synthases. *Matrix Biol.* **2014**, *35*, 8–13. [CrossRef] [PubMed]

95. Stuhlmeier, K.M.; Pollaschek, C. Differential effect of transforming growth factor beta (TGF-beta) on the genes encoding hyaluronan synthases and utilization of the p38 MAPK pathway in TGF-beta-induced hyaluronan synthase 1 activation. *J. Biol. Chem.* **2004**, *279*, 8753–8760. [CrossRef] [PubMed]

96. Vigetti, D.; Karousou, E.; Viola, M.; Deleonibus, S.; De Luca, G.; Passi, A. Hyaluronan: Biosynthesis and signaling. *Biochim. Biophys. Acta* **2014**, *1840*, 2452–2459. [CrossRef] [PubMed]

97. Zhang, Z.; Tao, D.; Zhang, P.; Liu, X.; Zhang, Y.; Cheng, J.; Yuan, H.; Liu, L.; Jiang, H. Hyaluronan synthase 2 expressed by cancer-associated fibroblasts promotes oral cancer invasion. *J. Exp. Clin. Cancer Res.* **2016**, *35*, 181. [CrossRef] [PubMed]

98. Li, Y.; Liang, J.; Yang, T.; Monterrosa Mena, J.; Huan, C.; Xie, T.; Kurkciyan, A.; Liu, N.; Jiang, D.; Noble, P.W. Hyaluronan synthase 2 regulates fibroblast senescence in pulmonary fibrosis. *Matrix Biol.* **2016**, *55*, 35–48. [CrossRef] [PubMed]

99. Zhang, H.; Tsang, J.Y.; Ni, Y.B.; Chan, S.K.; Chan, K.F.; Cheung, S.Y.; Tse, G.M. Hyaluronan synthase 2 is an adverse prognostic marker in androgen receptor-negative breast cancer. *J. Clin. Pathol.* **2016**, *69*, 1055–1062. [CrossRef] [PubMed]

100. Toole, B.P. Hyaluronan: From extracellular glue to pericellular cue. *Nat. Rev. Cancer* **2004**, *4*, 528–539. [CrossRef] [PubMed]

101. Stern, R.; Jedrzejas, M.J. Hyaluronidases: Their genomics, structures, and mechanisms of action. *Chem. Rev.* **2006**, *106*, 818–839. [CrossRef] [PubMed]

102. Csoka, A.B.; Frost, G.I.; Stern, R. The six hyaluronidase-like genes in the human and mouse genomes. *Matrix Biol.* **2001**, *20*, 499–508. [CrossRef]

103. Lokeshwar, V.B.; Rubinowicz, D.; Schroeder, G.L.; Forgacs, E.; Minna, J.D.; Block, N.L.; Nadji, M.; Lokeshwar, B.L. Stromal and epithelial expression of tumor markers hyaluronic acid and HYAL1 hyaluronidase in prostate cancer. *J. Biol. Chem.* **2001**, *276*, 11922–11932. [CrossRef] [PubMed]

104. Lepperdinger, G.; Strobl, B.; Kreil, G. HYAL2, a human gene expressed in many cells, encodes a lysosomal hyaluronidase with a novel type of specificity. *J. Biol. Chem.* **1998**, *273*, 22466–22470. [CrossRef] [PubMed]

105. Frost, G.I.; Csóka, A.B.; Wong, T.; Stern, R. Purification, cloning, and expression of human plasma hyaluronidase. *Biochem. Biophys. Res. Commun.* **1997**, *236*, 10–15. [CrossRef] [PubMed]

106. Csoka, A.B.; Frost, G.I.; Wong, T.; Stern, R. Purification and microsequencing of hyaluronidase isozymes from human urine. *FEBS Lett.* **1997**, *417*, 307–310. [CrossRef]
107. Stern, R. Hyaluronidases in cancer biology. *Semin. Cancer Biol.* **2008**, *18*, 275–280. [CrossRef] [PubMed]
108. Cherr, G.N.; Yudin, A.I.; Overstreet, J.W. The dual functions of GPI-anchored PH-20: Hyaluronidase and intracellular signaling. *Matrix Biol.* **2001**, *20*, 515–525. [CrossRef]
109. Soltés, L.; Mendichi, R.; Kogan, G.; Schiller, J.; Stankovska, M.; Arnhold, J. Degradative action of reactive oxygen species on hyaluronan. *Biomacromolecules* **2006**, *7*, 659–668. [CrossRef] [PubMed]
110. Stern, R.; Kogan, G.; Jedrzejas, M.J.; Soltés, L. The many ways to cleave hyaluronan. *Biotechnol. Adv.* **2007**, *25*, 537–557. [CrossRef] [PubMed]
111. Monzon, M.E.; Fregien, N.; Schmid, N.; Falcon, N.S.; Campos, M.; Casalino-Matsuda, S.M.; Forteza, R.M. Reactive oxygen species and hyaluronidase 2 regulate airway epithelial hyaluronan fragmentation. *J. Biol. Chem.* **2010**, *285*, 26126–26134. [CrossRef] [PubMed]
112. Schiller, J.; Arnhold, J.; Arnold, K. Contribution of reactive oxygen species to cartilage degradation in rheumatic diseases: Molecular pathways, diagnosis and potential therapeutic strategies. *Curr. Med. Chem.* **2003**, *10*, 2123–2145. [CrossRef] [PubMed]
113. Schiller, J.; Arnhold, J.; Arnold, K. Action of hypochlorous acid on polymeric components of cartilage. Use of 13C NMR spectroscopy. *Z. Naturforsch. C* **1995**, *50*, 721–728. [PubMed]
114. Jiang, D.; Liang, J.; Noble, P.W. Hyaluronan as an immune regulator in human diseases. *Physiol. Rev.* **2011**, *91*, 221–264. [CrossRef] [PubMed]
115. Tamer, T.M. Hyaluronan and synovial joint: Function, distribution and healing. *Interdiscip. Toxicol.* **2013**, *6*, 111–125. [CrossRef] [PubMed]
116. Wu, M.; Cao, M.; He, Y.; Liu, Y.; Yang, C.; Du, Y.; Wang, W.; Gao, F. A novel role of low molecular weight hyaluronan in breast cancer metastasis. *FASEB J.* **2015**, *29*, 1290–1298. [CrossRef] [PubMed]
117. Kim, M.Y.; Muto, J.; Gallo, R.L. Hyaluronic acid oligosaccharides suppress TLR3-dependent cytokine expression in a TLR4-dependent manner. *PLoS ONE* **2013**, *8*, e72421. [CrossRef] [PubMed]
118. Misra, S.; Toole, B.P.; Ghatak, S. Hyaluronan constitutively regulates activation of multiple receptor tyrosine kinases in epithelial and carcinoma cells. *J. Biol. Chem.* **2006**, *281*, 34936–34941. [CrossRef] [PubMed]
119. Toole, B.P.; Ghatak, S.; Misra, S. Hyaluronan oligosaccharides as a potential anticancer therapeutic. *Curr. Pharm. Biotechnol.* **2008**, *9*, 249–252. [CrossRef] [PubMed]
120. Campo, G.M.; Avenoso, A.; D'Ascola, A.; Prestipino, V.; Scuruchi, M.; Nastasi, G.; Calatroni, A.; Campo, S. 4-mer hyaluronan oligosaccharides stimulate in ammation response in synovial broblasts in part via TAK-1 and in part via p38-MAPK. *Curr. Med. Chem.* **2013**, *20*, 1162–1172. [CrossRef] [PubMed]
121. Yang, C.; Cao, M.; Liu, H.; He, Y.; Xu, J.; Du, Y.; Liu, Y.; Wang, W.; Cui, L.; Hu, J.; et al. The high and low molecular weight forms of hyaluronan have distinct effects on CD44 clustering. *J. Biol. Chem.* **2012**, *287*, 43094–43107. [CrossRef] [PubMed]
122. Turley, E.A.; Noble, P.W.; Bourguignon, LY. Signaling properties of hyaluronan receptors. *J. Biol. Chem.* **2002**, *277*, 4589–4592. [CrossRef] [PubMed]
123. Research, G.V. Hyaluronic Acid Market Size Worth USD 15.4 Billion by 2025 | CAGR: 8.8%. Available online: https://www.grandviewresearch.com/press-release/global-hyaluronic-acid-market (accessed on 8 March 2018).
124. Shiedlin, A.; Bigelow, R.; Christopher, W.; Arbabi, S.; Yang, L.; Maier, R.V.; Wainwright, N.; Childs, A.; Miller, R.J. Evaluation of hyaluronan from different sources: Streptococcus zooepidemicus, rooster comb, bovine vitreous, and human umbilical cord. *Biomacromolecules* **2004**, *5*, 2122–2127. [CrossRef] [PubMed]
125. Rangaswamy, V.; Jain, D. An efficient process for production and purification of hyaluronic acid from Streptococcus equi subsp. zooepidemicus. *Biotechnol. Lett.* **2008**, *30*, 493–496. [CrossRef] [PubMed]
126. Kim, J.H.; Yoo, S.J.; Oh, D.K.; Kweon, Y.G.; Park, D.W.; Lee, C.H.; Gil, G.H. Selection of a Streptococcus equi mutant and optimization of culture conditions for the production of high molecular weight hyaluronic acid. *Enzyme Microb. Technol.* **1996**, *19*, 440–445. [CrossRef]
127. Liu, L.; Liu, Y.; Li, J.; Du, G.; Chen, J. Microbial production of hyaluronic acid: Current state, challenges, and perspectives. *Microb. Cell Fact.* **2011**, *10*. [CrossRef] [PubMed]
128. Kaur, M.; Jayaraman, G. Hyaluronan production and molecular weight is enhanced in pathway-engineered strains of lactate dehydrogenase-deficient Lactococcus lactis. *Metab. Eng. Commun.* **2016**, *3*, 15–23. [CrossRef] [PubMed]

129. Chien, L.J.; Lee, C.K. Enhanced hyaluronic acid production in Bacillus subtilis by coexpressing bacterial hemoglobin. *Biotechnol. Prog.* **2007**, *23*, 1017–1022. [CrossRef] [PubMed]

130. Yu, H.; Stephanopoulos, G. Metabolic engineering of Escherichia coli for biosynthesis of hyaluronic acid. *Metab. Eng.* **2008**, *10*, 24–32. [CrossRef] [PubMed]

131. Cheng, F.; Gong, Q.; Yu, H.; Stephanopoulos, G. High-titer biosynthesis of hyaluronic acid by recombinant Corynebacterium glutamicum. *Biotechnol. J.* **2016**, *11*, 574–584. [CrossRef] [PubMed]

132. DeAngelis, P.L.; Achyuthan, A.M. Yeast-derived recombinant DG42 protein of Xenopus can synthesize hyaluronan in vitro. *J. Biol. Chem.* **1996**, *271*, 23657–23660. [CrossRef] [PubMed]

133. Jeong, E.; Shim, W.Y.; Kim, J.H. Metabolic engineering of Pichia pastoris for production of hyaluronic acid with high molecular weight. *J. Biotechnol.* **2014**, *185*, 28–36. [CrossRef] [PubMed]

134. Rakkhumkaew, N.; Shibatani, S.; Kawasaki, T.; Fujie, M.; Yamada, T. Hyaluronan synthesis in cultured tobacco cells (BY-2) expressing a chlorovirus enzyme: Cytological studies. *Biotechnol. Bioeng.* **2013**, *110*, 1174–1179. [CrossRef] [PubMed]

135. DeAngelis, P.L. Monodisperse hyaluronan polymers: Synthesis and potential applications. *Curr. Pharm. Biotechnol.* **2008**, *9*, 246–248. [CrossRef] [PubMed]

136. Schanté, C.E.; Zuber, G.; Herlin, C.; Vandamme, T.F. Chemical modifications of hyaluronic acid for the synthesis of derivatives for a broad range of biomedical applications. *Carbohydr. Polym.* **2011**, *85*, 469–489. [CrossRef]

137. Malson, T.; Lindqvist, B. Gels of Crosslinked Hyaluronic Acid for Use as a Vitreous Humor Substitute. International Publication No. WO1986000079 A1, 3 January 1986.

138. Shimojo, A.A.; Pires, A.M.; Lichy, R.; Rodrigues, A.A.; Santana, M.H. The crosslinking degree controls the mechanical, rheological, and swelling properties of hyaluronic acid microparticles. *J. Biomed. Mater. Res. A* **2015**, *103*, 730–737. [CrossRef] [PubMed]

139. Collins, M.N.; Birkinshaw, C. Physical properties of crosslinked hyaluronic acid hydrogels. *J. Mater. Sci. Mater. Med.* **2008**, *19*, 3335–3343. [CrossRef] [PubMed]

140. Bowman, S.; Awad, M.E.; Hamrick, M.W.; Hunter, M.; Fulzele, S. Recent advances in hyaluronic acid based therapy for osteoarthritis. *Clin. Transl. Med.* **2018**, *7*, 6. [CrossRef] [PubMed]

141. Nobile, V.; Buonocore, D.; Michelotti, A.; Marzatico, F. Anti-aging and filling efficacy of six types hyaluronic acid based dermo-cosmetic treatment: Double blind, randomized clinical trial of efficacy and safety. *J. Cosmet. Dermatol.* **2014**, *13*, 277–287. [CrossRef] [PubMed]

142. Kaur, K.; Singh, H.; Singh, M. Repair of tympanic membrane perforation by topical application of 1% sodium hyaluronate. *Indian J. Otolaryngol. Head Neck Surg.* **2006**, *58*, 241–244. [PubMed]

143. Gelardi, M.; Iannuzzi, L.; Quaranta, N. Intranasal sodium hyaluronate on the nasal cytology of patients with allergic and nonallergic rhinitis. *Int. Forum Allergy Rhinol.* **2013**, *3*, 807–813. [CrossRef] [PubMed]

144. Gelardi, M.; Guglielmi, A.V.; De Candia, N.; Maffezzoni, E.; Berardi, P.; Quaranta, N. Effect of sodium hyaluronate on mucociliary clearance after functional endoscopic sinus surgery. *Eur. Ann. Allergy Clin. Immunol.* **2013**, *45*, 103–108. [PubMed]

145. Maleki, A.; Kjøniksen, A.L.; Nyström, B. Characterization of the chemical degradation of hyaluronic acid during chemical gelation in the presence of different cross-linker agents. *Carbohydr. Res.* **2007**, *342*, 2776–2792. [CrossRef] [PubMed]

146. Bulpitt, P.; Aeschlimann, D. New strategy for chemical modification of hyaluronic acid: Preparation of functionalized derivatives and their use in the formation of novel biocompatible hydrogels. *J. Biomed. Mater. Res.* **1999**, *47*, 152–169. [CrossRef]

147. Lim, D.G.; Prim, R.E.; Kang, E.; Jeong, S.H. One-pot synthesis of dopamine-conjugated hyaluronic acid/polydopamine nanocomplexes to control protein drug release. *Int. J. Pharm.* **2018**, *542*, 288–296. [CrossRef] [PubMed]

148. Magnani, A.; Rappuoli, R.; Lamponi, S.; Barbucci, R. Novel polysaccharide hydrogels: Characterization and properties. *Polym. Adv. Technol.* **2000**, *11*, 488–495. [CrossRef]

149. Sigen, A.; Xu, Q.; McMichael, P.; Gao, Y.; Li, X.; Wang, X.; Greiser, U.; Zhou, D.; Wang, W. A facile one-pot synthesis of acrylated hyaluronic acid. *Chem. Commun.* **2018**, *54*, 1081–1084.

150. Felgueiras, H.P.; Wang, L.M.; Ren, K.F.; Querido, M.M.; Jin, Q.; Barbosa, M.A.; Ji, J.; Martins, M.C. Octadecyl chains immobilized onto hyaluronic acid coatings by thiol-ene "click chemistry" increase the surface antimicrobial properties and prevent platelet adhesion and activation to polyurethane. *ACS Appl. Mater. Interfaces* **2017**, *9*, 7979–7989. [CrossRef] [PubMed]

151. Smith, L.J.; Taimoory, S.M.; Tam, R.Y.; Baker, A.E.G.; Binth Mohammad, N.; Trant, J.F.; Shoichet, M.S. Diels-Alder Click-Cross-Linked Hydrogels with Increased Reactivity Enable 3D Cell Encapsulation. *Biomacromolecules* **2018**, *19*, 926–935. [CrossRef] [PubMed]

152. Fu, S.; Dong, H.; Deng, X.; Zhuo, R.; Zhong, Z. Injectable hyaluronic acid/poly(ethylene glycol) hydrogels crosslinked via strain-promoted azide-alkyne cycloaddition click reaction. *Carbohydr. Polym.* **2017**, *169*, 332–340. [CrossRef] [PubMed]

153. Shu, X.Z.; Liu, Y.; Luo, Y.; Roberts, M.C.; Prestwich, G.D. Disulfide cross-linked hyaluronan hydrogels. *Biomacromolecules* **2002**, *3*, 1304–1311. [CrossRef] [PubMed]

154. Bencherif, S.A.; Srinivasan, A.; Horkay, F.; Hollinger, J.O.; Matyjaszewski, K.; Washburn, N.R. Influence of the degree of methacrylation on hyaluronic acid hydrogels properties. *Biomaterials* **2008**, *29*, 1739–1749. [CrossRef] [PubMed]

155. Donnelly, P.E.; Chen, T.; Finch, A.; Brial, C.; Maher, S.A.; Torzilli, P.A. Photocrosslinked tyramine-substituted hyaluronate hydrogels with tunable mechanical properties improve immediate tissue-hydrogel interfacial strength in articular cartilage. *J. Biomater. Sci. Polym. Ed.* **2017**, *28*, 582–600. [CrossRef] [PubMed]

156. Yui, N.; Okano, T.; Sakurai, Y. Inflammation responsive degradation of crosslinked hyaluronic acid gels. *J. Control. Release* **1992**, *22*, 105–116.

157. Zhao, X. Process for the Production of Multiple Cross-Linked Hyaluronic Acid Derivatives. International Publication No. WO/2000/046253, 10 August 2000.

158. Collins, M.; Birkinshaw, C. Comparison of the effectiveness of four different crosslinking agents with hyaluronic acid hydrogel films for tissue-culture applications. *J. Appl. Polym. Sci.* **2007**, *104*, 3183–3191. [CrossRef]

159. Serban, M.; Yang, G.; Prestwich, G. Synthesis, characterization and chondroprotective properties of a hyaluronan thioethyl ether derivative. *Biomaterials* **2008**, *29*, 1388–1399. [CrossRef] [PubMed]

160. Crescenzi, V.; Francescangeli, A.; Taglienti, A.; Capitani, D.; Mannina, L. Synthesis and partial characterization of hydrogels obtained via glutaraldehyde crosslinking of acetylated chitosan and of hyaluronan derivatives. *Biomacromolecules* **2003**, *4*, 1045–1054. [CrossRef] [PubMed]

161. Tomihata, K.; Ikada, Y. Crosslinking of hyaluronic acid with glutaraldehyde. *J. Polym. Sci. Part A Polym. Chem.* **1997**, *35*, 3553–3559. [CrossRef]

162. Toemmeraas, K.; Eenschooten, C. Aryl/Alkyl Succinic Anhydride Hyaluronan Derivatives. International Publication No. WO/2007/033677, 29 March 2007.

163. Seidlits, S.K.; Khaing, Z.Z.; Petersen, R.R.; Nickels, J.D.; Vanscoy, J.E.; Shear, J.B.; Schmidt, C.E. The effects of hyaluronic acid hydrogels with tunable mechanical properties on neural progenitor cell differentiation. *Biomaterials* **2010**, *31*, 3930–3940. [CrossRef] [PubMed]

164. Pravata, L.; Braud, C.; Boustta, M.; El Ghzaoui, A.; Tømmeraas, K.; Guillaumie, F.; Schwach-Abdellaoui, K.; Vert, M. New amphiphilic lactic acid oligomer–hyaluronan conjugates: Synthesis and physicochemical characterization. *Biomacromolecules* **2008**, *9*, 340–348. [CrossRef] [PubMed]

165. Mlcochová, P.; Bystrický, S.; Steiner, B.; Machová, E.; Koós, M.; Velebný, V.; Krcmár, M. Synthesis and characterization of new biodegradable hyaluronan alkyl derivatives. *Biopolymers* **2006**, *82*, 74–79. [CrossRef] [PubMed]

166. Della Valle, F.; Romeo, A. Esters of Hyaluronic Acid. U.S. Patent 4,851,521, 25 July 1989.

167. Huin-Amargier, C.; Marchal, P.; Payan, E.; Netter, P.; Dellacherie, E. New physically and chemically crosslinked hyaluronate (HA)-based hydrogels for cartilage repair. *J. Biomed. Mater. Res. A* **2006**, *76*, 416–424. [CrossRef] [PubMed]

168. Hirano, K.; Sakai, S.; Ishikawa, T.; Avci, F.Y.; Linhardt, R.J.; Toida, T. Preparation of the methyl ester of hyaluronan and its enzymatic degradation. *Carbohydr. Res.* **2005**, *340*, 2297–2304. [CrossRef] [PubMed]

169. Prata, J.E.; Barth, T.A.; Bencherif, S.A.; Washburn, N.R. Complex fluids based on methacrylated hyaluronic acid. *Biomacromolecules* **2010**, *11*, 769–775. [CrossRef] [PubMed]

170. Benedetti, L.; Cortivo, R.; Berti, T.; Berti, A.; Pea, F.; Mazzo, M.; Moras, M.; Abatangelo, G. Biocompatibility and biodegradation of different hyaluronan derivatives (Hyaff) implanted in rats. *Biomaterials* **1993**, *14*, 1154–1160. [CrossRef]

171. Bellini, D.; Topai, A. Amides of Hyaluronic Acid and the Derivatives Thereof and a Process for Their Preparation. International Application No. PCT/IB1999/001254, 13 January 2000.

172. Kaczmarek, B.; Sionkowska, A.; Kozlowska, J.; Osyczka, A.M. New composite materials prepared by calcium phosphate precipitation in chitosan/collagen/hyaluronic acid sponge cross-linked by EDC/NHS. *Int. J. Biol. Macromol.* **2018**, *107*, 247–253. [CrossRef] [PubMed]

173. Kirk, J.F.; Ritter, G.; Finger, I.; Sankar, D.; Reddy, J.D.; Talton, J.D.; Nataraj, C.; Narisawa, S.; Millán, J.L.; Cobb, R.R. Mechanical and biocompatible characterization of a cross-linked collagen-hyaluronic acid wound dressing. *Biomatter* **2013**, *3*, pii: E25633. [CrossRef] [PubMed]

174. Bergman, K.; Elvingson, C.; Hilborn, J.; Svensk, G.; Bowden, T. Hyaluronic acid derivatives prepared in aqueous media by triazine-activated amidation. *Biomacromolecules* **2007**, *8*, 2190–2195. [CrossRef] [PubMed]

175. D'Este, M.; Eglin, D.; Alini, M. A systematic analysis of DMTMM vs EDC/NHS for ligation of amines to hyaluronan in water. *Carbohydr. Polym.* **2014**, *108*, 239–246. [CrossRef] [PubMed]

176. Crescenzi, V.; Francescangeli, A.; Segre, A.; Capitani, D.; Mannina, L.; Renier, D.; Bellini, D. NMR structural study of hydrogels based on partially deacetylated hyaluronan. *Macromol. Biosci.* **2002**, *2*, 272–279. [CrossRef]

177. Bhattacharya, D.; Svechkarev, D.; Souchek, J.J.; Hill, T.K.; Taylor, M.A.; Natarajan, A.; Mohs, A.M. Impact of structurally modifying hyaluronic acid on CD44 interaction. *J. Mater. Chem. B* **2017**, *5*, 8183–8192. [CrossRef] [PubMed]

178. Kong, J.H.; Oh, E.J.; Chae, S.Y.; Lee, K.C.; Hahn, S.K. Long acting hyaluronate—Exendin 4 conjugate for the treatment of type 2 diabetes. *Biomaterials* **2010**, *31*, 4121–4128. [CrossRef] [PubMed]

179. Mero, A.; Campisi, M. Hyaluronic acid bioconjugates for the delivery of bioactive molecules. *Polymers* **2014**, *6*, 346–369. [CrossRef]

180. Mangano, K.; Vergalito, F.; Mammana, S.; Mariano, A.; De Pasquale, R.; Meloscia, A.; Bartollino, S.; Guerra, G.; Nicoletti, F.; Di Marco, R. Evaluation of hyaluronic acid-P40 conjugated cream in a mouse model of dermatitis induced by oxazolone. *Exp. Ther. Med.* **2017**, *14*, 2439–2444. [CrossRef] [PubMed]

181. Manju, S.; Sreenivasan, K. Conjugation of curcumin onto hyaluronic acid enhances its aqueous solubility and stability. *J. Colloid Interface Sci.* **2011**, *359*, 318–325. [CrossRef] [PubMed]

182. Sharma, M.; Sahu, K.; Singh, S.P.; Jain, B. Wound healing activity of curcumin conjugated to hyaluronic acid: In vitro and in vivo evaluation. *Artif. Cells Nanomed. Biotechnol.* **2018**, *46*, 1009–1017. [CrossRef] [PubMed]

183. Chen, Y.; Peng, F.; Song, X.; Wu, J.; Yao, W.; Gao, X. Conjugation of paclitaxel to C-6 hexanediamine-modified hyaluronic acid for targeted drug delivery to enhance antitumor efficacy. *Carbohydr. Polym.* **2018**, *181*, 150–158. [CrossRef] [PubMed]

184. Nguyen, V.D.; Zheng, S.; Han, J.; Le, V.H.; Park, J.O.; Park, S. Nanohybrid magnetic liposome functionalized with hyaluronic acid for enhanced cellular uptake and near-infrared-triggered drug release. *Colloids Surf. B. Biointerfaces* **2017**, *154*, 104–114. [CrossRef] [PubMed]

185. Hayward, S.L.; Wilson, C.L.; Kidambi, S. Hyaluronic acid-conjugated liposome nanoparticles for targeted delivery to CD44 overexpressing glioblastoma cells. *Oncotarget* **2016**, *7*, 34158–34171. [CrossRef] [PubMed]

186. Han, N.K.; Shin, D.H.; Kim, J.S.; Weon, K.Y.; Jang, C.Y.; Kim, J.S. Hyaluronan-conjugated liposomes encapsulating gemcitabine for breast cancer stem cells. *Int. J. Nanomed.* **2016**, *11*, 1413–1425. [CrossRef] [PubMed]

187. Chi, Y.; Yin, X.; Sun, K.; Feng, S.; Liu, J.; Chen, D.; Guo, C.; Wu, Z. Redox-sensitive and hyaluronic acid functionalized liposomes for cytoplasmic drug delivery to osteosarcoma in animal models. *J. Control. Release* **2017**, *261*, 113–125. [CrossRef] [PubMed]

188. Negi, L.M.; Jaggi, M.; Joshi, V.; Ronodip, K.; Talegaonkar, S. Hyaluronan coated liposomes as the intravenous platform for delivery of imatinib mesylate in MDR colon cancer. *Int. J. Biol. Macromol.* **2015**, *73*, 222–235. [CrossRef] [PubMed]

189. Moustafa, M.A.; Elnaggar, Y.S.R.; El-Refaie, W.M.; Abdallah, O.Y. Hyalugel-integrated liposomes as a novel ocular nanosized delivery system of fluconazole with promising prolonged effect. *Int. J. Pharm.* **2017**, *534*, 14–24. [CrossRef] [PubMed]

190. Manconi, M.; Manca, M.L.; Valenti, D.; Escribano, E.; Hillaireau, H.; Fadda, A.M.; Fattal, E. Chitosan and hyaluronan coated liposomes for pulmonary administration of curcumin. *Int. J. Pharm.* **2017**, *525*, 203–210. [CrossRef] [PubMed]

191. Leite Nascimento, T.; Hillaireau, H.; Vergnaud, J.; Rivano, M.; Deloménie, C.; Courilleau, D.; Arpicco, S.; Suk, J.S.; Hanes, J.; Fattal, E. Hyaluronic acid-conjugated lipoplexes for targeted delivery of siRNA in a murine metastatic lung cancer model. *Int. J. Pharm.* **2016**, *514*, 103–111. [CrossRef] [PubMed]

192. Maudens, P.; Meyer, S.; Seemayer, C.A.; Jordan, O.; Allémann, E. Self-assembled thermoresponsive nanostructures of hyaluronic acid conjugates for osteoarthritis therapy. *Nanoscale* **2018**, *10*, 1845–1854. [CrossRef] [PubMed]

193. Kim, H.; Park, H.T.; Tae, Y.M.; Kong, W.H.; Sung, D.K.; Hwang, B.W.; Kim, K.S.; Kim, Y.K.; Hahn, S.K. Bioimaging and pulmonary applications of self-assembled Flt1 peptide-hyaluronic acid conjugate nanoparticles. *Biomaterials* **2013**, *34*, 8478–8490. [CrossRef] [PubMed]

194. Pandey, M.; Choudhury, H.; Gunasegaran, T.A.P.; Nathan, S.S.; Md, S.; Gorain, B.; Tripathy, M.; Hussain, Z. Hyaluronic acid-modified betamethasone encapsulated polymeric nanoparticles: Fabrication, characterisation, in vitro release kinetics, and dermal targeting. *Drug Deliv. Transl. Res.* **2018**. [CrossRef] [PubMed]

195. Pereira, G.G.; Detoni, C.B.; Balducci, A.G.; Rondelli, V.; Colombo, P.; Guterres, S.S.; Sonvico, F. Hyaluronate nanoparticles included in polymer films for the prolonged release of vitamin E for the management of skin wounds. *Eur. J. Pharm. Sci.* **2016**, *83*, 203–211. [CrossRef] [PubMed]

196. Wang, T.; Hou, J.; Su, C.; Zhao, L.; Shi, Y. Hyaluronic acid-coated chitosan nanoparticles induce ROS-mediated tumor cell apoptosis and enhance antitumor efficiency by targeted drug delivery via CD44. *J. Nanobiotechnol.* **2017**, *15*, 7. [CrossRef] [PubMed]

197. Cho, H.J.; Yoon, H.Y.; Koo, H.; Ko, S.H.; Shim, J.S.; Lee, J.H.; Kim, K.; Kwon, I.C.; Kim, D.D. Self-assembled nanoparticles based on hyaluronic acid-ceramide (HA-CE) and Pluronic® for tumor-targeted delivery of docetaxel. *Biomaterials* **2011**, *32*, 7181–7190. [CrossRef] [PubMed]

198. Parashar, P.; Rathor, M.; Dwivedi, M.; Saraf, S.A. Hyaluronic acid decorated naringenin nanoparticles: Appraisal of chemopreventive and curative potential for lung cancer. *Pharmaceutics* **2018**, *10*. [CrossRef] [PubMed]

199. Hwang, S.M.; Kim, D.D.; Chung, S.J.; Shim, C.K. Delivery of ofloxacin to the lung and alveolar macrophages via hyaluronan microspheres for the treatment of tuberculosis. *J. Control. Release* **2008**, *129*, 100–106. [CrossRef] [PubMed]

200. Li, Y.; Han, M.; Liu, T.; Cun, D.; Fang, L.; Yang, M. Inhaled hyaluronic acid microparticles extended pulmonary retention and suppressed systemic exposure of a short-acting bronchodilator. *Carbohydr. Polym.* **2017**, *172*, 197–204. [CrossRef] [PubMed]

201. Fatnassi, M.; Jacquart, S.; Brouillet, F.; Rey, C.; Combes, C.; Girod Fullana, S. Optimization of spray-dried hyaluronic acid microspheres to formulate drug-loaded bone substitute materials. *Powder Technol.* **2014**, *255*, 44–51. [CrossRef]

202. Simsolo, E.E.; Eroğlu, İ.; Tanrıverdi, S.T.; Özer, Ö. Formulation and evaluation of organogels containing hyaluronan microparticles for topical delivery of caffeine. *AAPS PharmSciTech* **2018**, *19*, 1367–1376. [CrossRef] [PubMed]

203. Esposito, E.; Menegatti, E.; Cortesi, R. Hyaluronan-based microspheres as tools for drug delivery: A comparative study. *Int. J. Pharm.* **2005**, *288*, 35–49. [CrossRef] [PubMed]

204. Koivusalo, L.; Karvinen, J.; Sorsa, E.; Jönkkäri, I.; Väliaho, J.; Kallio, P.; Ilmarinen, T.; Miettinen, S.; Skottman, H.; Kellomäki, M. Hydrazone crosslinked hyaluronan-based hydrogels for therapeutic delivery of adipose stem cells to treat corneal defects. *Mater. Sci. Eng. C Mater. Biol. Appl.* **2018**, *85*, 68–78. [CrossRef] [PubMed]

205. Luo, Y.; Kirker, K.R.; Prestwich, G.D. Cross-linked hyaluronic acid hydrogel films: New biomaterials for drug delivery. *J. Control. Release* **2000**, *69*, 169–184. [CrossRef]

206. Kroin, J.S.; Kc, R.; Li, X.; Hamilton, J.L.; Das, V.; van Wijnen, A.J.; Dall, O.M.; Shelly, D.A.; Kenworth, T.; Im, H.J. Intraarticular slow-release triamcinolone acetate reduces allodynia in an experimental mouse knee osteoarthritis model. *Gene* **2016**, *591*, 1–5. [CrossRef] [PubMed]

207. Pirard, D.; Vereecken, P.; Mélot, C.; Heenen, M. Three percent diclofenac in 2.5% hyaluronan gel in the treatment of actinic keratoses: A meta-analysis of the recent studies. *Arch. Dermatol. Res.* **2005**, *297*, 185–189. [CrossRef] [PubMed]

208. Gao, Y.; Foster, R.; Yang, X.; Feng, Y.; Shen, J.K.; Mankin, H.J.; Hornicek, F.J.; Amiji, M.M.; Duan, Z. Up-regulation of CD44 in the development of metastasis, recurrence and drug resistance of ovarian cancer. *Oncotarget* **2015**, *6*, 9313–9326. [CrossRef] [PubMed]

209. Buffa, R.; Šedová, P.; Basarabová, I.; Bobula, T.; Procházková, P.; Vágnerová, H.; Dolečková, I.; Moravčíková, S.; Hejlová, L.; Velebný, V. A new unsaturated derivative of hyaluronic acid—Synthesis, analysis and applications. *Carbohydr. Polym.* **2017**, *163*, 247–253. [CrossRef] [PubMed]

210. Aya, K.L.; Stern, R. Hyaluronan in wound healing: Rediscovering a major player. *Wound Repair Regen.* **2014**, *22*, 579–593. [CrossRef] [PubMed]

211. Tagliagambe, M.; Elstrom, T.A.; Ward, D.B. Hyaluronic Acid Sodium Salt 0.2% Gel in the Treatment of a Recalcitrant Distal Leg Ulcer: A Case Report. *J. Clin. Aesthet. Dermatol.* **2017**, *10*, 49–51. [PubMed]

212. Neuman, M.G.; Nanau, R.M.; Oruña-Sanchez, L.; Coto, G. Hyaluronic acid and wound healing. *J. Pharm. Pharm. Sci.* **2015**, *18*, 53–60. [CrossRef] [PubMed]

213. Rueda López, J.; Segovia Gómez, T.; Guerrero Palmero, A.; Bermejo Martínez, M.; Muñoz Bueno, A.M. Hyaluronic acid: A new trend to cure skin injuries an observational study. *Rev. Enferm.* **2005**, *28*, 53–57. [PubMed]

214. Shi, L.; Zhao, Y.; Xie, Q.; Fan, C.; Hilborn, J.; Dai, J.; Ossipov, D.A. Moldable Hyaluronan Hydrogel Enabled by Dynamic Metal-Bisphosphonate Coordination Chemistry for Wound Healing. *Adv. Healthc. Mater.* **2018**, *7*. [CrossRef] [PubMed]

215. Li, H.; Xue, Y.; Jia, B.; Bai, Y.; Zuo, Y.; Wang, S.; Zhao, Y.; Yang, W.; Tang, H. The preparation of hyaluronic acid grafted pullulan polymers and their use in the formation of novel biocompatible wound healing film. *Carbohydr. Polym.* **2018**, *188*, 92–100. [CrossRef] [PubMed]

216. Berce, C.; Muresan, M.S.; Soritau, O.; Petrushev, B.; Tefas, L.; Rigo, I.; Ungureanu, G.; Catoi, C.; Irimie, A.; Tomuleasa, C. Cutaneous wound healing using polymeric surgical dressings based on chitosan, sodium hyaluronate and resveratrol. A preclinical experimental study. *Colloids Surf. B Biointerfaces* **2018**, *163*, 155–166. [CrossRef] [PubMed]

217. Sanad, R.A.; Abdel-Bar, H.M. Chitosan-hyaluronic acid composite sponge scaffold enriched with Andrographolide loaded lipid nanoparticles for enhanced wound healing. *Carbohydr. Polym.* **2017**, *173*, 441–450. [CrossRef] [PubMed]

218. Wirostko, B.; Mann, B.K.; Williams, D.L.; Prestwich, G.D. Ophthalmic Uses of a Thiol-Modified Hyaluronan-Based Hydrogel. Adv. *Wound Care* **2014**, *3*, 708–716. [CrossRef] [PubMed]

219. Cassano, M.; Russo, G.M.; Granieri, C.; Cassano, P. Cytofunctional changes in nasal ciliated cells in patients treated with hyaluronate after nasal surgery. *Am. J. Rhinol. Allergy* **2016**, *30*, 83–88. [CrossRef] [PubMed]

220. Dahiya, P.; Kamal, R. Hyaluronic acid: A boon in periodontal therapy. *N. Am. J. Med. Sci.* **2013**, *5*, 309–315. [CrossRef] [PubMed]

221. Neumayer, T.; Prinz, A.; Findl, O. Effect of a new cohesive ophthalmic viscosurgical device on corneal protection and intraocular pressure in small-incision cataract surgery. *J. Cataract Refract. Surg.* **2008**, *34*, 1362–1366. [CrossRef] [PubMed]

222. Vandermeer, G.; Chamy, Y.; Pisella, P.J. Comparison of objective optical quality measured by double-pass aberrometry in patients with moderate dry eye: Normal saline vs. artificial tears: A pilot study. *J. Fr. Ophtalmol.* **2018**, *41*, e51–e57. [CrossRef] [PubMed]

223. Carracedo, G.; Villa-Collar, C.; Martin-Gil, A.; Serramito, M.; Santamaría, L. Comparison Between Viscous Teardrops and Saline Solution to Fill Orthokeratology Contact Lenses Before Overnight Wear. *Eye Contact Lens.* **2017**. [CrossRef] [PubMed]

224. Johnson, M.E.; Murphy, P.J.; Boulton, M. Effectiveness of sodium hyaluronate eyedrops in the treatment of dry eye. *Graefe's Arch. Clin. Exp. Ophthalmol.* **2006**, *244*, 109–112. [CrossRef] [PubMed]

225. Aragona, P.; Papa, V.; Micali, A.; Santocono, M.; Milazzo, G. Long term treatment with sodium hyaluronate–containing artificial tears reduces ocular surface damage in patients with dry eye. *Br. J. Ophthalmol.* **2002**, *86*, 181–184. [CrossRef] [PubMed]

226. Ho, W.T.; Chiang, T.H.; Chang, S.W.; Chen, Y.H.; Hu, F.R.; Wang, I.J. Enhanced corneal wound healing with hyaluronic acid and high-potassium artificial tears. *Clin. Exp. Optom.* **2013**, *96*, 536–541. [CrossRef] [PubMed]

227. Laffleur, F.; Dachs, S. Development of novel mucoadhesive hyaluronic acid derivate as lubricant for the treatment of dry eye syndrome. *Ther. Deliv.* **2015**, *6*, 1211–1219. [CrossRef] [PubMed]

228. Ghosh, P.; Guidolin, D. Potential mechanism of action of intra-articular hyaluronan therapy in osteoarthritis: Are the effects molecular weight dependent? *Semin. Arthritis Rheum.* **2002**, *32*, 10–37. [CrossRef] [PubMed]

229. Brown, T.J.; Laurent, U.B.; Fraser, J.R. Turnover of hyaluronan in synovial joints: Elimination of labelled hyaluronan from the knee joint of the rabbit. *Exp. Physiol.* **1991**, *76*, 125–134. [CrossRef] [PubMed]

230. Greenberg, D.D.; Stoker, A.; Kane, S.; Cockrell, M.; Cook, J.L. Biochemical effects of two different hyaluronic acid products in a co-culture model of osteoarthritis. *Osteoarthr. Cartil.* **2006**, *14*, 814–822. [CrossRef] [PubMed]

231. Sun, S.F.; Hsu, C.W.; Lin, H.S.; Liou, I.H.; Chen, Y.H.; Hung, C.L. Comparison of single intra-articular injection of novel hyaluronan (HYA-JOINT Plus) with synvisc-one for knee osteoarthritis: A randomized, controlled, double-blind trial of efficacy and safety. *J. Bone Joint Surg. Am.* **2017**, *99*, 462–471. [CrossRef] [PubMed]

232. Monzon, M.E.; Casalino-Matsuda, S.M.; Forteza, R.M. Identification of glycosaminoglycans in human airway secretions. *Am. J. Respir. Cell Mol. Biol.* **2006**, *34*, 135–141. [CrossRef] [PubMed]

233. Garantziotis, S.; Brezina, M.; Castelnuovo, P.; Drago, L. The role of hyaluronan in the pathobiology and treatment of respiratory disease. *Am. J. Physiol. Lung Cell Mol. Physiol.* **2016**, *310*, L785–L795. [CrossRef] [PubMed]

234. Gerdin, B.; Hällgren, R. Dynamic role of hyaluronan (HYA) in connective tissue activation and inflammation. *J. Intern. Med.* **1997**, *242*, 49–55. [CrossRef] [PubMed]

235. Furnari, M.L.; Termini, L.; Traverso, G.; Barrale, S.; Bonaccorso, M.R.; Damiani, G.; Piparo, C.L.; Collura, M. Nebulized hypertonic saline containing hyaluronic acid improves tolerability in patients with cystic fibrosis and lung disease compared with nebulized hypertonic saline alone: A prospective, randomized, double-blind, controlled study. *Ther. Adv. Respir. Dis.* **2012**, *6*, 315–322. [CrossRef] [PubMed]

236. Gavina, M.; Luciani, A.; Villella, V.R.; Esposito, S.; Ferrari, E.; Bressani, I.; Casale, A.; Bruscia, E.M.; Maiuri, L.; Raia, V. Nebulized hyaluronan ameliorates lung inflammation in cystic fibrosis mice. *Pediatr. Pulmonol.* **2013**, *48*, 761–771. [CrossRef] [PubMed]

237. Petrigni, G.; Allegra, L. Aerosolised hyaluronic acid prevents exercise-induced bronchoconstriction, suggesting novel hypotheses on the correction of matrix defects in asthma. *Pulm. Pharmacol. Ther.* **2006**, *19*, 166–171. [CrossRef] [PubMed]

238. Turino, G.M.; Ma, S.; Lin, Y.Y.; Cantor, J.O. The therapeutic potential of hyaluronan in COPD. *Chest* **2017**. [CrossRef] [PubMed]

239. Nenna, R.; Papasso, S.; Battaglia, M.; De Angelis, D.; Petrarca, L.; Felder, D.; Salvadei, S.; Berardi, R.; Roberti, M.; Papoff, P.; et al. 7% hypertonic saline and hyaluronic acid and in the treatment of infants mild-moderate bronchiolitis. *Eur. Respir. J.* **2011**, *38*, 1717.

240. Goddard, J.C.; Janssen, D.A.W. Intravesical hyaluronic acid and chondroitin sulfate for recurrent urinary tract infections: Systematic review and meta-analysis. *Int. Urogynecol. J.* **2017**. [CrossRef] [PubMed]

241. Ząbkowski, T.; Jurkiewicz, B.; Saracyn, M. Treatment of recurrent bacterial cystitis by intravesical instillations of hyaluronic acid. *Urol. J.* **2015**, *12*, 2192–2195. [PubMed]

242. American Society of Plastic Surgeons. 2017 Plastic Surgery Statistics Report. Available online: https://www.plasticsurgery.org/documents/News/Statistics/2017/plastic-surgery-statistics-report-2017.pdf (accessed on 8 March 2018).

243. Muhn, C.; Rosen, N.; Solish, N.; Bertucci, V.; Lupin, M.; Dansereau, A.; Weksberg, F.; Remington, B.K.; Swift, A. The evolving role of hyaluronic acid fillers for facial volume restoration and contouring: A Canadian overview. *Clin. Cosmet. Investig. Dermatol.* **2012**, *5*, 147–158. [CrossRef] [PubMed]

244. Janiš, R.; Pata, V.; Egner, P.; Pavlačková, J.; Zapletalová, A.; Kejlová, K. Comparison of metrological techniques for evaluation of the impact of a cosmetic product containing hyaluronic acid on the properties of skin surface. *Biointerphases* **2017**, *12*, 021006. [CrossRef] [PubMed]

245. Trommer, H.; Wartewig, S.; Böttcher, R.; Pöppl, A.; Hoentsch, J.; Ozegowski, J.H.; Neubert, R.H. The effects of hyaluronan and its fragments on lipid models exposed to UV irradiation. *Int. J. Pharm.* **2003**, *254*, 223–234. [CrossRef]

246. Hašová, M.; Crhák, T.; Safránková, B.; Dvořáková, J.; Muthný, T.; Velebný, V.; Kubala, L. Hyaluronan minimizes effects of UV irradiation on human keratinocytes. *Arch. Dermatol. Res.* **2011**, *303*, 277–284. [CrossRef] [PubMed]

247. Drealos, Z.D. Nutrition and enhancing youthful-appearing skin. *Clin. Dermatol.* **2010**, *28*, 400–408. [CrossRef] [PubMed]

248. Kimura, M.; Maeshima, T.; Kubota, T.; Kurihara, H.; Masuda, Y.; Nomura, Y. Adsorption of orally administerd hyaluronan. *J. Med. Food* **2016**, *19*, 1172–1179. [CrossRef] [PubMed]

249. Zgheib, C.; Xu, J.; Liechty, K.W. Targeting Inflammatory Cytokines and Extracellular Matrix Composition to Promote Wound Regeneration. *Adv. Wound Care* **2014**, *3*, 344–355. [CrossRef] [PubMed]

250. Sato, T.; Iwaso, H. An effectiveness study of hyaluronic acid [Hyabest® (J)] in the treatment of osteoarthritis of the knee on the patients in the United States. *J. New Rem. Clin.* **2009**, *58*, 260–269.

251. Balogh, L.; Polyak, A.; Mathe, D.; Kiraly, R.; Thuroczy, J.; Terez, M.; Janoki, G.; Ting, Y.; Bucci, L.R.; Schauss, A.G. An effectiveness study of hyaluronic acid [Hyabest® (J)] in the treatment of osteoarthritis of the knee on the patients in the United States. *J. Agric. Food Chem.* **2008**, *56*, 10582–10593. [CrossRef] [PubMed]

252. Oe, M.; Mitsugi, K.; Odanaka, W.; Yoshida, H.; Matsuoka, R.; Seino, S.; Kanemitsu, T.; Masuda, Y. Dietary hyaluronic acid migrates into the skin of rats. *Sci. World J.* **2014**, *2014*, 378024. [CrossRef] [PubMed]

253. Solà, R.; Valls, R.M.; Martorell, I.; Giralt, M.; Pedret, A.; Taltavull, N.; Romeu, M.; Rodríguez, À.; Moriña, D.; Lopez de Frutos, V.; Montero, M.; et al. A low-fat yoghurt supplemented with a rooster comb extract on muscle joint function in adults with mild knee pain: A rondomized, double blind, parallel, placebo-controlled, clinical trial of efficacy. *Food Funct.* **2015**, *6*, 3531–3539. [CrossRef] [PubMed]

254. Göllner, I.; Voss, W.; von Hehn, U.; Kammerer, S. Ingestion of an Oral Hyaluronan Solution Improves Skin Hydration, Wrinkle Reduction, Elasticity, and Skin Roughness: Results of a Clinical Study. *J. Evid. Based Complement. Altern. Med.* **2017**, *22*, 816–823. [CrossRef] [PubMed]

255. Kawada, C.; Yoshida, T.; Yoshida, H.; Matsuoka, R.; Sakamoto, W.; Odanaka, W.; Sato, T.; Yamasaki, T.; Kanemitsu, T.; Masuda, Y.; et al. Ingested hyaluronan moisturizes dry skin. *Nutr. J.* **2014**, *13*, 70. [CrossRef] [PubMed]

256. Kawada, C.; Yoshida, T.; Yoshida, H.; Sakamoto, W.; Odanaka, W.; Sato, T.; Yamasaki, T.; Kanemitsu, T.; Masuda, Y.; Urushibata, O. Ingestion of hyaluronans (molecular weights 800 k and 300 k) improves dry skin conditions: A randomized, double blind, controlled study. *J. Clin. Biochem. Nutr.* **2015**, *56*, 66–73. [CrossRef] [PubMed]

257. Kawada, C.; Kimura, M.; Masuda, Y.; Nomura, Y. Oral administration of hyaluronan prevents skin dryness and epidermal thickening in ultraviolet irradiated hairless mice. *J. Photochem. Photobiol. B* **2015**, *153*, 215–221. [CrossRef] [PubMed]

258. Oe, M.; Tashiro, T.; Yoshida, H.; Nishiyama, H.; Masuda, Y.; Maruyama, K.; Koikeda, T.; Maruya, R.; Fukui, N. Oral hyaluronan relieves knee pain: A review. *Nutr. J.* **2016**, *15*, 11. [CrossRef] [PubMed]

259. Larson, B. 3D Cell Culture: A Review of Current Techniques. 2015. Available online: http://mktg.biotek.com/news/2015/Fall/featured-application.html (accessed on 14 June 2018).

260. Gurski, L.A.; Jha, A.K.; Zhang, C.; Jia, X.; Farach-Carson, M.C. Hyaluronic acid-based hydrogels as 3D matrices for in vitro evaluation of chemotherapeutic drugs using poorly adherent prostate cancer cells. *Biomaterials* **2009**, *30*, 6076–6085. [CrossRef] [PubMed]

261. Jeffery, A.F.; Churchward, M.A.; Mushahwar, V.K.; Todd, K.G.; Elias, A.L. Hyaluronic acid-based 3D culture model for in vitro testing of electrode biocompatibility. *Biomacromolecules* **2014**, *15*, 2157–2165. [CrossRef] [PubMed]

262. Ouyang, L.; Highley, C.B.; Rodell, C.B.; Sun, W.; Burdick, J.A. 3D Printing of Shear-Thinning Hyaluronic Acid Hydrogels with Secondary Cross-Linking. *ACS Biomater. Sci. Eng.* **2016**, *2*, 1743–1751. [CrossRef]

263. Fallacara, A.; Busato, L.; Pozzoli, M.; Ghadiri, M.; Ong, H.X.; Young, P.M.; Manfredini, S.; Traini, D. Combination of urea-crosslinked Hyaluronic acid and sodium ascorbyl phosphate for the treatment of inflammatory lung diseases: An in vitro study. *Eur. J. Pharm. Sci.* **2018**, *120*, 96–106. [CrossRef] [PubMed]

 *polymers*

Article

Preparation, Physicochemical Properties and Hemocompatibility of Biodegradable Chitooligosaccharide-Based Polyurethane

Weiwei Xu, Minghui Xiao, Litong Yuan, Jun Zhang and Zhaosheng Hou *

College of Chemistry, Chemical Engineering and Materials Science, Shandong Normal University, Jinan 250014, China; 17853135516@163.com (W.X.); xiaominghui98@163.com (M.X.); m17853135556@163.com (L.Y.); zj971127@163.com (J.Z.)
* Correspondence: houzs@sdnu.edu.cn

Received: 6 May 2018; Accepted: 21 May 2018; Published: 24 May 2018

Abstract: The purpose of this study was to develop a process to achieve biodegradable chitooligosaccharide-based polyurethane (CPU) with improved hemocompatibility and mechanical properties. A series of CPUs with varying chitooligosaccharide (COS) content were prepared according to the conventional two-step method. First, the prepolymer was synthesized from poly(ε-caprolactone) (PCL) and uniform-size diurethane diisocyanates (HBH). Then, the prepolymer was chain-extended by COS in N,N-dimethylformamide (DMF) to obtain the weak-crosslinked CPU, and the corresponding films were obtained from the DMF solution by the solvent evaporation method. The uniform-size hard segments and slight crosslinking of CPU were beneficial for enhancing the mechanical properties, which were one of the essential requirements for long-term implant biomaterials. The chemical structure was characterized by FT-IR, and the influence of COS content in CPU on the physicochemical properties and hemocompatibility was extensively researched. The thermal stability studies indicated that the CPU films had lower initial decomposition temperature and higher maximum decomposition temperature than pure polyurethane (CPU-1.0) film. The ultimate stress, initial modulus, and surface hydrophilicity increased with the increment of COS content, while the strain at break and water absorption decreased, which was due to the increment of crosslinking density. The results of in vitro degradation signified that the degradation rate increased with the increasing content of COS in CPU, demonstrating that the degradation rate could be controlled by adjusting COS content. The surface hemocompatibility was examined by protein adsorption and platelet adhesion tests. It was found that the CPU films had improved resistance to protein adsorption and possessed good resistance to platelet adhesion. The slow degradation rate and good hemocompatibility of the CPUs showed great potential in blood-contacting devices. In addition, many active amino and hydroxyl groups contained in the structure of CPU could carry out further modification, which made it an excellent candidate for wide application in biomedical field.

Keywords: chitooligosaccharide; polyurethane; biodegradability; physicochemical properties; hemocompatibility

---

1. Introduction

Polyurethanes (PUs) are becoming more and more important as engineering materials because of their excellent abrasion resistance and the properties of both rubber and plastics [1–3]. The hard-segment-rich and soft-segment-rich domains existing in PUs contribute to the specific microphase-separated structure, which give them unique mechanical properties [4]. Due to the excellent physic-mechanical properties and good biocompatibility, PUs have been used in many biomedical engineering areas, including blood vessels, cardioids, artificial skins, cartilages, joints, and catheters [5–7]. But, when PUs are used as long-term blood-contacting materials, surface-induced thrombosis, protein fouling, and poor cytocompatibility are three popular problems which are difficult to conquer [8,9]. To achieve improved biocompatibility of synthetic polymers, natural biopolymers were used to prepare novel hemocompatible biocomposites [10,11]. Much attention has been paid to producing a nonspecific protein repelling surface by surface modification via various kinds of methods and creating highly effective non-thrombogenic devices. A preferred strategy is to immobilize natural biopolymers that shield the surface, thus introducing a high activation barrier to repel proteins [12–15]. But the biopolymers can fall out slowly from the surface in the uage process, which limits the application as long-term implant materials. Currently, much effort has been applied to use natural material in the design and preparation of new biomaterial, and bulk-modified PU by natural biopolymers has become a new frontier [16,17]. Among the natural biopolymers, polysaccharides, which are readily available, inexpensive, and biodegradable, appear to be good candidates for this purpose.

Chitosan (CH), the linear cationic (1,4)-2-amino-2-deoxy-β-D-glucan produced from the natural parent chitin by partial deacetylation, is the second most abundant polysaccharide in nature and has been utilized in the biomedical field due to its biological properties, such as nontoxicity, thermal stability, cytocompatibility, and hemocompatibility [18]. Several researches have been reported regarding chitosan-polyurethane copolymers [19–21], which possess improved thermal stability and mechanical properties. In some of their works, the swollen CH, which is obtained by dispersion in glacial acetic acid/*N,N*-dimethylformamide (DMF) mixtures [22], is used as an extender in the copolymerization because CH cannot be dissolved in organic solvent; others use the waterborne PU (WPU) to react with CH in glacial acetic acid [23,24]. However, the CH-modified PU materials obtained by these methods are limited by the processing difficulties, because the forms of these materials are either gel or solid. On the other hand, CH exhibits other drawbacks of poor flexibility and degradability, which are related to chemical and physical characteristics such as the high molecular weight and crystallinity [25].

As depolymerized product by chemical and enzymatic hydrolysis of CH, chitooligosaccharide (COS) consisting of 2–10 glucosamine units bounded via β-1,4-glycoside linkages has attracted more and more attention recently, because the latter has a shorter chain length and can be easily soluble in water and in some organic solvent, such as dimethylsulfoxide (DMSO) and DMF. In addition, there are several papers reported that COS can be absorbed readily through the intestine, quickly getting into the blood flow [26,27]. On the other hand, the reactive amino and hydroxyl groups of COS make it easy to prepare COS-based modified biomaterials. Because of its unusual properties, COS shows great potential to be a biopolymer to improve the performance of PUs [28–31]. Base on the good water solubility of COS, Jia groups prepared novel hemocompatible WPUs using COS as an extender via the self-emulsion polymerization method [17,32]. The COS-based WPU emulsion showed satisfactory freeze/thaw stability, and the films cast from the emulsions exhibited excellent mechanical properties and good anticoagulating character. However, only a little work has been published to describe the synthesis of COS-based PU in organic solvent [33], which can provide a novel way to design and prepare bulk-modified PU biomaterials by COS and exploit the application of PU in biomedical field.

In this article, a series of block COS-based PUs (CPU) with varying COS content were prepared in organic solvent via the conventional two-step method. First, the prepolymer was synthesized from poly(ε-caprolactone) (PCL) and uniform-size diurethane diisocyanates (HBH). Then, the prepolymer was chain-extended by COS in DMF to obtain the CPU with low crosslinking density, and the corresponding films were prepared from the DMF solution by the solvent evaporation method. The uniform-size hard segments and slight crosslinking of CPU are beneficial for enhancing the mechanical properties, which are the essential requirement for long-term implant biomaterials. The influence of COS content in CPU on the physicochemical properties of the films, including thermal properties, mechanical properties, surface hydrophilicity, swellability, and in vitro hydrolytic biodegradability, were researched. Surface hemocompatibility of the films was examined by protein adsorption and platelet adhesion tests. Moreover, many active amino and hydroxyl groups contained in the structure of CPU could carry out further modification, which made it an excellent candidate for wide application in biomedical field.

2. Materials and Methods

2.1. Materials

COS (number-average molecular weight: 3000 g/mol; degree of deacetylation: 92%) was supplied by Qingdao Yunzhou Biochemistry Co., Ltd. (Qingdao, China) and dried for 4 h at 50 °C under vacuum prior to use. PCL (number-average molecular weight: 2000 g/mol) was obtained from Shenzhen Polymtek Biomaterial Co., Ltd. (Shenzhen, China) and used without further purification. 1,6-Hexanediisocyanate (HDI) and dibutyltin dilaurate (DBTDL) were purchased from Sigma-Aldrich Chemical Co. (St. Louis, MO, USA) and used as received. 1,4-Butanediol (BDO, Aladdin Reagent Co., Shanghai, China) were dried over 4Å molecular sieves and redistilled before use. DMF (AR grade, Aladdin Reagent Co., Shanghai, China) was dried with phosphorus pentoxide and distilled under reduced pressure prior to use. Phosphate buffer saline (PBS, pH = 7.4) was supplied by Beijing Chemical Reagent Co., Ltd. (Beijing, China) and used as received. Other reagents were AR grade and purified by standard methods.

2.2. Synthesis of CPU

The basic formulations are given in Table 1. A typical procedure was described as below: A predetermined amount of diurethane diisocyanates (1,6-hexanediisocyanate-1,4-butanediol-1,6-hexanediisocyanate, HBH), which was synthesized according to the previous literature [34], and PCL was charged into a three-necked flask equipped with a mechanical stirrer under dried nitrogen atmosphere. DMF was added and the mixture was stirred at room temperature to get a homogeneous solution (~0.4 g/mL). After two drops of DBTDL was added to the solution (0.3 wt % of PCL and HBH), the reaction was carried out at 80 °C for about 2.5 h until the isocyanate group content in the system reaching the theoretical value, which was determined using the standard di-n-butylamine back titration method [35,36]. Then the system (prepolymer solution) was cooled to 25 °C, and the COS solution (DMF, 0.2 g/mL) was added in one portion under vigorous mechanical stirring. When the reaction mixture became viscous, small amounts of DMF were re-added to keep the system homogeneous. The reaction mixture was allowed to proceed at 25 °C for about 1.5 h until the NCO peak (~2270 cm^{-1}) in the FT-IR spectrum disappeared to obtain the CPU solution. The reaction was carried out according to the general reaction scheme as shown in Figure 1, and the CPUs was named as CPU-X (X: the molar ratio of n_{-NCO}:n_{-OH}).

Table 1. The basic formulations and chitooligosaccharide (COS) content of chitooligosaccharide-based polyurethane (CPU).

Samples	PCL/g	HBH/g	COS/g	COS Content/wt % *	n_{-OH}:n_{-NCO}:n_{-NH2} **
CPU-1.0	8.0	1.70	0	0	1:1:0
CPU-1.4	8.0	2.39	1.14	9.89	1:1.4:0.8
CPU-1.7	8.0	2.90	2.0	15.5	1:1.7:1.4
CPU-2.0	8.0	3.41	2.86	20.0	1:2.0:2.0

* COS content in CPU; ** the molar ratio of –OH in PCL, –NCO in HBH and –NH$_2$ in COS.

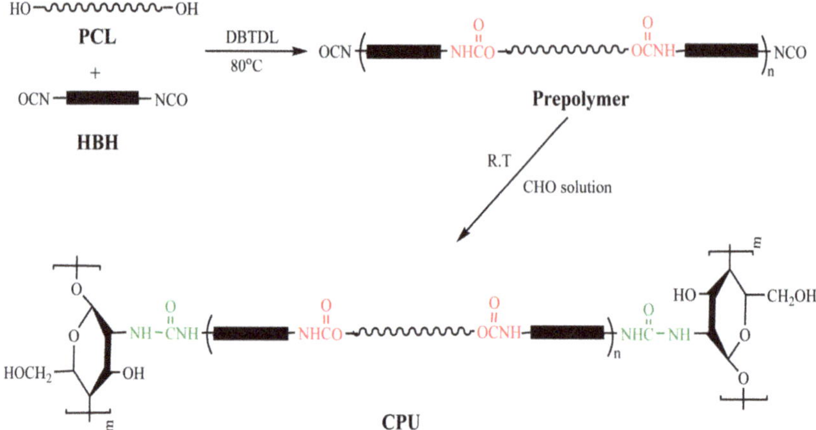

Figure 1. General reaction scheme of CPU composites.

2.3. Preparation of CPU Films

The CPU solution was first diluted with DMF to ~4.5 g/100 mL, and then the diluted solution was cast into a Teflon mold. Most of solvent was first removed by natural volatilizing at 50 °C for four days, and then the last traces of solvent was removed under reduced pressure for one day to obtain the semitransparent pale-brown films with 0.40 ± 0.02 mm thickness. Finally the films were punched into discs with ~10 mm in diameter for measurement.

2.4. Characterization

FT-IR: the Fourier transform infrared (FT-IR) spectrophotometer used was an Alpha infrared spectrometer (Bruker, Rheinstetten, Germany) equipped with a Bruker platinum ATR accessory at room temperature. The spectra covered the infrared region 4000–400 cm^{-1} with the resolution of 4 cm^{-1}. The scans were collected on COS powder, prepolymer, and CPU films.

Thermal properties: A differential scanning calorimeter (DSC) (DSC2910, Universal, New Brunswick, NJ, USA) was employed to study the thermal transition behavior of the polymers. The samples (1~1.3 mg) were first heated up to 150 °C at a heating rate of 10 °C/min to erase the thermal history, then cooled to −70 °C at 5 °C/min, and finally heated to 150 °C at 10 °C/min. The measurements were taken under a continuous nitrogen purge (30 mL/min), and the reported thermal transition temperatures were collected during the second heating cycle. Thermogravimetric analysis (TGA) was recorded on a TGA 2050 analyzer (Universal, New Brunswick, NJ, USA). The instrument was calibrated using a pure calcium oxalate sample before analysis. About 8~10 mg of sample was subjected to TGA scans, which were performed from ambient to 800 °C with a heating rate of 20 °C/min under nitrogen atmosphere with a gas flow rate of 50 mL/min.

Crystallization behavior: The crystallization behaviors of CPU films were measured by X-ray powder diffraction (XRD) analysis. The measurements were conducted by a Max 2200PC power X-ray diffractometer (Rigaku Corp., Tokyo, Japan) with 40 kV and 20 mA using Cu Kα (1.54051 Å) radiation. The sample holder containing samples were scanned from 5° to 55° with a scanning rate of 2θ = 0.02°.

Mechanical properties: Tensile strength tests were carried out using a single-column tensile test machine (Model HY939C, Dongguan Hengyu, Ltd., Dongguan, China) at room temperature with a crosshead speed of 50 mm/min. The films were punched into Dumbbell-shaped specimens using a punch cutter with a punching die of 12 mm width and 75 mm length, the neck width and length were 4.0 and 30 mm, respectively. Property values reported here represent averaged results of at least five samples.

Water contact angle: The contact angle measurement was used to evaluate the surface hydrophilicity of the films. The sessile static water contact angle measurements were carried on a drop shape analysis system (CAM 200, KSV Instruments, Helsinki, Finland). The ultrapure water was dripped onto the sample surface at room temperature, and in oder to ensure that the droplet did not soak into the compact, the tests were performed within 10 s. Three different points were tested for each sample and six readings were averaged for each film.

Water absorption: The amount of water that each film absorbed was adopted to quantify the swellability of the CPU films. Each film disc (~10 mm diameter) was immersed in 10 mL deionized water which maintained the temperature of 37 ± 0.1 °C until reaching the water absorption equilibrium (~48 h). Then the discs were removed from water, and the surplus surface water was gently wiped off with a filter paper and weighed. The sample mass change resulting from the water uptake expressed in percent was calculated according to the formula: Water absorption (%) = $(W_t - W_o)/W_o \times 100$, where W_o and W_t are the weights of dry and wet samples, respectively. Each sample was tested at least five times and the results were averaged.

In vitro degradation: In vitro degradation studies of the films were performed in PBS (pH = 7.4) through the weight loss. The film discs (~10 mm diameter) were placed into an individual sealed bottle containing 10 mL PBS in a biochemical incubator at 37 ± 0.1 °C. At given time intervals, the samples were removed from the buffer, rinsed three times with distilled water and dried to a constant weight at 35 °C under vacuum. Post-degradation weight was measured and mass loss of the films was obtained using the following formula: Weight loss ratio (%) = $(W_o - W_r)/W_o \times 100$, where W_o and W_r mean the original dry weight and the rest dry weight after degradation for a predetermined time, respectively. The assessments were conducted for 12 months or until the films lost mechanical properties and became fragments. Three samples were tested, and the average value was taken.

Surface morphologies: A cold field emission scanning electron microscope (FE-SEM, Hitachi SU8010, Tokyo, Japan) was used to investigate the surface morphologies of the films after degradation for a fixed time. The dried discs, which were first mounted on aluminum stubs with conductive graphite-filled tapes, were coated with a gold layer under vacuum and then used for SEM observation.

Protein adsorption: A Bradford protein assay with bovine serum albumin (BSA) as the model protein was used to measure the amount of albumin adsorbed onto the surface. In order to achieve complete hydration, the film discs (~10 mm diameter) were immersed into PBS (pH = 7.4) for about 24 h until equilibrating, and then were filled with 1.0 mL of BSA solution (45 µg/mL, the same as the concentration of normal plasma) at 37 °C for 1 h. After absorption, the discs were taken out and first rinsed with PBS for three times to remove the unbound BSA. Then the adsorbed protein on the surface was desorbed with sodium dodecylsulfonate aqueous solutions (1 wt %) at 37 °C with agitating at 100 rpm for 1 h. A micro-Bradford protein assay kit (Sangon Biotech Co., Ltd., Shanghai, China) with a multiwell microplate reader (Multiskan Mk3-Thermolabsystems, Thermo Fisher Scientific, Inc., Waltham, MA, USA) was used to test the concentration of the adsorbed BSA at 595 nm. The final adsorbed protein quantity could be obtained referring to the standard curve of optical density against BSA concentration. Three independent measurements were performed, and values relative to the control (PBS) were collected.

Platelet adhesion: The interaction between the blood and film surface was assayed by platelet adhesion experiment. Platelet-rich plasma (PRP) was obtained from the fresh rabbit blood (Shandong Success Biological Technology Co., Ltd., Qingzhou, China) containing sodium citrate as an anticoagulant by centrifugation from blood at 2000 rpm for 20 min at 4 °C. The film discs (~10 mm diameter) were first equilibrated with PBS (pH = 7.4) for 12 h, and then were taken out and incubated with 1.0 mL PRP which was pre-warmed to 37 °C. After incubation for 2 h at 37 °C, the discs were rinsed three times with PBS by mild shaking to remove nonadherent and weakly adhered platelets. The platelets adhering to the surface were fixed with a glutaraldehyde solution (2.5 v/v %) in PBS buffer for 30 min at room temperature. After thorough washing with PBS, the discs were dehydrated by treating with gradual ethanol/water solution from 50% to 100% ethanol (v:v) with a step of 10% for

30 min in each step and allowed to dry on a clean bench at room temperature. The platelet-attached surfaces were coated with gold prior to this, and different fields were randomly observed by FE-SEM.

3. Results and Discussion

3.1. FT-IR

FT-IR spectroscopy has been extensively used in PU research, since it presents an easy method of obtaining direct information on chemical changes. The FT-IR spectra of COS, prepolymer, and CPU-1.7 are shown in Figure 2 (CPUs with different COS content have the similar spectra except for the intensity of the peak). In the spectrum of COS powder (Figure 2a), the broad characteristic peak at about 3200~3420 cm^{-1} was attributed to the stretching vibrations of –OH and –NH$_2$, and the absorption peaks at 2882 and 1035 cm^{-1} corresponded to the saturated stretching of –CH$_2$ and cyclic ether C–O–C, respectively [37]. The weak absorption bands observed at 1664 and 1549 cm^{-1} belonged to the characteristic absorption peaks of amide I and amide II, which was attributed to the residual amide linkage in COS. In the spectrum of prepolymer (Figure 2b), an obvious absorption peak at 2265 cm^{-1} belonged to the characteristic absorption of terminal –NCO of the prepolymer. The absorbance in the region near 3325 cm^{-1} indicated that most of the N–H groups are hydrogen bonded [38]. The other absorption bands at 1726, 1673, 1535, and 1150 cm^{-1} were attributed to the characteristic stretching frequencies of ester C=O, amide I, amide II, and ester C–O–C, respectively. The absorption peak of –NCO disappeared completely in the spectrum of CPU-1.7 (Figure 2c), meaning that the prepolymer was chain-extended with COS. As the paper reported [18], the ureido will be formed because the –NH$_2$ groups have much higher reactivity with –NCO than –OH groups in COS. Meanwhile, the relative intensity of –NH– (~3322 cm^{-1}), amide I (~1680 cm^{-1}), and amide II (~1535 cm^{-1}) increased obviously compared with that of ester C=O, which should be due to the formation of new ureido between COS and the prepolymer. In addition, all the other characteristic absorptions of COS and prepolymers also appeared in the spectrum. The results from FT-IR spectra show the successful preparation of CPU.

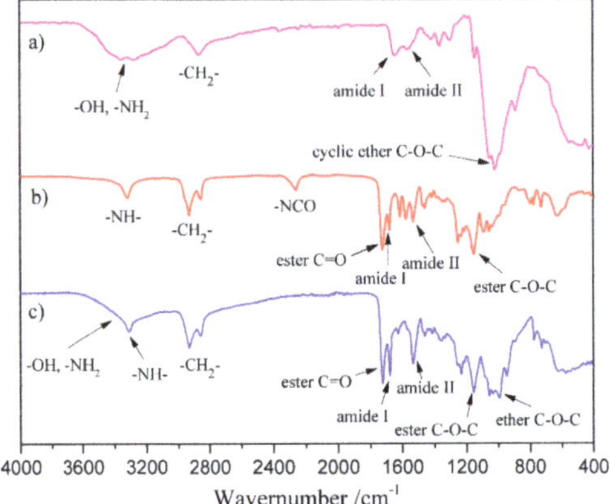

Figure 2. FT-IR spectra of (**a**) COS; (**b**) prepolymer and (**c**) CPU-1.7.

3.2. Thermal Transition

The thermal transitions of polymeric materials, such as glass transition temperature (T_g), melting temperature (T_m) and melting enthalpy (ΔH_f), are always be determined by DSC [39]. The DSC thermograms of the COS, PCL, and CPU with different COS content are displayed in Figure 3 and the corresponding thermal transition temperature is listed in Table 2. There was no obvious T_g observed in the thermogram of COS (Figure 3a), which should be due to the low molecular weight (~3000 g/mol). A broad melting endothermic peak (T_m) around 45–130 °C with ΔH_f of 52 J/g signified that COS was a noncrystalline polymer [40]. The T_g of PCL was observed at −58.7 °C (Figure 3b), which had been reported in a previous study [41]. The T_m observed at ~61.2 °C with ΔH_f of 61.8 J/g was assigned to the melting transition of crystallized segments, which demonstrated the high crystallinity of PCL. Two glass transition temperatures (T_{g1} and T_{g2}) were observed in the thermograms of CPU films with different COS content (Figure 3c–f). The first glass transition point (T_{g1}) at a low temperature of ~−20 °C was attributed to the soft segments. Another glass transition area (T_{g2}) at a high temperature was observed within broad temperatures of from 46 to 64 °C. Because the polarity of urethane (–OCONH–) or/and ureido (–NHCONH–) groups in hard segments is higher than that of ester groups in soft segments, the hard segments are hardly miscible with the soft segments, resulting in micro-phase separation and the appearance of two clear T_g for soft and hard domains [42]. The broad temperature range (T_{g2}) probably corresponded to the relaxation of mixed amorphous intermediate phase (soft and hard domains) [43]. In addition, one endothermic peak with broader temperature (T_m) was found (Table 2), and the ΔH_f increased from 48 to 133 J/g with the increasing content of COS (0~20 wt %) and hard segments (17.5~23.9 wt %) in CPU. The broad endothermic peak should be attributed to the melting transition of COS segments and hard domains. There was no obvious melting endothermic peak of PCL segments in the thermograms, which is probably because the multiple H-bonds between urethane/ureido and ester groups and crosslinking restrict the movement of PCL segments as the crystallization of the soft segment decreases [44].

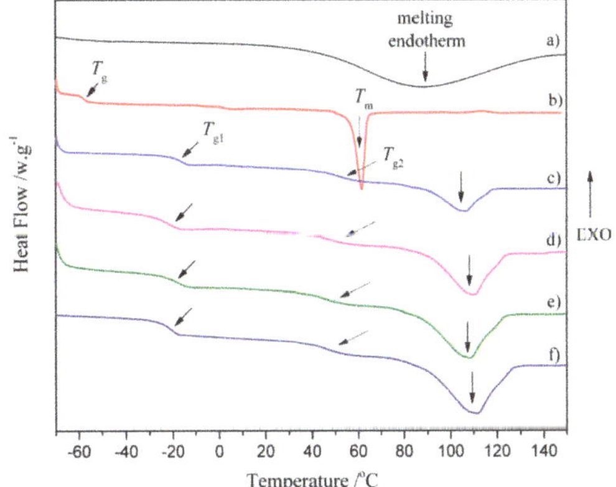

Figure 3. Differential scanning calorimeter DSC thermograms of (**a**) COS; (**b**) poly(ε-caprolactone) PCL; (**c**) CPU-1.0; (**d**) CPU-1.3; (**e**) CPU-1.7 and (**f**) CPU-2.0.

Table 2. The thermal transition temperatures of COS, PCL, and CPUs.

Samples	COS	PCL	CPU-1.0	CPU-1.4	CPU-1.7	CPU-2.0
T_{g1} (°C)	-	−58.7	−17.4	−21.0	−18.3	−20.2
T_{g2} (°C)	-	-	48~63	46~61	46~62.5	45.5~63
T_m (°C)	45~130	61.2	90~118	83~123	82~125	80~129
ΔH_f (J/g)	52	61.8	48.2	78.3	108.9	133.4

3.3. Thermal Stability

The thermal stability of materials is often evaluated by TGA, and from the results, it can deduce the mechanism by which a material loses weight as a result of controlled heating [45]. Figure 4 shows the TGA curves of COS powder and CPU films with different COS content. It could be seen that two consecutive weight loss steps were observed in the COS. The first stage of weight loss (~4.6 wt %) between 30 and 110 °C was responsible for the water loss in COS, indicating its hygroscopic nature. The weight loss of about 45 wt % in the second step with a rapid decomposition between 155 and 330 °C was ascribed to the complex disintegration processes including saccharide rings and macromolecule chains of COS. Approximately 30 wt % remained as residue at the end of the measurement, which indicated that COS had high thermal stability at higher temperature. The thermal degradation of COS, the same as CH, started with the amino groups and formed an unsaturated structure [46]. Only one weight loss step was observed in the curve of pure PU (CPU-1.0), and it showed higher initial decomposition temperature (245 °C) and lower remaining weight (0.6 wt %) compared with COS. With the COS content increasing from 0 to 20 wt % (CPU-1.0~CPU-2.0), the maximum degradation temperature of CPU films increased from 304 to 378 °C, which should be attributed to the increment of chemical crosslinking density in the total structure. Obviously, the lower initial decomposition temperature and higher remaining weight of CPU (CPU-1.4~CPU-2.0) than pure PU (CPU-1.0) were attributed to the COS segments in CPU.

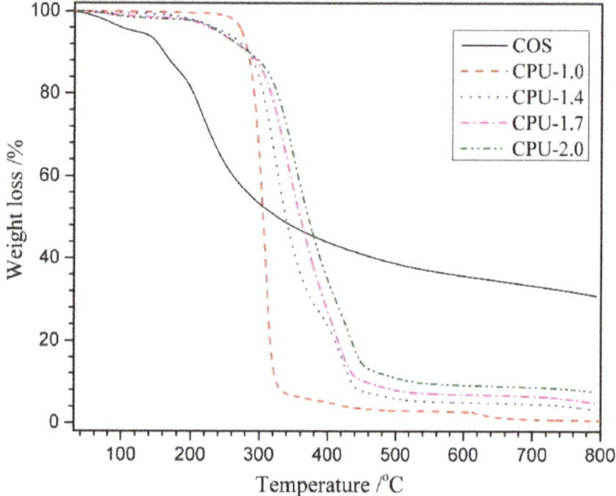

Figure 4. Thermogravimetric analysis (TGA) curves of COS powder and CPU films with different COS content.

3.4. Crystallization Behavior

The crystallization behaviors of the COS powder and CPU films with different COS content were investigated by XRD analysis, and the scattering patterns are shown in Figure 5. A broad diffuse peak appearing in the scattering pattern of COS powder (Figure 5a) signified an amorphous structure, which was consistent with the result of DSC. The pure PU (CPU-1.0, Figure 5b) exhibited a clear diffuse peak with a maximum at $2\theta = {\sim}23.7°$, indicating that pure PU was a hemicrystalline polymer. The crystal composition arised from the crystalline soft domains and uniform-size hard regions formed by the structural symmetry. With the increment of COS content in CPU (CPU-1.4~CPU-2.0, Figure 5c–e), the intensity of diffraction peaks ($2\theta = {\sim}20.7°$ and 23.7°) increased obviously, which should correspond to the increasing content of uniform-size hard segment. The new diffraction peak ($2\theta = {\sim}20.7°$) may be assigned to the ureido formed between COS and the prepolymer, indicating that new crystalline zones were formed. Meanwhile, no sharp diffraction peaks were observed in their scattering patterns. It can be deduced that, with the addition of COS, the CPU forms more crosslinking points which makes it more difficult for COS to react with the prepolymer due to the steric effect, and the unreacted COS is physically mixed with CPU, resulting in slightly blunt peaks in the scattering patterns.

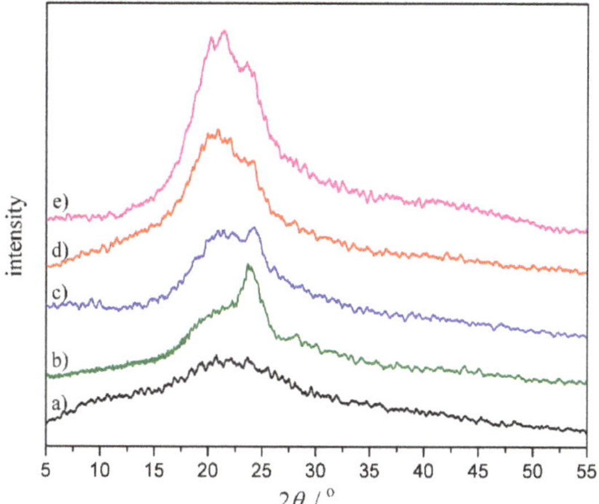

Figure 5. XRD patterns of (**a**) COS powder; (**b**) CPU-1.0; (**c**) CPU-1.4; (**d**) CPU-1.7 and (**e**) CPU-2.0 films.

3.5. Mechanical Properties

Mechanical properties are an important quality for biocompatible polymers in soft tissue engineering [47]. The typical stress–strain behaviors of the CPU films with COS content from 0 to 20 wt % are presented in Figure 6, and the characteristic values derived from these curves, including ultimate stress, strain at break, and initial modulus, are shown in Table 3 (pure COS film was too brittle to be obtained because of the low molecular weight). All the films exhibited a similar obvious yield point and manifested as the normal elastomers, which displayed a smooth transition from the elastic to plastic deformation regions in stress–strain behaviors [48]. The pure PU (CPU-1.0) showed an ultimate stress of 24.1 MPa with a strain at break of 774%. The excellent mechanical properties should be related to the uniformity of the hard segments (HBH), which reinforces the hard segments to form hard domains and serves as reinforcing material in a soft segment matrix. In addition, the compact physical-linking network structure, which is formed by the multiple H-bonds existing not only among urethane groups but also between urethane and ester groups, provides an additional energy dissipation mechanism [43]. With the COS content increasing from 9.9 to 20 wt % (CPU-1.4~CPU-2.0), the ultimate stress increased gradually from 30.9 to 35.3 MPa and strain at break decreased from 671% to 462%, as outlined in Table 3. The following two reasons may be explain the results: first, the increased urethane and ureido groups can form more intermolecular hydrogen bonds which give more physical crosslinking points; second, the higher chemical crosslinking density enhances the ultimate stress, especially in a rubbery state. In addition, the initial modulus increased from 25.5 to 53.8 MPa with the COS content increasing from 0 to 20 wt % (Table 3). From these results, it is suggested that the chemical crosslinking density is very important to the mechanical properties of materials and the mechanical properties of CPU can be controlled by adjusting the crosslinking density.

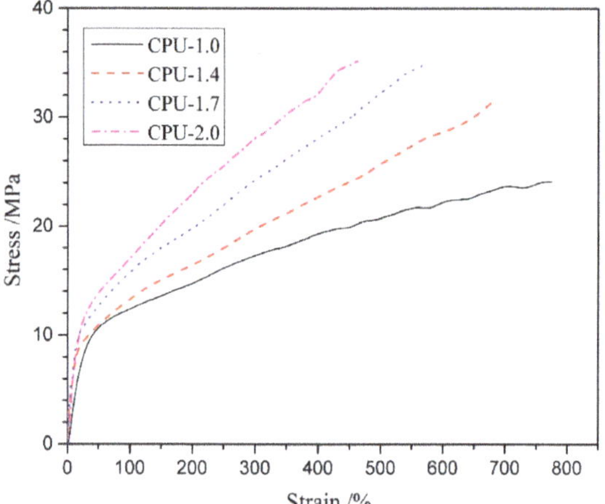

Figure 6. Stress–strain behaviors of CPU films with different COS content.

Table 3. Mechanical properties of CPU films.

Films	Strain at Break (%)	Ultimate Stress (MPa)	Yield Stress (MPa)	Yield Strain (%)	Initial Modulus (MPa)
CPU-1.0	774 ± 17	24.1 ± 2.2	10.6 ± 1.6	41.3 ± 3.5	25.5
CPU-1.4	671 ± 15	30.9 ± 1.8	9.3 ± 1.1	22.6 ± 1.2	41.2
CPU-1.7	569 ± 15	34.9 ± 1.6	10.2 ± 1.1	19.7 ± 1.0	51.7
CPU-2.0	462 ± 13	35.3 ± 1.6	11.9 ± 1.2	22.1 ± 1.3	53.8

3.6. Surface Hydrophilicity and Swellability

The surface hydrophilicity and swellability, which are commonly related to the protein adsorption, platelet adhesion, and biodegradability, are important parameters for biomaterials in many medical applications [49]. The surface hydrophilicity and swellability of the CPU films with different COS content were evaluated by measuring the water contact angle and water absorption, and the results are displayed in Figure 7. The pure PU (CPU-1.0) exhibited a characteristically high water contact angle of 92°, indicating a hydrophobic surface. When the COS content in CPU increased from 9.9 to 20 wt %, the surface hydrophilicity increased gradually with the water contact angle decreasing from 71° to 45.3°, which was ascribed to the introduction of hydrophilic (unreacted) amino groups and hydroxyl groups in COS segments. The result showed that the COS content had outstanding effect on the surface hydrophilicity. The low water absorption (9.5 wt %) of CPU-1.0 indicated that pure PU had a bulk hydrophobic structure, which was consistent with the results for the water contact angle. While CPU-1.4 with lower COS content exhibited a higher water absorption (53 wt %), and with the increment of COS content in CPU (CPU-1.4~CPU-2.0), the water absorption decreased sharply. Obviously, the water absorption is closely related to two factors: hydrophilic chain segment and crosslinking density. The low COS content results in few crosslinking points and the water absorption is mainly affected by hydrophilic COS segments. In addition, the introduction of COS can somewhat destroy the well-defined structure. Thus, water molecules can pass through the film easily, resulting in high water absorption. When the COS content is higher, more crosslinking points are formed, which limits the movement and relaxation of the chains in the films and restricts water to get into the matrix. Consequently, higher crosslinking density can make the films more difficult to swell in water.

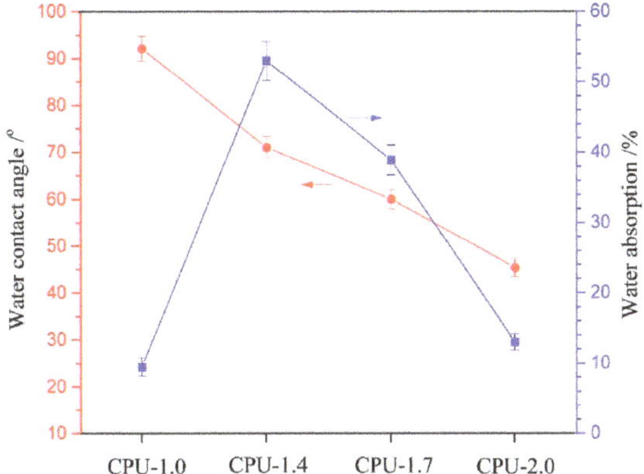

Figure 7. Water contact angle and water absorption of CPU films with different COS content.

3.7. In Vitro Degradation

The degradation behavior has a crucial impact on the performance of implant biomaterials. The degradation behaviors were examined in vitro in PBS at 37 °C, and the percentage weight loss of the CPU films with time are shown in Figure 8. The pure PU (CPU-1.0) exhibited a slow degradation rate with less than 2 wt % after 6 months, and the film had no obvious change except that the transparency decreased. Only 11 wt % weight loss was observed until the end of the test. It is evident that the degradation is mainly caused by hydrolysis of ester groups, the low surface and bulk hydrophilicity and a more compact network structure formed by hydrogen bonds hinder water to approach the ester groups, resulting in slow hydrolytic degradation rate. When the COS was introduced into CPU (CPU-1.4~CPU-2.0), the degradation rate increased obviously after a slow weight loss during two months. The degradation rate increased with the increment of COS content in CPU, which was not in agreement with the results of swellability mentioned above. On the one hand, chemical crosslinked network can somewhat destroy the ordering of polymer structure and reduce the mobility of the chain, which allow the ester groups to be easily exposed to water and result in an increased susceptibility to degradation [38]. On the other hand, the process of degradation of crosslinked materials can be accelerated by blending with polymers susceptible to degradation [50]. In the case of CPU with higher COS content (especially CPU-2.0), the unreacted COS being physically mixed with CPU, as described in XRD, acts as a plasticizer and makes the material more ductile. Then, it was easy for chain scission to take place through hydrolysis of ester bonds. The results are consistent with that of COS-based WPU [16], and manifests that the degradation rate of CPU can be controlled by adjusting the COS content.

The hydrolytic degradation process can be demonstrated directly by morphological changes in film surface. Figure 9 shows the typical surface morphologies of CPU-1.4 after different degradation periods (predegradation and 3, 6, 10 and 12 months' postdegradation). The non-degraded film (Figure 9a) was pale-brown semitransparent and exhibited a smooth surface. After three and six months of degradation (Figure 9b,c), the surface became rougher and rougher, and then turned into many irregular hollows at ten months (Figure 9d) which should be due to the loss of unreacted COS. At the end of the measurement, large cavities appeared (Figure 9e), indicating that the film gradually lost its mechanical properties.

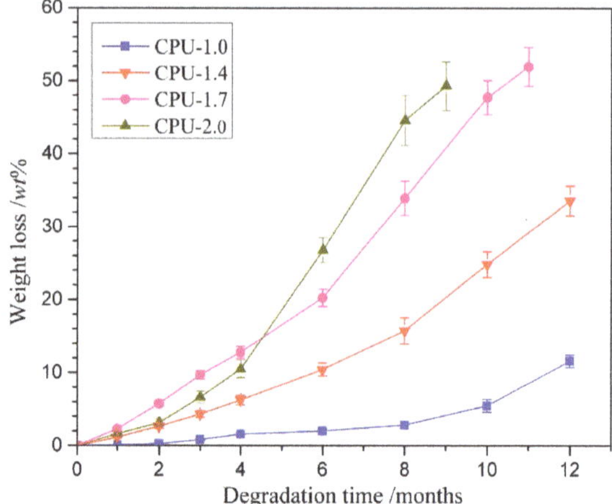

Figure 8. Degradation behaviors of CPU films with different COS content in PBS (pH: 7.4) at 37 ± 0.1 °C.

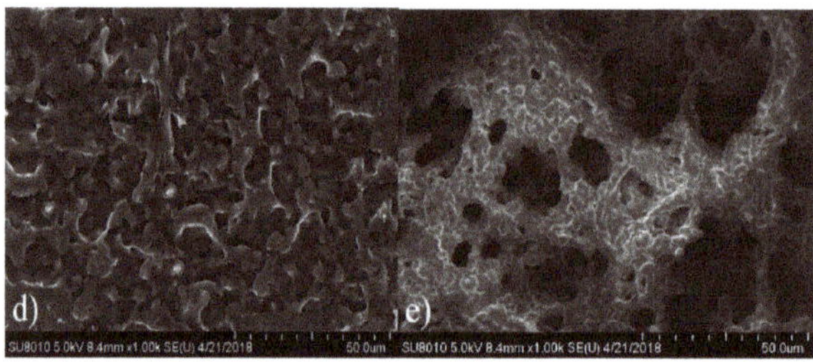

Figure 9. Surface morphologies of CPU-1.4 film in PBS (pH: 7.4) at 37 ± 0.1 °C after (**a**) 0; (**b**) 3; (**c**) 6; (**d**) 10; and (**e**) 12 months' degradation.

3.8. Protein Adsorption

One of the most important measurements to evaluate the hemocompatibility of the implantable materials is plasma protein adsorption [51]. The adsorption of BSA on the CPU surfaces was examined to investigate the effects of the COS content on the interaction between the surface and the proteins. The amount of BSA absorbed on the film surface is exhibited in Figure 10. Pure PU (CPU-1.0) exhibited high adsorption of BSA protein (0.91 μg/cm^2), and the adsorbed amount of protein decreased after the introduction of COS into the films. With the COS content increased from 9.9 to 20 wt % (CPU-1.4~CPU-2.0), the amount of adsorbed protein decreased gradually from 0.43 to 0.24 μg/cm^2. It was attributed to the hydrophilic amino groups and hydroxyl groups of COS segments on the interface, which could bond with the water molecules to form a hydrated layer and reduce the interaction with protein, leading to a repulsive force to protein. Our experimental data were consistent with the previous reports [32,52,53]: the amount of BSA protein adsorbed on the film surface decreased while the surface hydrophilicity increased. The result is also in accordance with the surface hydrophilicity mentioned above. The lower protein adsorption capacity of CPU films indicated that they had better hemocompatibility than pure PU films.

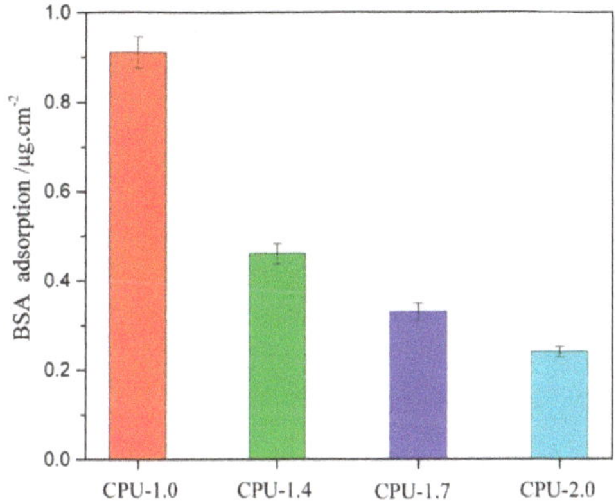

Figure 10. Amount of BSA adsorbed on the surface of CPU films with different COS content at 37 ± 0.5 °C.

3.9. Platelet Adhesion

Platelet adhesion on the film surface is another important test for evaluation of the hemocompatibility of the implantable materials. Obviously, the less interaction between the platelet and the material surface means the lower probability of thrombus [15]. The morphologies of the platelets adherent on the surfaces of CPU films were assessed by SEM observation, and the representative micrographs are given in Figure 11. The surface distribution of platelets on pure PU surface (CPU-1.0, Figure 11a) was non-random and aggregated to some extent, presenting the highly activated state. As shown in the Figure 11b–d, the platelet adhesion was obviously reduced on the surface of COS-based CPU films. The amount of adhesive platelet on the surface decreased gradually with the increment of COS content, and no obvious gathering of platelets was found, which proved a better anti-platelet adhesion surface [54]. One possible explanation is that the hydrophilic surface of films suppresses platelet adhesion. The trend of platelet adhesion was consistent with that of protein adsorption.

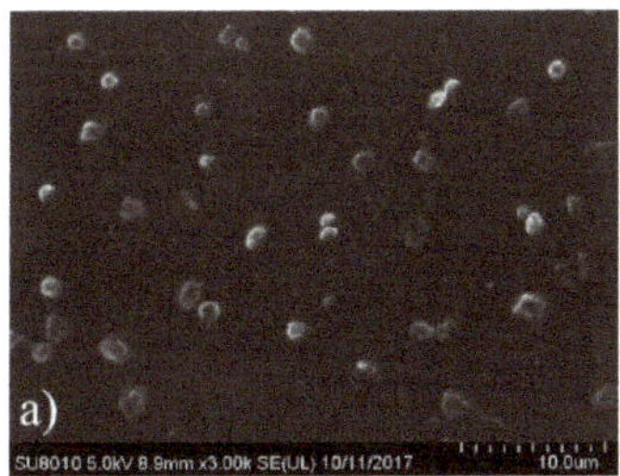
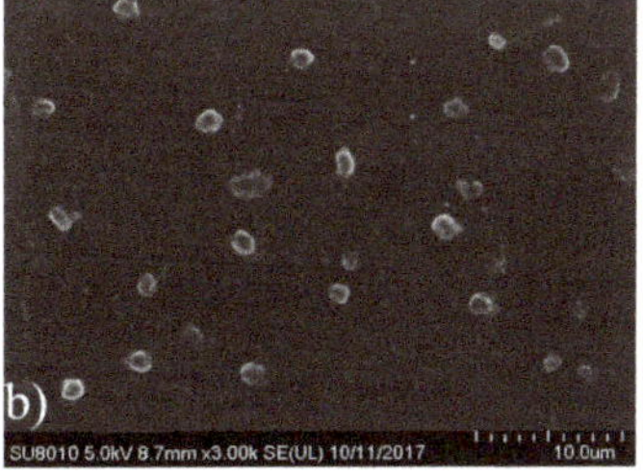

Figure 11. *Cont.*

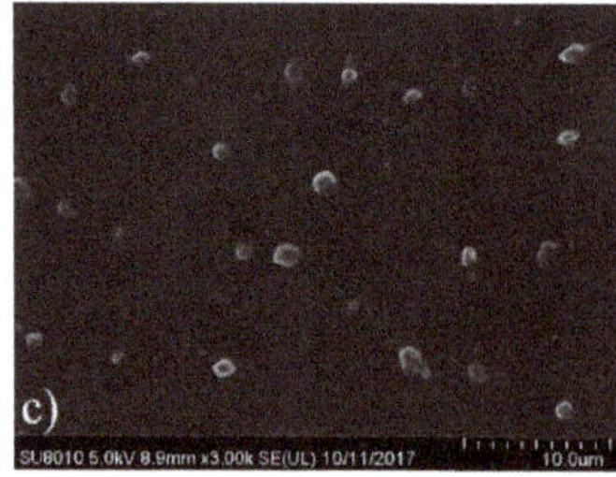 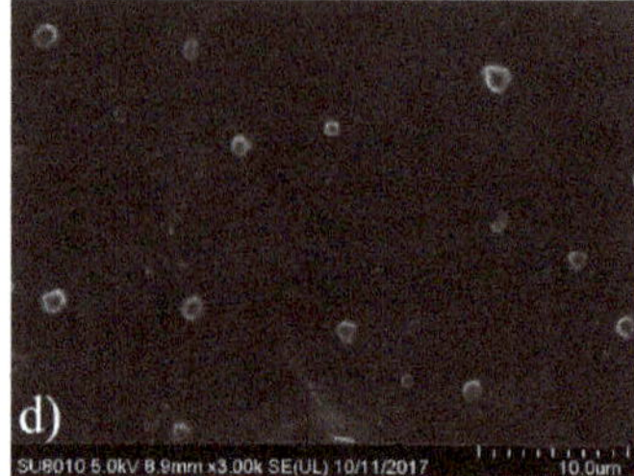

Figure 11. Representative SEM micrographs of platelet adhesion on the surface of (**a**) CPU-1.0; (**b**) CPU-1.4; (**c**) CPU-1.7 and (**d**) CPU-2.0 films.

4. Conclusions

In this work, a series of biodegradable COS-based PUs (CPU) with uniform-size hard segments were prepared in DMF via the conventional two-step method, in which COS was employed as a chain extender. The corresponding films were obtained by the solvent evaporation method. The chemical structure was characterized by FT-IR, and the influence of COS content in CPU on the physicochemical properties and hemocompatibility was extensively researched. The thermal stability studies indicated that the CPU films had a lower initial decomposition temperature and higher maximum decomposition temperature than pure PU (CPU-1.0) film. The ultimate stress, initial modulus, and surface hydrophilicity increased with the increment of COS content, and, due to the increment of crosslinking density, the strain at break and water absorption decreased. In vitro degradation studies showed that the degradation rate increased with the increasing content of COS in CPU, demonstrating that the degradation rate could be controlled by adjusting COS content. The surface hemocompatibility was evaluated by protein adsorption and platelet adhesion tests, and the results demonstrated that the CPU surface had improved resistance to protein adsorption and possessed good resistance to platelet adhesion. The slow degradation and good hemocompatibility of the CPUs display great potential in blood-contacting devices. Moreover, many active amino and hydroxyl groups contained in the structure of CPU could carry out further modification, which made it an excellent candidate for wide application in biomedical field.

Author Contributions: Z.H. and W.X. conceived and designed the experiments; W.X., M.X., L.Y. and J.Z. performed the experiments; Z.H. and W.X. analyzed the data and wrote the paper.

Funding: This research was funded by [Shandong Provincial Natural Science Foundation, China] grant number [ZR2018MEM024]; and [National Undergraduate Training Programs for Innovation and Entrepreneurship, China] grant number [201710445057].

References

1. Janik, H.; Marzec, M. A review: Fabrication of porous polyurethane scaffolds. *Mat. Sci. Eng. C Mater.* **2015**, *48*, 586–591. [CrossRef] [PubMed]
2. Noreen, A.; Zia, K.M.; Zuber, M.; Tabasum, S.; Zahoor, A.F. Bio-based polyurethane: An efficient and environment friendly coating systems: A review. *Prog. Org. Coat.* **2016**, *91*, 25–32. [CrossRef]
3. John, K.R.S. The use of polyurethane materials in the surgery of the spine: a review. *Spine J.* **2014**, *14*, 3038–3047. [CrossRef] [PubMed]
4. Chen, H.; Jiang, X.; He, L.; Zhang, T.; Xu, M.; Yu, X. Novel biocompatible waterborne polyurethane using L-lysine as an extender. *J. Appl. Polym. Sci.* **2002**, *84*, 2474–2480. [CrossRef]
5. Spaans, C.J.; Belgraver, V.W.; Rienstra, O.; Groot, J.H.D.; Veth, R.P.; Pennings, A.J. Solvent-free fabrication of micro-porous polyurethane amide and polyurethane-urea scaffolds for repair and replacement of the knee-joint meniscus. *Biomaterials* **2000**, *21*, 2453–2460. [CrossRef]

6. Shen, Z.; Kang, C.; Chen, J.; Ye, D.; Qiu, S.; Guo, S.; Zhu, Y. Surface modification of polyurethane towards promoting the ex vivo cytocompatibility and in vivo biocompatibility for hypopharyngeal tissue engineering. *J. Biomater. Appl.* **2013**, *28*, 607–616. [CrossRef] [PubMed]

7. Bochyńska, A.I.; Hammink, G.; Grijpma, D.W.; Buma, P. Tissue adhesives for meniscus tear repair: An overview of current advances and prospects for future clinical solutions. *J. Mater. Sci. Mater. Med.* **2016**, *27*, 85–102. [CrossRef] [PubMed]

8. Li, D.; Chen, H.; Glenn, M.W.; Brash, J.L. Lysine-PEG-modified polyurethane as a fibrinolytic surface: Effect of PEG chain length on protein interactions, platelet interactions and clot lysis. *Acta Biomater.* **2009**, *5*, 1864–1871. [CrossRef] [PubMed]

9. Puskas, J.E.; Chen, Y. Biomedical application of commercial polymers and novel polyisobutylene-based thermoplastic elastomers for soft tissue replacement. *Biomacromolecule* **2004**, *5*, 1141–1154. [CrossRef] [PubMed]

10. Tao, Y.; Hasan, A.; Deeb, G.; Hu, C.; Han, H. Rheological and mechanical behavior of silk fibroin reinforced waterborne polyurethane. *Polymers* **2016**, *8*, 94. [CrossRef]

11. Dong, C.; Lv, Y. Application of collagen scaffold in tissue engineering: Recent advances and new perspectives. *Polymers* **2016**, *8*, 42. [CrossRef]

12. Adipurnama, I.; Yang, M.C.; Ciach, T.; Butruk-Raszeja, B. Surface modification and endothelialization of polyurethane for vascular tissue engineering applications: A review. *Biomater. Sci.* **2017**, *5*, 22–37. [CrossRef] [PubMed]

13. Lin, W.C.; Tseng, C.H.; Yang, M.C. In-vitro hemocompatibility evaluation of a thermoplastic polyurethane membrane with surface-immobilized water-soluble chitosan and heparin. *Macromol. Biosci.* **2010**, *5*, 1013–1021. [CrossRef] [PubMed]

14. Zhang, Q.; Liao, J.F.; Shi, X.H.; Qiu, Y.G.; Chen, H.J. Surface biocompatible construction of polyurethane by heparinization. *J. Polym. Res.* **2015**, *22*, 68–79. [CrossRef]

15. Ren, Z.; Chen, G.; Wei, Z.; Lin, S.; Qi, M. Hemocompatibility evaluation of polyurethane film with surface-grafted poly(ethylene glycol) and carboxymethyl-chitosan. *J. Appl. Polym. Sci.* **2013**, *127*, 308–315. [CrossRef]

16. Xu, D.; Meng, Z.; Han, M.; Xi, K.; Jia, X.; Yu, X.; Chen, Q. Novel blood-compatible waterborne polyurethane using chitosan as an extender. *J. Appl. Polym. Sci.* **2008**, *109*, 240–246. [CrossRef]

17. Tan, A.C.W.; Polo-Cambronell, B.J.; Provaggi, E.; Ardila-Suárez, C.; Ramirez-Caballero, G.E.; Baldovino-Medrano, V.G.; Kalaskar, D.M. Design and development of low cost polyurethane biopolymer based on castor oil and glycerol for biomedical applications. *Biopolymers* **2018**, *109*, e23078. [CrossRef] [PubMed]

18. Pandey, A.R.; Singh, U.S., Momin, M.; Bhavsar, C. Chitosan: Application in tissue engineering and skin grafting. *J. Polym. Res.* **2017**, *24*, 125–146. [CrossRef]

19. Shih, C.; Chen, C.; Huang, K. Adsorption of color dyestuffs on polyurethane-chitosan blends. *J. Appl. Polym. Sci.* **2004**, *91*, 3991–3998. [CrossRef]

20. Shih, C.; Huang, K. Synthesis of a polyurethane-chitosan blended polymer and a compound process for shrink-proof and antimicrobial woolen fabrics. *J. Appl. Polym. Sci.* **2003**, *88*, 2356–2363. [CrossRef]

21. Silva, S.S.; Menezes, S.M.C.; Garcia, R.B. Synthesis and characterization of polyurethane-g-chitosan. *Eur. Polym. J.* **2003**, *39*, 1515–1519. [CrossRef]

22. Zhu, Y.B.; Gao, C.Y.; He, T.; Shen, J.C. Endothelium regeneration on luminal surface of polyurethane vascular scaffold modified with diamine and covalently grafted with gelatin. *Biomaterials* **2004**, *25*, 423–430. [CrossRef]

23. El-Sayed, A.A.; Gabry, L.K.E.; Allam, O.G. Application of prepared waterborne polyurethane extended with chitosan to impart antibacterial properties to acrylic fabrics. *J. Mater. Sci. Mater. Med.* **2010**, *21*, 507–514. [CrossRef] [PubMed]

24. Nikje, M.M.A.; Tehrani, Z.M. Synthesis and characterization of waterborne polyurethane-chitosan nanocomposites. *Polym. Plast. Technol.* **2010**, *49*, 812–817. [CrossRef]

25. Lee, K.Y.; Ha, W.S.; Park, W.H. Blood compatibility and biodegradability of partially *N*-acylated chitosan derivatives. *Biomaterials* **1995**, *16*, 1211–1216. [CrossRef]

26. Chae, S.Y.; Jang, M.K.; Nah, J.W. Influence of molecular weight on oral absorption of water soluble chitosans. *J. Control. Release* **2005**, *102*, 383–394. [CrossRef] [PubMed]

27. Xu, Q.; Wang, W.; Yang, W.; Du, Y.; Song, L. Chitosan oligosaccharide inhibits EGF-induced cell growth possibly through blockade of epidermal growth factor receptor/mitogen-activated protein kinase pathway. *Int. J. Biol. Macromol.* **2017**, *98*, 502–505. [CrossRef] [PubMed]

28. Jeon, Y.J.; Kim, S.K. Production of chitooligosaccharides using an ultrafiltration membrane reactor and their antibacterial activity. *Carbohydr. Polym.* **2000**, *41*, 133–141. [CrossRef]

29. Kim, K.Y.; Kwon, S.L.; Park, J.H.; Chung, H.; Jeong, S.Y.; Kwon, I.C. Physicochemical characterizations of self-assembled nanoparticles of glycol chitosan–deoxycholic acid conjugates. *Biomacromolecules* **2005**, *6*, 1154–1158. [CrossRef] [PubMed]

30. Kwon, S.; Park, J.H.; Chung, H.; Kwon, I.C.; Jeong, S.Y.; Kim, I.S. Physicochemical characteristics of self-assembled nanoparticles based on glycol chitosan bearing 5β-cholanic acid. *Langmuir* **2003**, *19*, 10188–10193. [CrossRef]

31. Yokasan, R.; Matsusaki, M.; Akashi, M.; Chirachanchai, S. Controlled hydrophobic/hydrophilic chitosan: Colloidal phenomena and nanosphere formation. *Colloid Polym. Sci.* **2004**, *282*, 337–342. [CrossRef]

32. Xu, D.; Wu, K.; Zhang, Q.; Hu, H.; Xi, K.; Chen, Q.; Yu, X.; Chen, J.; Jia, X. Synthesis and biocompatibility of anionic polyurethane nanoparticles coated with adsorbed chitosan. *Polymer* **2010**, *51*, 1926–1933. [CrossRef]

33. Usman, A.; Zia, K.M.; Zuber, M.; Tabasum, S.; Rehman, S.; Zia, F. Chitin and chitosan based polyurethanes: A review of recent advances and prospective biomedical applications. *Int. J. Biol. Macromol.* **2016**, *86*, 630–645. [CrossRef] [PubMed]

34. Qu, W.Q.; Xia, Y.R.; Jiang, L.J.; Zhang, L.W.; Hou, Z.S. Synthesis and characterization of a new biodegradable polyurethanes with good mechanical properties. *Chin. Chem. Lett.* **2016**, *27*, 135–138. [CrossRef]

35. Hou, Z.; Qu, W.; Kan, C. Synthesis and properties of triethoxysilane-terminated anionic polyurethane and its waterborne dispersions. *J. Polym. Res.* **2015**, *22*, 111–119. [CrossRef]

36. Pei, D.; Wang, J.; Mu, Y.; Wan, X. A simple and low-cost synthesis of antibacterial polyurethane with high mechanical and antibacterial properties. *Macromol. Chem. Phys.* **2017**, *218*, 1700203. [CrossRef]

37. Ajitha, P.; Vijayalakshmi, K.; Saranya, M.; Gomathi, T.; Rani, K.; Sudha, P.N.; Anil, S. Removal of toxic heavy metal lead (II) using chitosan oligosaccharide-graft-maleic anhydride/polyvinyl alcohol/silk fibroin composite. *Int. J. Biol. Macromol.* **2017**, *104*, 1469–1482.

38. Barrioni, B.R.; Carvalho, S.M.D.; Oréfice, R.L.; Oliveira, A.A.R.D.; Pereira, M.D.M. Synthesis and characterization of biodegradable polyurethane films based on HDI with hydrolyzable crosslinked bonds and a homogeneous structure for biomedical applications. *Mater. Sci. Eng. C* **2015**, *52*, 22–30. [CrossRef] [PubMed]

39. Xu, J.; Teng, H.; Hou, Z.; Gu, C.; Zhu, L. Comb-like polysiloxanes with oligo(oxyethylene) and sulfonate groups in side chains for solvent-free dimethoxysilyl-terminated polypropylene oxide waterborne emulsions. *Colloid Polym. Sci.* **2018**, *296*, 157–163. [CrossRef]

40. Li, F.H.; Sun, Y.; Li, S.X.; Ma, S.J. Synthesis and characterization of thermoplastic biomaterial based on acylated chitosan oligosaccharide. *Appl. Mech. Mater.* **2012**, *117–119*, 1433–1436. [CrossRef]

41. Li, F.H.; Li, S.X.; Jiang, T.; Sun, Y. Syntheses and characterization of chitosan oligosaccharide-graft-polycaprolactone copolymer I thermal and spherulite morphology studies. *Adv. Mater. Res.* **2011**, *183–185*, 155–160. [CrossRef]

42. Król, P. Synthesis methods, chemical structures and phase structures of linear polyurethanes. Properties and applications of linear polyurethanes in polyurethane elastomers, copolymers and ionomers. *Prog. Mater. Sci.* **2007**, *52*, 915–1015. [CrossRef]

43. Yin, S.; Xia, Y.; Jia, Q.; Hou, Z.; Zhang, N. Preparation and properties of biomedical segmented polyurethanes based on poly(ether ester) and uniform-size diurethane diisocyanates. *J. Biomater. Sci. Polym. Ed.* **2017**, *28*, 119–138. [CrossRef] [PubMed]

44. Barikani, M.; Zia, K.M.; Bhatti, I.A.; Zuber, M.; Bhatti, H.N. Molecular engineering and properties of chitin based shape memory polyurethanes. *Carbohydr. Polym.* **2008**, *74*, 621–626. [CrossRef]

45. Liu, X.; Xia, W.; Jiang, Q.; Xu, Y.; Yu, P. Synthesis, characterization, and antimicrobial activity of kojic acid grafted chitosan oligosaccharide. *J. Agric. Food Chem.* **2014**, *62*, 297–303. [CrossRef] [PubMed]

46. Teng, S.H.; Lee, E.J.; Yoon, B.H.; Shin, D.S.; Kim, H.E.; Oh, J.S. Chitosan/nanohydroxyapatite composite membranes via dynamic filtration for guided bone regeneration. *J. Biomed. Mater. Res. A* **2009**, *88*, 569–580. [CrossRef] [PubMed]

47. Haut, R.C. Biomechanics of soft tissue. In *Accidental Injury*; Nahum, A.M., Melvin, J.W., Eds.; Springer: New York, NY, USA, 2002; pp. 228–253.

48. Caracciolo, P.C.; Queiroz, A.A.D.; Higa, Q.Z.; Buffa, F.; Abraham, G.A. Segmented poly(esterurethane urea)s from novel urea-diol chain extenders: Synthesis, characterization and in vitro biological properties. *Acta Biomater.* **2008**, *4*, 976–988. [CrossRef] [PubMed]

49. Takami, K.; Matsuno, R.; Ishihara, K. Synthesis of polyurethanes by polyaddition using diol compounds with methacrylate-derived functional groups. *Polymer* **2011**, *52*, 5445–5451. [CrossRef]

50. Brzeska, J.; Morawska, M.; Heimowska, A.; Sikorska, W.; Tercjak, A.; Kowalczuk, M.; Rutkowska, M. Degradability of cross-linked polyurethanes/chitosan composites. *Polimery* **2017**, *62*, 567–575. [CrossRef]

51. Kwon, M.J.; Bae, J.H.; Kim, J.J.; Na, K.; Lee, E.S. Long acting porous microparticle for pulmonary protein delivery. *Int. J. Pharm.* **2007**, *333*, 5–9. [CrossRef] [PubMed]

52. Zhang, Z.; Chen, S.; Chang, Y.; Jiang, S. Surface grafted sulfobetaine polymers via atom transfer radical polymerization as superlow fouling coatings. *J. Phys. Chem. B* **2006**, *110*, 10799–10804. [CrossRef] [PubMed]

53. Tangpasuthadol, V.; Pongchaisirikul, N.; Hoven, V.P. Surface modification of chitosan films. Effects of hydrophobicity on protein adsorption. *Carbohydr. Res.* **2003**, *338*, 937–942. [CrossRef]

54. Yu, S.H.; Mi, F.L.; Shyu, S.S.; Tsai, C.H.; Peng, C.K.; Lai, J.Y. Miscibility, mechanical characteristic and platelet adhesion of 6-*O*-carboxymethylchitosan/polyurethane semi-IPN membranes. *J. Membr. Sci.* **2006**, *276*, 68–80. [CrossRef]

Article

A Novel Delivery System for the Controlled Release of Antimicrobial Peptides: Citropin 1.1 and Temporin A

Urszula Piotrowska [1,2,*], Ewa Oledzka [1], Anna Zgadzaj [3], Marta Bauer [4] and Marcin Sobczak [1,2]

[1] Department of Biomaterials Chemistry, Chair of Inorganic and Analytical Chemistry, Faculty of Pharmacy with the Laboratory Medicine Division, Medical University of Warsaw, Banacha 1 St., 02-097 Warsaw, Poland; eoledzka@wum.edu.pl (E.O.); marcin.sobczak@e-mail.com (M.S.)

[2] Department of Organic Chemistry and Biochemistry, Faculty of Materials Science and Design, Kazimierz Pulaski University of Technology and Humanities in Radom, 27 Chrobrego St., 26-600 Radom, Poland

[3] Department of Environmental Health Science, Faculty of Pharmacy with the Laboratory Medicine Division, Medical University of Warsaw, 1 Banacha St., 02-097 Warsaw, Poland; azgadzaj@wum.edu.pl

[4] Department of Inorganic Chemistry, Faculty of Pharmacy with the Laboratory Medicine Division, Medical University of Gdansk, Al. Gen. J. Hallera 107 St., 80-416 Gdansk, Poland; bauerm@gumed.edu.pl

* Correspondence: upiotrowska@wum.edu.pl or piotrowska_urszula@wp.pl; Tel.: +48-225-720-755; Fax: +48-225-720-784

Received: 16 April 2018; Accepted: 30 April 2018; Published: 2 May 2018

Abstract: Antimicrobial peptides (AMPs) are prospective therapeutic options for treating multiple-strain infections. However, clinical and commercial development of AMPs has some limitations due to their limited stability, low bioavailability, and potential hemotoxicity. The purpose of this study was to develop new polymeric carriers as highly controlled release devices for amphibian peptides citropin 1.1 (CIT) and temporin A (TEMP). The release rate of the active pharmaceutical ingredients (APIs) was strongly dependent on the API characteristics and the matrix microstructure. In the current work, we investigated the effect of the polymer microstructure on in vitro release kinetics of AMPs. Non-contact laser profilometry, scanning electron microscopy (SEM), and differential scanning calorimetry (DSC) were used to determine the structural changes during matrix degradation. Moreover, geno- and cytotoxicity of the synthesized new carriers were evaluated. The in vitro release study of AMPs from the obtained non-toxic matrices shows that peptides were released with near-zero-order kinetics. The peptide "burst release" effect was not observed. New devices have reached the therapeutic concentration of AMPs within 24 h and maintained it for 28 days. Hence, our results suggest that these polymeric devices could be potentially used as therapeutic options for the treatment of local infections.

Keywords: antimicrobial peptides; biodegradable polymers; biocompatible polymers; drug delivery systems; controlled release; citropin; temporin; ionic liquids

---

1. Introduction

Our life expectancy has increased as the quality of medical care is constantly improving and new therapies are being developed all the time. However, microorganism resistance to antibiotics has become an important challenge in modern medicine due to the global uncontrolled use of antibiotics [1]. After the golden age of antibiotic discovery, only two new classes of antibiotics have been marketed [2,3].

Antimicrobial peptides (AMPs), also called host defense peptides or cationic peptide antibiotics, are a prospective therapeutic option for treating multiple-strain infections. AMPs represent a diverse class of naturally occurring biologically active molecules with a broad spectrum of activity. They possess activity against viruses, both Gram-positive and Gram-negative bacteria (with endotoxin-neutralizing activity), fungi, and protozoa. Moreover, AMPs display anti-inflammatory, immunomodulatory, antitumor, angiogenic, and wound healing properties [4–6].

However, from the therapeutic point of view, clinical and commercial development of AMPs as low-molecular-weight active pharmaceutical ingredients (APIs) still has some limitations. AMPs have limited stability, low bioavailability, and potential hemotoxicity. In order to overcome the above-mentioned issues, peptides could be loaded into polymeric drug delivery systems (DDSs). Non-toxic polymeric matrices with appropriate microstructure would be able to release APIs with optimal pharmacokinetics, improving the efficacy and toxicological safety of the therapy [7]. Novel DDSs, which are characterized by highly controlled drug release profiles, are currently being demanded by both the pharmaceutical industry and medical practitioners.

Recent studies focused on the development of polymeric carriers for AMPs or short lipopeptides. The tested matrices were poly (lactic acid-co-castor oil) [8], poly (ester-anhydride) [8], poly(ε-caprolactone) (PCL) [9], copolymer of ε-caprolactone and trimethylene carbonate [9] and polyphosphoesters [10]. New systems were reported to be effective against *Enterococcus faecalis* [8], *Bacillus anthracis* [9], *Enterococcus hirae* [9], *Staphylococcus aureus* [9,10], *Escherichia coli* [10] and *Pseudomonas aeruginosa* [10]. However, the previous studies did not take into account the determination of the kinetics and mechanisms of the peptide release from the new systems. Moreover, there are no studies based on released amount of peptides with reference to therapeutic index (TI). AMPs due to their high hydrophobicity are able to interact not only with bacterial cell membranes, but also with red blood cells, so they show hemolytic activity in certain concentrations. Therefore, it is very important to achieve a therapeutic concentration of peptides in the target tissues. The TI for AMPs defined by Chen et al. [11] is a ratio that compares the minimal hemolytic concentration (MHC) that induces erythrocyte lysis and the minimal inhibitory concentration (MIC) that inhibits the visible growth of the bacterium being investigated. Larger TI values indicate simultaneously greater antimicrobial specificity of the peptide molecule.

In this study, new polymeric matrices were synthesized by enzymatic ring opening polymerization (eROP) of ε-caprolactone (CL) in order to obtain non-toxic devices for highly controlled and prolonged release of amphibian AMPs: citropin 1.1 (CIT) and temporin A (TEMP).

CIT, a 16–amino acid (GLFDVIKKVASVIGGL-NH$_2$) cationic peptide (+2 at pH 7) is produced by the tree frog *Litoria citropa* [12]. It has a high content of hydrophobic residues (56%) in its molecule and an average molecular mass of 1397 Da. CIT is one of the simplest amphibian peptides with a broad spectrum of biological activity reported to date. The solution structure of CIT is an alpha helix with well-defined hydrophobic/hydrophilic regions. CIT has shown antimicrobial activity toward Gram-positive bacteria (*Staphylococcus aureus*, including methicillin-resistant *Staphylococcus aureus*, *Rhodococcus equi*, *Bacillus subtilis*) with antibiofilm properties [13,14]. Moreover, it has antifungal and antitumor activity [15,16].

TEMP is a short, 13–amino acid peptide (FLPLIGRVLSGIL-NH$_2$) isolated from skin secretions of the European red frog *Rana temporaria*. It has a positive net charge (+2 at pH 7) and a higher content of hydrophobic residues (61%) than CIT. Average molecular mass of TEMP is 1615 Da. It has broad antimicrobial activity against both Gram-positive and Gram-negative bacteria [17]. We chose ionic liquids (ILs) as an alternative to the volatile organic solvents during poly(ε-caprolactone) PCL synthesis [18]. ILs are commonly known as solvents that enhance enzyme stability and activity. Moreover, they affect the microstructure of the polymers [19]. The amount of drug released was mainly dependent on the kind of polyester, its number average molecular weight (M_n), dispersity (M_w/M_n), or microstructure. However, our main intention was to check how the kinetics of peptide release might depend on the microstructural characteristics of the polyester matrices.

2. Materials and Methods

2.1. Materials

CL ($\geq$97%; Aldrich, Poznan, Poland) was dried and distilled over CaH_2 at reduced pressure before use. Immobilized lipase B from *Candida antarctica* (CALB) (catalog #L4777) was purchased from Sigma, Poland. 1-Butyl-3-methylimidazolium bis(trifluoromethylsulfonyl)imide ([bmim][NTf$_2$] $\geq$98%; Aldrich, Poznan, Poland) and 1-butyl-3-methylimidazolium hexafluorophosphate ([bmim][PF$_6$] $\geq$ 98%, Merck, Poland) were used as received. CIT ($\geq$95%) and TEMP ($\geq$95%) (Lipopharm.pl, Poland) were used as received. Dichloromethane (DCM), toluene, and methanol (Avantor, Gliwice, Poland) were used as received. Phosphate buffered saline (PBS) was purchased from Sigma (Poznan, Poland) and was also used as received. Dulbecco's modified Eagle's medium (DMEM) was purchased from Gibco (Thermo Fisher Scientific, Waltham, MA, USA).

2.2. Synthesis of Polymeric Matrices

In a typical experiment, the substrate (3.0 g of CL) was mixed with solvent (3.0 mL) under argon atmosphere, followed by the addition of 300 mg CALB, according to our previously described method [20]. Finally, the obtained PCL was precipitated in cold methanol, filtered, and dried under vacuum at room temperature for 4 days.

2.3. Spectroscopy Data of PCL

^{1}H-NMR (CDCl$_3$, δ, ppm): 4.07 [2H, t, $-CH_2CH_2CH_2CH_2\mathbf{CH_2}OC(O)-$], 3.66 [2H, t, $-\mathbf{CH_2}OH$], 2.32 [2H, t, $-CH_2CH_2CH_2CH_2\mathbf{CH_2}COO-$], 1.67 [4H, m, $-CH_2\mathbf{CH_2}CH_2\mathbf{CH_2}CH_2COO-$], 1.39 [2H, m, $-CH_2CH_2\mathbf{CH_2}CH_2CH_2COO-$]; ^{13}C-NMR (CDCl$_3$, δ, ppm): 173.7 [$-\mathbf{C}(O)O-$], 64.3 [$-CH_2CH_2CH_2CH_2\mathbf{C}H_2OC(O)-$], 62.7 [$-\mathbf{C}H_2OH$] 34.3 [$-CH_2CH_2CH_2CH_2\mathbf{C}H_2COO-$], 28.5 [$-CH_2CH_2CH_2\mathbf{C}H_2CH_2OC(O)-$], 25.7 [$-CH_2CH_2CH_2\mathbf{C}H_2CH_2COO-$], 24.7 [$-CH_2CH_2\mathbf{C}H_2CH_2CH_2COO-$]; FT-IR (KBr, cm^{-1}): 3440 (νO-H), 2938 ($\nu_{as}CH_2$), 2866 ($\nu_{as}CH_3$), 1727 (νC=O), 1245 (νC-O).

2.4. Preparation of Poly(ε-Caprolactone) Devices of CIT and TEMP

The obtained PCL matrices were dissolved in DCM under argon atmosphere. Then, the CIT or TEMP was slowly added to the vigorously stirred PCL solution. The samples were dried in vacuo at room temperature until a constant weight was reached. The tablet disc (13 mm diameter, 1 mm thick) was obtained using a hydraulic press (Specac, London, UK) at 98 kN. The mean weight of the developed devices was about 200 mg, corresponding to approximately 5 mg of peptide.

2.5. Genotoxicity Assay of Polymeric Matrices

The *umu*-test was performed with *Salmonella typhimurium* TA1535/pSK1002 in 96-well microplates according to the ISO 13829 protocol with and without metabolic activation, as we described before [21].

2.6. Neutral Red Uptake (NRU) Assay of Polymeric Matrices

Cytotoxicity assays were performed on BALB/c 3T3, clone A31 mammalian fibroblasts (American Type Culture Collection) on the basis of the ISO 10993-5:2009 guideline: Biological evaluation of medical devices—Part 5: Tests for in vitro cytotoxicity. Cells were seeded into 96-well microplates (10^4 cells·mL^{-1}) in DMEM culture medium (supplemented with 10% calf bovine serum, 100 IU/mL penicillin, and 0.1 mg·mL^{-1} streptomycin) and incubated for 24 h (5% CO_2, 37 °C, >90% humidity). Subsequently wells were examined under a microscope, and culture medium was replaced with the tested extracts (prepared as for the *umu*-test) mixed with the treatment medium (1:1). All extracts were tested in twofold dilution series for 24 h (4 data points for each one). The next day, cells were washed with PBS and treated with the neutral red medium for 3 h and then with desorbing fixative (ethanol and acetic acid water solution). The amount of neutral red accumulated by cells was evaluated colorimetrically at 540 nm. PBS and SLS were used as negative and positive controls, respectively. The cell viability was calculated relative to the negative control.

2.7. Tests for In Vitro Cytotoxicity of Polymeric Matrices

For the cytotoxicity test, agar cells were seeded (1.5×10^5 cells·mL^{-1}, DMEM supplemented as for the NRU test) into 6-well plates and maintained in culture for 24 h. The next day, each plate was examined under a microscope, and the culture medium was replaced with fresh DMEM with 1% agar. Round or square sterile samples (about 1 cm diameter/side length) of all tested materials were placed carefully on the solidified agar layer in each well. As negative and positive controls, polyethylene foil and latex were used, respectively. In each test one well seeded with cells was left with clear surface as a control of the cell culture. After 24 h, cells were examined under a microscope to determine the cytotoxic effect before and after carefully removing the specimens from the agar.

2.8. In Vitro Studies of Citropin 1.1 and Temporin A Release from Polymeric Devices

The polymeric devices were immersed in 0.1 M PBS (10 mL) at 37 °C and gently shaken. The sample solutions were withdrawn for analysis at selected time intervals and replaced with new buffer solution. The quantity of peptide was analyzed by high-performance liquid chromatography (HPLC). The samples were prepared in triplicate.

2.9. Measurements

The polymerization products were characterized by means of ^{1}H and ^{13}C NMR (Varian 300 MHz, Palo Alto, CA, USA) at room temperature, with CDCl$_3$ as solvent. Fourier transform infrared (FT-IR) spectra were measured from KBr pellets (Perkin Elmer spectrometer, Perkin Elmer, Warsaw, Poland).

SEC traces were recorded using an Agilent 1100 Series isocratic pump, a degasser, an autosampler thermostatic box for columns, and a set of TSKgel columns (2 × PLGel 5 microns MIXED-C) at 30 °C. A Wyatt Optilab rEX interferometric refractometer and a MALS DAWN EOS laser photometer (Wyatt Technology Corp., Santa Barbara, CA, USA) were used as detectors. Methylene chloride was used as the eluent, at a flow rate of 0.8 mL·min^{-1}. M_n and M_w/M_n were calculated from the experimental traces using the Wyatt ASTRA v 4.90.07 program.

M_n and M_w/M_n of the obtained polyesters were determined by SEC/MALS with a system composed of an Agilent 1100 chromatograph series with an isocratic pump, an autosampler, a degasser, and a thermostatic box for columns. Refractive index (OPTILAB rex, Wyatt Technology Corporation, Santa Barbara, CA, USA) and MALS (DAWN EOS, Wyatt Technology Corporation, Santa Barbara, CA, USA) were applied as detectors and the TSKgel HXL columns were used in series for separations. DCM was used as eluent at a flow rate of 0.8 mL·min^{-1}.

The matrix-assisted laser desorption/ionization mass spectrometry (MALDI-ToF-MS) experiments were performed on an Axima-Performance TOF spectrometer (Shimadzu Biotech, Manchester, UK), equipped with a nitrogen laser (337 nm) using a dithranol (1,8-dihydroxy-9(10H)-anthracenone) matrix.

Thermogravimetric analyses (TGA) were performed using TA Instruments Q50 (New Castle, DE, USA) at a heating rate of 10 °C·min^{-1} under nitrogen flow (60 mL·min^{-1}). Differential scanning calorimetry (DSC) measurements were performed using a TA Instruments DSC Q20 (New Castle, DE, USA) with aluminum pans at a heating rate of 10 °C min^{-1} under nitrogen flow (50 mL·min^{-1}). The degree of crystallinity (X_c) was calculated from the peak enthalpies normalized to the actual weight fraction of polymer using Equation (1):

$$X_c = [\Delta H_f\,(T_m)] \times [\Delta H_{f0}\,(T_{0m})]^{-1} \tag{1}$$

where $\Delta H_f\,(T_m)$ is the enthalpy of fusion at the melting point and $\Delta H_{f0}(T_{0m})$ is the heat of fusion of 100% crystalline PCL (139.5 J·g^{-1}).

Morphological assessment of the samples was carried out with a FEI Quanta 250 FEG scanning electron microscope (SEM) (FEI Inc., Eindhoven, The Netherlands).

Surfaces of tablet matrices were determined with the help of a Contact 3D surface profiler (Taylor Hobson TalySurf CCI Lite, Ametek, UK).

HPLC was performed using a Waters apparatus with the use of an Agilent ZORBAX column (1.8 um, SB-C18, 4.6 × 50 mm). The mobile phase consisted of water (0.1% formic acid) and acetonitryl (0.1% formic acid) at a flow rate of 1 ml·min^{-1}.

2.10. Mathematical Models for Peptide Release Studies

The release data points were subjected to zero-order and first-order kinetics and Higuchi and Korsmeyer–Peppas models [22–24].

Zero-order model:

$$F = kt \tag{2}$$

First-order model:

$$\log F = \log F_0 - \frac{kt}{2.303} \tag{3}$$

Korsmeyer–Peppas model:

$$F = kt^n \ (F < 0.6) \tag{4}$$

where F is the fraction of peptide released up to time (t), F_0 is the initial concentration of peptide, k is the constant of the mathematical models, and n is the exponent of the Korsmeyer–Peppas model.

3. Results and Discussion

3.1. Characterization of PCL Matrices

The purpose of our work was to obtain nontoxic PCLs as devices for highly controlled release of CIT and TEMP. Biodegradable polyesters were synthesized using CALB as a biocatalyst. Lipase was used to avoid the difficulty of completely removing the residual of the conventional metal catalyst from the final product. This paper is a continuation of our earlier investigations [20]. The polymerization was carried out at 60 and 80 °C for 7 days in two ILs: [bmim][NTf$_2$] and [bmim][PF$_6$]. Moreover, the process was performed without solvent to compare the effect of ILs on PCL characterization.

As is commonly known, many factors can influence the release rate of API from polymeric matrices. It is mainly dependent on the kind of polymer and its M_n, M_w/M_n, and microstructure. However, our main intention was to check how peptide release kinetics could be dependent on the polymeric matrix microstructure, thus polyester matrices were synthesized and characterized (Table 1).

Table 1. Characterization of poly(ε-caprolactone) (PCL) matrices synthesized in the presence of *Candida antarctica* (CALB) in ionic liquids (ILs).

No.	Solvent	Temp. [°C]	M_n [Da] [a]	M_w/M_n [b]	Yield [%]	Conv. [%] [c]	T_c [°C] [d]	T_m [°C] [d]	ΔH_f [J·g^{-1}]	X_c [%]	T_{deg} [°C]	MC [%] [e]
PCL-1	[bmim][NTf$_2$]	80	6100	1.8	78	95	34.7	53.1	70.4	50.5	405.4	35
PCL-2	[bmim][PF$_6$]	80	5700	1.5	87	92	28.5	50.6	78.4	56.3	410.3	19
PCL-3	[bmim][NTf$_2$]	60	3600	2.0	88	90	32.1	50.4	73.7	52.8	406.9	23
PCL-4	[bmim][PF$_6$]	60	2100	1.4	85	94	30.4	47.3	82.7	59.2	411.7	0
PCL-5	–	80	1000	2.4	61	73	31.6	49.1	75.7	54.2	407.9	0
PCL-6	–	60	2700	1.2	78	87	32.5	50.4	84.6	60.1	414.5	0

[a] Determined by SEC using the correction coefficient $M_{n(SEC)} = 0.56 \cdot M_{n(SEC\ raw\ data)}$; [b] determined by SEC; [c] determined by ^{1}H NMR [20]; [d] onset temperature; [e] determined by MALDI-ToF-MS.

The chemical structure of the obtained homopolymers was confirmed by ^{13}C, ^{1}H NMR (Figure 1) and IR studies (Experimental section).

The M_n of PCLs synthesized in ILs was determined by SEC and was in the range of 2100–6100 Da. The M_w/M_n indices were about 1.5 for homopolymers synthesized in [bmim][PF$_6$] and about 1.9 for homopolymers synthesized in [bmim][NTf$_2$]. The reaction yields ranged from 78 to 88%. The results obtained in bulk conditions confirm our earlier assumptions that ILs improve enzyme stability at higher temperatures and lead to obtaining polyesters with higher M_n, lower M_w/M_n, and a higher conversion.

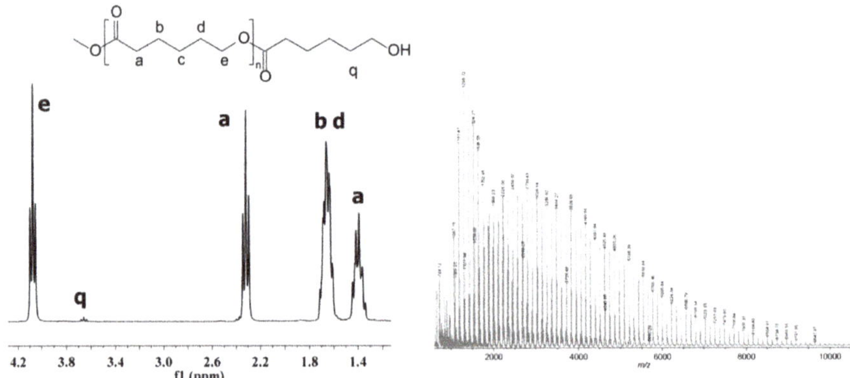

Figure 1. ^{1}H NMR and MALDI-ToF mass spectrum of synthesized PCL-1 analyzed in CDCl$_3$. The reaction was carried out using CALB and [bmim][NTf$_2$] as solvent at 80 °C for 7 days.

The thermal properties of the synthesized matrices determined by the DSC method showed that PCLs synthesized in [bmim][NTf$_2$] were characterized by a lower degree of crystallinity (X_c) (50.5–52.8%) as compared to PCLs obtained in [bmim][PF$_6$] under the same experimental conditions (56.3–59.2%). The low X_c values for PCLs is related to the high content of amorphous domains of cyclic macromolecules in the polymer structure that are formed as side products during eROP [25]. Their presence is the main factor determining the degradation rate of the polyester matrices. The amorphous domains in the PCL chain are mobile and have less packing, while the linear chains in the crystalline domains are rigid and take up less space. As a result, macrocycles show greater exposure of ester bonds to water during hydrolysis, so they degrade faster [26,27]. The content of cyclic oligomers can be controlled by selecting the appropriate reaction conditions. Analysis of the PCL structure determined by MALDI-ToF-MS and DSC showed that the highest content of macrocycles (35%) and the highest proportion of the amorphous phase (about 49.5%) was noted for PCL-1 synthesized in [bmim][NTf$_2$] at 80 °C.

To evaluate the biocompatibility of the matrices, which is extremely important for medical devices like internal implants or urinary and venous catheters, we performed toxicity assays on the selected PCL samples synthesized in different ILs. For the first bioassay, the *umu*-test was used to evaluate the genotoxic potential of the tested materials. The genotoxic potential of the sample is presented as the induction ratio (IR). Results with IR $\geq$ 1.5 are considered genotoxic. Additionally, bacteria growth (G) is evaluated by a measurement of optical density to determine the toxicity of the tested samples.

All tested samples (Table 2) were not toxic for *Salmonella typhimurium* TA1535 (G > 0.5). The IR rises as a result of different types of DNA damage that induces the umuC gene in the SOS system linked in this bacterial strain to the synthesis of β-galactosidase. The IR ratio of both samples did not differ from the results obtained for negative or solvent control, therefore none of the tested PCLs matrices exhibited genotoxic activity (IR < 1.5). Similar negative results were evaluated with and without metabolic activation which disclaims the activity of tested samples as a procarcinogens in this bioassay.

Table 2. The *umu*-test results for the highest concentrations of tested extracts (0.67 mg·mL^{-1}).

Sample	−S9 [a]		+S9 [b]	
	G ± SD	IR ± SD	G ± SD	IR ± SD
PCL-1	0.90 ± 0.08	0.97 ± 0.14	0.87 ± 0.03	0.94 ± 0.07
PCL-2	0.92 ± 0.05	1.02 ± 0.09	0.88 ± 0.01	0.90 ± 0.08
Solvent control	1.01 ± 0.09	0.94 ± 0.15	0.88 ± 0.01	1.03 ± 0.06
Negative control	1.00 ± 0.05	1.00 ± 0.01	1.00 ± 0.05	1.00 ± 0.08

[a] version without metabolic activation; [b] version with metabolic activation.

In the next step, two in vitro cytotoxicity assays were performed using the BALB/c mouse 3T3 fibroblast cell line. The results revealed that in the NRU assay, BALB/c 3T3 cell viability was not decreased by any of the tested samples (Table 3).

Table 3. Results of the NRU test for the highest concentrations of tested extracts (0.5 mg·mL^{-1}).

Sample	Cell Viability ± SD (%)
PCL-1	96 ± 2
PCL-2	95 ± 1
Solvent control	100 ± 5

The percentage of viable cells did not differ significantly from the negative control. Moreover, the agar overlay assay also did not reveal any cytotoxic potential of any tested polyesters after 24 h of contact through the agar layer. The results show that there were no changes in the integrity of culture monolayer or general morphology of cells in comparison with the culture control (Figure 2). Moreover, there was no detectable zone around or under the specimen.

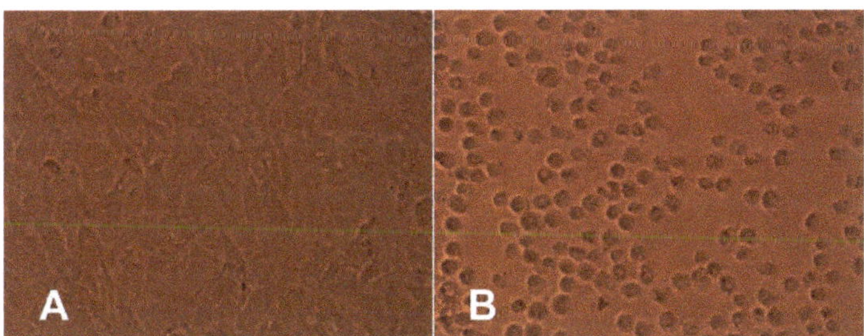

Figure 2. Sample PCL-1: (**A**) no detectable zone around or under specimen; (**B**) positive control zone with degenerated cells extending up to 0.7 cm around the specimen. Magnification × 200.

3.2. In Vitro Kinetics Release of CIP and TEMP from New Devices

The kinetic release of CIP or TEMP from the selected synthesized matrices (PCL-1-CIT and PCL-1-TEMP, obtained from the PCL-1 matrix; PCL-2-CIT and PCL-2-TEMP, obtained from the PCL-2 matrix) was determined at pH 7.4 and 37 °C over 28 days (Figure 3). The ordinate of the plot was calculated based on the cumulative amount of peptide released [28].

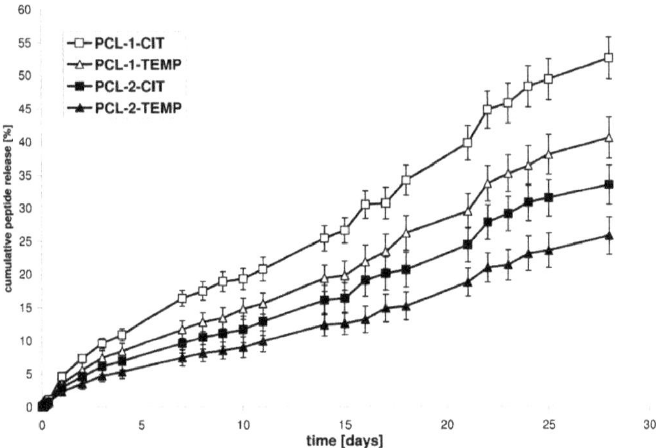

Figure 3. Cumulative release of peptides from PCL-1-CIT, PCL-1-TEMP, PCL-2-CIT, and PCL-2-TEMP over 28 days (each point represents mean ± SD of three points).

It was found that the difference in the release rate observed for the PCL devices was attributed to the difference in the X_c of the PCL matrices. The rate of in vitro peptide release increased as the X_c of the matrices decreased. The percentage of CIT released after 28 days of incubation was about 52.6% for PCL-1-CIT (obtained from PCL-1, X_c = 50.5%) and 33.5% for PCL-2-CIT (obtained from PCL-2, X_c = 56.3%). Similarly, 40.6% of TEMP was released from PCL-1-TEMP (obtained from PCL-1, X_c = 50.5%) and 25.8% from PCL-2-TEMP (obtained from PCL-2, X_c = 56.3%). It was also found that CIT and TEMP were released in a rather regular and continuous way. Furthermore, CIT was released more quickly than TEMP. For instance, the amount of peptide released from PCL-1-CIT was about 16.4, 25.4, 39.8, and 52.6% after 7, 14, 21, and 28 days of incubation, respectively, whereas for PCL-1-TEMP these values were about 11.7, 19.4, 29.5, and 40.6% after 7, 14, 21, and 28 days of incubation, respectively. These differences were probably related to differences in the hydrophobic nature of the peptides. The amount of hydrophobic residue was 61 and 56% for TEMP and CIT, respectively. As a result, TEMP interact stronger with the hydrophobic polymeric matrix. Moreover, a high content of non-polar amino acids in the TEMPs structure (85%) provide to a poor solvation in water as compared to CIT (75% of the non-polar hydrophobic residues).

The release data points were subjected to zero-order and first-order kinetics and the Korsmeyer–Peppas model in order to evaluate the kinetics and mechanisms of peptide release from the obtained polyester matrices (Table 4).

Table 4. Analysis data of peptide release from polymeric matrices.

No.	Zero-Order Model	First-Order Model	Korsmeyer–Peppas Model	
	R^2	R^2	R^2	n
PCL-1-CIT	0.9877	0.9574	0.9797	>0.89
PCL-1-TEMP	0.9853	0.9603	0.9742	>0.89
PCL- 2-CIT	0.9855	0.9644	0.9732	>0.89
PCL-2-TEMP	0.9841	0.9656	0.9722	>0.89

As is commonly known, according to the Korsmeyer–Peppas model, for the diffusion-degradation–controlled peptide release system, the exponent value (n) varies between 0.45 and 0.89 (anomalous, non-Fickian). When n was close to 0.45, diffusion (Fickian diffusion) dominated in the process. When *n* was close to 0.89 (zero-order release), degradation controlled the release [22–24].

Our work shows that the obtained devices demonstrated a rather controlled release profile. High R^2 values (from 0.9841 to 0.9877) were obtained for the near-zero-order kinetics model. The R^2 values of the Korsmeyer–Peppas model were also high (from 0.9722 to 0.9797). The controlled drug release profiles were obtained with no significant burst release. This suggests that AMP release from the new PCLs matrices is a highly controlled process. Moreover, Table 4 shows that the exponent *n* in the Korsmeyer–Peppas model was approximately above 0.89. This is likely due to a rather degradation-controlled mechanism. The lower peptide release rate from PLA-2 than PLA-1 matrices is a logical consequence of the lower matrices' crystallinity, because the degradation process is initiated in the amorphous phase [26].

3.3. Analysis of Matrix Surfaces After Degradation

Noncontact laser profilometry and SEM were used to determine the structural changes during the matrix degradation process. Figure 4 presents SEM micrographs and profilometer images of the microstructure of the PCL tablets after 28 d of incubation in PBS solution at 37 °C and pH 7.4.

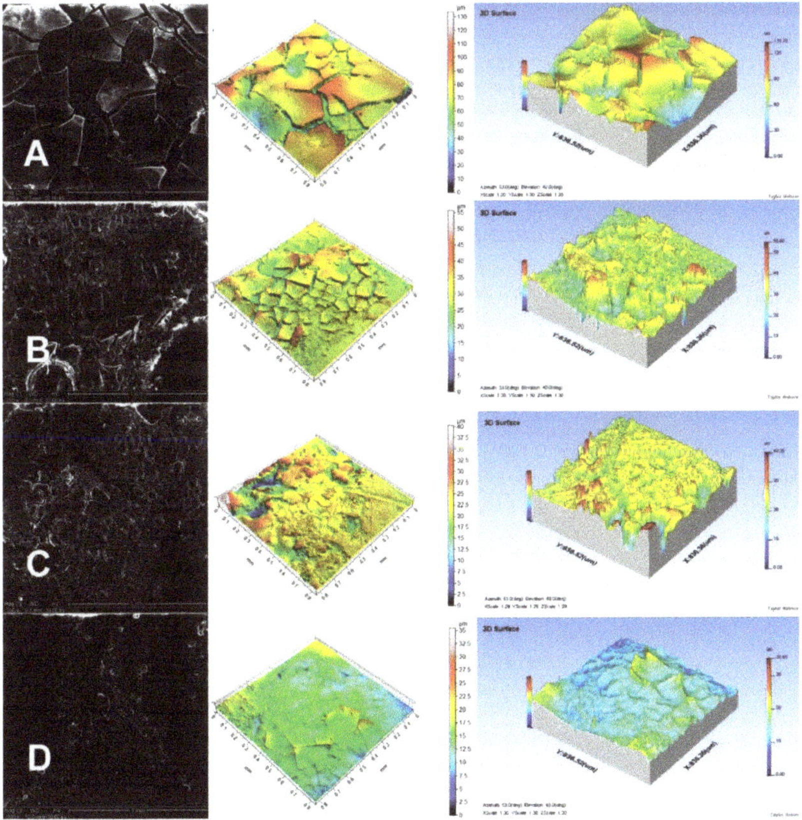

Figure 4. SEM micrographs and profilometer images showing the microstructure of PCL tablets after 28 days in PBS solution at 37 °C and pH 7.4: (**A**) PCL-1-CIT; (**B**) PCL-1-TEMP; (**C**) PCL-2-CIT; (**D**) PCL-2-TEMP.

We could easily see an increase in the depth of holes visible on the tablet surface in the following order: PCL-2-TEMP > PCL-2-CIT > PCL-1-TEMP > PCL-1-CIT. This occurrence confirms our previous assumption that PCL tablets synthesized in [bmim][NTf$_2$] degraded faster than PCLs synthesized in [bmim][PF$_6$] due to the lower crystallinity.

Table 5 shows average roughness (*Ra*) of the tablets and mass loss after incubation in PBS. It was found that the difference in the release rate observed for the PCL devices was also attributed to the difference in *Ra* values of the PCL matrices. Profilometer analysis indicated that the highest *Ra* value (5.34) was obtained for PCL-1-CIT. For PCL-1-TEMP and PCL-2-CIT samples, the Ra values were 2.56 and 1.35, respectively. The lowest Ra value was obtained for the PCL-2-TEMP sample (about 0.32). Moreover, percentage mass loss was higher for PCL-1 than PCL-2 devices (54.2–22.8% and 16.4–7.9%, respectively), which is consistent with our previous report [20].

Table 5. Parameters of surface roughness and mass loss after 28 days in PBS solution at 37 °C and pH 7.4.

No.	Ra ± SD	Mass Loss ± SD (%)
PCL-1-CIT	5.34 ± 1.60	54.2 ± 2.2
PCL-1-TEMP	2.56 ± 0.74	22.8 ± 1.7
PCL-2-CIT	1.35 ± 0.75	16.4 ± 0.9
PCL-2-TEMP	0.32 ± 0.10	7.9 ± 0.5

As it is commonly known, PCL is water-insoluble polymer. However, it is hydrolytically unstable and is degraded by hydrolysis of ester bonds. The process first occurs in the amorphous regions and is followed by a slower degradation in crystalline regions. Various studies have revealed that PCLs degradation is a result of several simultaneously occurring processes. These include water uptake, swelling, ester hydrolysis, diffusion of oligomers and degradation products, and local pH drop [29,30]. However, completely explanation of the mechanism which involves several factors is not yet possible.

3.4. Evaluation of Therapeutic Effect of New AMP Devices

From the therapeutic point of view, it is very important to achieve therapeutic concentration of APIs in the target tissue. In the current work, we also investigated the daily release rate of AMPs after 1, 7, 14, 21, and 28 days (Figure 5) considering their reported MIC and MHC values [13,31]. We focused on the anti-infectious effect of AMPs in infections caused by major pathogens in surgical wound infections, *Staphylococcus aureus*, *Enterococcus faecalis*, *Escheria coli*, *Klebsiella pneumoniae*, and *Pseudomonas aeruginosa* [32].

CIT, due to its higher TI value (8.8), shows greater antimicrobial specificity as compared to TEMP (TI = 4.5). This indicates that CIT could be a promising antimicrobial compound for polymeric devices in wound infections. However, regardless the type of API, the type of matrix is a key factor for effective therapy. It was noted that PCL-2 was the best matrix for CIT. New PCL-2-CIT devices reach therapeutic concentration of AMPs (about 110–190 µg·mL^{-1}) within 24 h and maintain it for the next 28 days. For TEMP, the most effective matrix appeared to be PCL-1, with peptide daily release of about 130–190 µg·mL^{-1} for 28 days.

For PCL-1, the peptide maximum daily concentration was dependent on the type of APIs. In the case of TEMP, maximum concentration was reached after 14 days of incubation in PBS at pH 7.4 at 37 °C. Interestingly, the maximum CIT daily release concentration was reached after 7 and 21 days (about 280 µg·mL^{-1}), and this value exceeded the MHC values (about 256 µg·mL^{-1} [31]). For PCL-2, the highest daily concentration of both peptides was noted after 21 days.

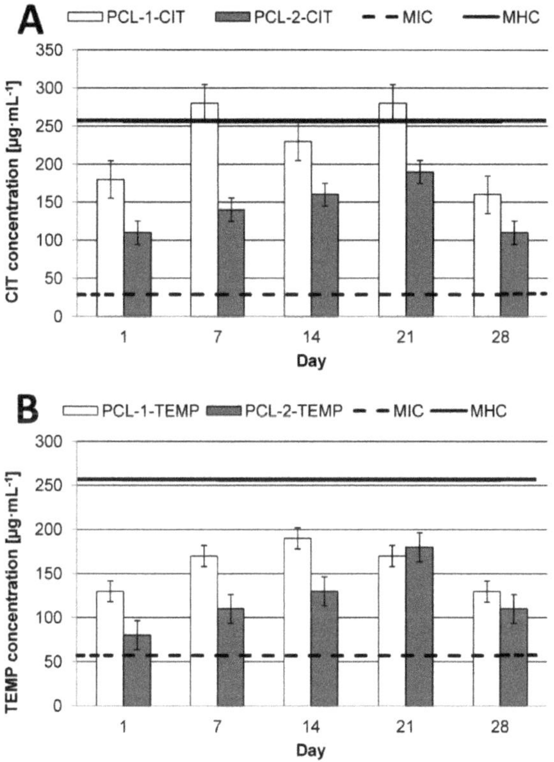

Figure 5. Peptide daily concentration after 1, 7, 14, 21, and 28 days of incubation in PBS solution at 37 °C and pH 7.4: (**A**) CIT; (**B**) TEMP. Geometric values of minimal inhibitory concentration (MIC) and minimal hemolytic concentration (MHC) were taken from the literature [13,31].

4. Conclusions

In the present work, new nontoxic polymeric matrices for controlled CIT and TEMP release were obtained. PCLs with different crystallinity were successfully synthesized by eROP in two different ILs, [bmim][NTf$_2$] and [bmim][PF$_6$]. The in vitro release study demonstrated that CIT and TEMP release rates were strongly dependent on the crystallinity of the polymeric matrices. However, peptide characteristics were also important. CIT, a less hydrophobic peptide, was released more quickly than the more hydrophobic TEMP. We found that the peptides were released with high control, according to the degradation mechanism with near-zero-order kinetics. The peptide "burst release" was not observed during the degradation process. New devices reached a therapeutic concentration of AMPs within 24 h and maintained it for 28 days. From a broader perspective, this study suggests that synthesized polyester matrices may contribute to a potential application as medium- and long-term controlled CIT or TEMP delivery systems, offering fast antibacterial effects on local infections in implantology.

Author Contributions: U.P. conceived, designed, and directed the studies, performed the synthesis and physicochemical characterization of the materials and new devices, and wrote the manuscript. M.S., E.O., and M.B. performed the peptide release studies (tests and their interpretation) and helped write the paper. A.Z. performed geno- and cytotoxicity assays (tests and their interpretation). All authors read and approved the final manuscript.

Acknowledgments: This work was financially supported by the National Science Centre (Poland), PRELUDIUM 6 research scheme, decision No. DEC-2013/11/N/NZ7/02342 ("Synthesis, physico-chemical and biological studies on macromolecular carriers of antimicrobial peptides") and funded with the assistance of the Ministry of Science and Higher Education from subsidies for statutory activity, project No. 3086/35/P ("Development of formulations and technologies for the manufacture of innovate cosmetics, pharmacy supplies, household and industrial chemicals"). The authors are indebted to Violetta Kowalska from the Medical University of Warsaw and Dominik Czerwonka from the Kazimierz Pulaski University of Technology and Humanities in Radom for the performance of the NMR (V.K.) and roughness (D.C.) measurements.

Conflicts of Interest: The authors declare no conflict of interest.

References

1. Gomes, B.; Augusto, M.T.; Felício, M.R.; Hollmann, A.; Franco, O.L.; Gonçalves, S.; Santos, N.C. Designing improved active peptides for therapeutic approaches against infectious diseases. *Biotechnol. Adv.* **2018**, *36*, 415–429. [CrossRef] [PubMed]

2. Coates, A.R.; Halls, G.; Hu, Y. Novel classes of antibiotics or more of the same? *Br. J. Pharmacol.* **2011**, *163*, 184–194. [CrossRef] [PubMed]

3. Hu, Y.; Shamaei-Tousi, A.; Liu, Y.; Coates, A. A new approach for the discovery of antibiotics by targeting non-multiplying bacteria: A novel topical antibiotic for staphylococcal infections. *PLoS ONE* **2010**, *5*, e11818. [CrossRef] [PubMed]

4. Jenssen, H.; Hamill, P.; Hancock, R.E. Peptide antimicrobial agents. *Clin. Microbiol. Rev.* **2006**, *19*, 491–511. [CrossRef] [PubMed]

5. Zasloff, M. Antimicrobial peptides of multicellular organisms. *Nature* **2002**, *415*, 389–395. [CrossRef] [PubMed]

6. Piotrowska, U.; Sobczak, M.; Oledzka, E. Current state of a dual behaviour of antimicrobial peptides—Therapeutic agents and promising delivery vectors. *Chem. Biol. Drug Des.* **2017**, *90*, 1079–1093. [CrossRef] [PubMed]

7. Zhao, Y.N.; Xu, X.; Wen, N.; Song, R.; Meng, Q.; Guan, Y.; Cheng, S.; Cao, D.; Dong, Y.; Qie, J.; et al. A Drug Carrier for Sustained Zero-Order Release of Peptide Therapeutics. *Sci. Rep.* **2017**, *7*, 5524. [CrossRef] [PubMed]

8. Eckhard, L.H.; Sol, A.; Abtew, E.; Shai, Y.; Domb, A.J.; Bachrach, G.; Beyth, N. Biohybrid polymer-antimicrobial peptide medium against *Enterococcus faecalis*. *PLoS ONE* **2014**, *9*, e109413. [CrossRef] [PubMed]

9. Sobczak, M.; Kamysz, W.; Tyszkiewicz, W.; Dębek, C.; Kozłowski, R.; Olędzka, E.; Piotrowska, U.; Nałęcz-Jawecki, G.; Plichta, A.; Grzywacz, D. Biodegradable macromolecular conjugates of citropin: Synthesis, characterization and in vitro efficiency study. *React. Funct. Polym.* **2014**, *83*, 54–61. [CrossRef]

10. Pranantyo, D.; Xu, L.Q.; Kang, E.-T.; Mya, M.K.; Chan-Park, M.B. Conjugation of Polyphosphoester and Antimicrobial Peptide for Enhanced Bactericidal Activity and Biocompatibility. *Biomacromolecules* **2016**, *17*, 4037–4044. [CrossRef] [PubMed]

11. Chen, Y.; Mant, C.T.; Farmer, S.W.; Hancock, R.E.; Vasil, M.L.; Hodges, R.S. Rational design of α-helical antimicrobial peptides with enhanced activities and specificity/therapeutic index. *J. Biol. Chem.* **2005**, *280*, 12316–12329. [CrossRef] [PubMed]

12. Sikorska, E.; Greber, K.; Rodziewicz-Motowidło, S.; Szultka, L.; Łukasiak, J.; Kamysz, W. Synthesis and antimicrobial activity of truncated fragments and analogs of citropin 1.1: The solution structure of the SDS micelle-bound citropin-like peptides. *J. Struct. Biol.* **2009**, *168*, 250–258. [CrossRef] [PubMed]

13. Giacometti, A.; Cirioni, O.; Kamysz, W.; Silvestri, C.; Del Prete, M.S.; Licci, A.; D'amato, G.; Łukasiak, J.; Scalise, G. In vitro activity and killing effect of citropin 1.1 against Gram-positive pathogens causing skin and soft tissue infections. *Antimicrob. Agents Chemother.* **2005**, *49*, 2507–2509. [CrossRef] [PubMed]

14. Giacometti, A.; Cirioni, O.; Kamysz, W.; Silvestri, C.; Del Prete, M.S.; Licci, A.; D'amato, G.; Łukasiak, J.; Scalise, G. In vitro activity of citropin 1.1 alone and in combination with clinically used antimicrobial agents against *Rhodococcus equi*. *Antimicrob. Agents Chemother.* **2005**, *56*, 410–412. [CrossRef] [PubMed]

15. Doyle, J.; Brinkworth, C.S.; Wegener, K.L.; Carver, J.A.; Llewellyn, L.E.; Olver, I.N.; Bowie, J.H.; Wabnitz, P.A.; Tyler, M.J. nNOS inhibition, antimicrobial and anticancer activity of the amphibian skin peptide, citropin 1.1 and synthetic modifications. *Eur. J. Biochem.* **2003**, *270*, 1141–1153. [CrossRef] [PubMed]

16. Mulder, K.C.; Lima, L.A.; Miranda, V.J.; Dias, S.C.; Franco, O.L. Current scenario of peptide-based drugs: The key roles of cationic antitumor and antiviral peptides. *Front. Microbiol.* **2013**, *4*, 1–23. [CrossRef] [PubMed]

17. Wade, D.; Silberring, J.; Soliymani, R.; Heikkinen, S.; Kilpeläinen, I.; Lankinen, H.; Kuusela, P. Antibacterial activities of temporin A analogs. *FEBS Lett.* **2000**, *479*, 6–9. [CrossRef]

18. Vekariya, R.L. A review of ionic liquids: Applications towards catalytic organic transformations. *J. Mol. Liq.* **2016**, *227*, 44–60. [CrossRef]

19. Piotrowska, U.; Sobczak, M.; Oledzka, E. Characterization of aliphatic polyesters synthesized via enzymatic ring-opening polymerization in ionic liquids. *Molecules* **2017**, *22*, 923. [CrossRef] [PubMed]

20. Piotrowska, U.; Sobczak, M.; Oledzka, E.; Combes, C. Effect of ionic liquids on the structural, thermal, and in vitro degradation properties of poly(epsilon-caprolactone) synthesized in the presence of *Candida antarctica* lipase B. *J. Appl. Polym. Sci.* **2016**, *133*, 10. [CrossRef]

21. Żółtowska, K.; Piotrowska, U.; Oledzka, E.; Kuras, M.; Zgadzaj, A.; Sobczak, M. Biodegradable Poly (ester-urethane) Carriers Exhibiting Controlled Release of Epirubicin. *Pharm. Res.* **2017**, *34*, 780–792. [CrossRef] [PubMed]

22. Alexis, F. Factors affecting the degradation and drug-release mechanism of poly(lactic acid) and poly[(lactic acid)-co-(glycolic acid)]. *Polym. Int.* **2005**, *54*, 36–46. [CrossRef]

23. Siepmann, J.; Göpferich, A. Mathematical modeling of bioerodible, polymeric drug delivery systems. *Adv. Drug Deliv. Rev.* **2001**, *48*, 229–247. [CrossRef]

24. Dash, S.; Murthy, P.N.; Nath, L.; Chowdhury, P. Kinetic modeling on drug release from controlled drug delivery systems. *Acta Pol. Pharm.* **2010**, *67*, 217–223. [PubMed]

25. Kobayashi, S. Lipase-catalyzed polyester synthesis—A green polymer chemistry. *Proc. Jpn. Acad. Ser. B Phys. Biol. Sci.* **2010**, *86*, 338–365. [CrossRef] [PubMed]

26. Bosworth, L.A.; Downes, S. Physicochemical characterisation of degrading polycaprolactone scaffolds. *Polym. Degrad. Stab.* **2010**, *95*, 2269–2276. [CrossRef]

27. Woodruff, M.A.; Hutmacher, D.W. The return of a forgotten polymer—Polycaprolactone in the 21st century. *Prog. Polym. Sci.* **2010**, *35*, 1217–1256. [CrossRef]

28. Żółtowska, K.; Piotrowska, U.; Oledzka, E.; Luchowska, U.; Sobczak, M.; Bocho-Janiszewska, A. Development of biodegradable polyesters with various microstructures for highly controlled release of epirubicin and cyclophosphamide. *Eur. J. Pharm. Sci.* **2017**, *96*, 440–448. [CrossRef] [PubMed]

29. Göpferich, A. Mechanism of polymer degradation and erosion. *Biomaterials* **1996**, *17*, 103–114. [CrossRef]

30. Avgoustakis, K.; Nixon, J.R. Biodegradable controlled tablets 3. Effect of polymer characteristics on drug release. *Int. J. Pharm.* **1993**, *99*, 247–252. [CrossRef]

31. Neubauer, D.; Jaśkiewicz, M.; Migoń, D.; Bauer, M.; Sikora, K.; Sikorska, E.; Kamysz, E.; Kamysz, W. Retro analog concept: Comparative study on physico-chemical and biological properties of selected antimicrobial peptides. *Amino Acids* **2017**, *49*, 1755–1771. [CrossRef] [PubMed]

32. Giacometti, A.; Cirioni, O.; Schimizzi, A.M.; Del Prete, M.S.; Barchiesi, F.; D'errico, M.M.; Petrelli, E.; Scalise, G. Epidemiology and microbiology of surgical wound infections. *J. Clin. Microbiol.* **2000**, *38*, 918–922. [PubMed]

MDPI

St. Alban-Anlage 66

4052 Basel

Switzerland

Tel. +41 61 683 77 34

Fax +41 61 302 89 18

www.mdpi.com

Polymers Editorial Office

E-mail: polymers@mdpi.com

www.mdpi.com/journal/polymers